PHYSICS AND HUMAN AFFAIRS

Art Hobson
University of Arkansas

1807 **WJ** 1982
175 YEARS OF PUBLISHING

John Wiley & Sons
NEW YORK • CHICHESTER • BRISBANE • TORONTO • SINGAPORE

Cover illustration: Tom Huffman
Text design: Sue Taube
Illustrations: John Balbalis
    with the assistance of the
    Wiley Illustration Dept.
Editorial supervisor: Deborah Herbert
Manuscript editor: Jenifer McLean
Production: Lilly Kaufman

Library of Congress Cataloging in Publication Data:

Hobson, Art, 1934–
    Physics and human affairs.

    Includes bibliographical references and index.
    1. Physics.   2. Physics—Social aspects.
3. Physics—Philosophy.   I. Title.   84- 7869

QC23.H74   530   81-11407
ISBN 0-471-04746-5   AACR2

Printed in the United States of America

10 9 8 7 6 5 4 3 2 1

For Ziva and David,
a couple of nice kids.
May their world be peaceful.

ACKNOWLEDGMENTS

I am grateful to Michael Lieber, Charles Richardson, John Meason, and Milton McLain, Jr. of the University of Arkansas faculty for their helpful comments on portions of the manuscript. Professors Karl F. Kuhn, Edward G. Grimsal, Charles Long, and Edsel W. Winter all reviewed the manuscript and made many helpful suggestions for improvements. The Wiley staff has been a delight to work with, from Donald Deneck's enthusiasm for the initial manuscript through Robert McConnin's editorial guidance and including the perceptive editorial corrections of Jenifer McLean. I thank Terry Bremer for help in preparing the manuscript and Beverly Daven for her careful typing of both the first and the second "final" versions and for deciphering the thousands of penciled-in alterations. This book evolved during five years of teaching Physics and Human Affairs at the University of Arkansas. I especially thank all the students whose vitality and interest in these topics encouraged me to write this book and who gave me useful feedback during class testing of several preliminary versions of the manuscript. And thanks to my special friends Debra, John, and Bootie, who helped me more than they knew.

A. H.

TO THE STUDENT

In an age dominated by science and technology it is important that nonscientists help us come to grips with the delights and the dilemmas posed by our expanding knowledge of the natural world. In an age of fragmented and narrowed fields of study it is important that some individuals bridge the interdisciplinary gaps. Today, when scientific knowledge is bursting the confines of human imagination, it is important that the poets among us experience these new songs. Thus I have written this book for the generalist in our sometimes overspecialized society.

Physics and Human Affairs is about physics, of course, but it is also about the many-sided human implications of physics. I hope that this book will make some of the wonderful insights of physics accessible and exciting to you. Most of all, I hope that this book will help you integrate science-related ideas into your own life and that it will encourage you to bring your intelligence and your energy to bear on the many science-related issues of our time.

If you are a student of the arts, the social sciences, education, or business, you probably have little need and less desire for a quantitative, technical physics course. You'll probably never need to solve the mathematical problems emphasized in technical courses. This book approaches physics qualitatively, emphasizing concepts and implications rather than techniques. Although we will use arithmetic and proportions, we won't need algebraic manipulations and formulas.

Exercises are interspersed with the reading material. They are meant to be answered during the reading. Most exercises are an integral part of the text. Their purpose is to give you a chance to check your understanding and to emphasize certain points in the book. Check your answers with my answers. Answers to these interspersed exercises are given either in the back of the book or, when stated, in the text immediately following the exercise. No answers are provided to the Further Exercises given at the end of each chapter.

You will also find Thought Questions throughout the text. Most of these questions have no specific correct answers. They are open-ended, sometimes controversial, usually a matter of opinion. They are meant to help you develop your own views. Think about them as you read. You won't find any list of answers to the Thought Questions because there are many possible answers. I give my own views on a few of these questions in the text.

Each chapter contains a Checklist of key concepts to help you review. Also at the end of each chapter you'll find a list of Recommended Readings, with a short description of each. These books and articles are nonmathematical and are written for nonscientists. Try them—you might like them!

Happy learning!

A. H.

PREFACE

This is a textbook for a one-semester (or one-quarter) physics course for nonscientists. It could also be used in a physical science course, as it contains a significant amount of astronomy and some chemistry. My goal in writing it is to promote understanding and enthusiasm about physics and its social and philosophical implications. Most important, my goal is to encourage students to integrate science-related ideas into their ways of thinking about the world and to provide some of the tools needed to make intelligent decisions in a science-based culture. I have tried to achieve these goals by presenting physics in a language comprehensible to nonscientists and by dealing directly with many of the physics-related topics that are important or interesting to nonscientists. These interdisciplinary topics include the methods and validity of science; extraterrestrial life; space colonies; the energy crisis; carbon dioxide heating of the atmosphere; atmospheric ozone depletion; philosophical questions raised by quantum mechanics and high-energy physics; nuclear war and the arms race; nuclear power, and cosmology.

Now, it is impossible to cover all of the usual physics topics *plus* the above interdisciplinary material in a one-semester course. My solution to this problem is twofold:

1. I have omitted the *techniques* of physics and have focused instead on *concepts*. Nonscientists have no need for the problem-solving mathematical methods that occupy so much time in more technical courses. Thus this book uses only simple arithmetic and proportions, with no algebra or trigonometry.
2. Although the book treats each of the standard divisions of physics (classical mechanics, thermodynamics, electromagnetism, light, relativity, quantum mechanics, nuclear physics), it explores only the most significant or interesting topics within each division. My choice of topics is strongly influenced by my desire for students to develop an enthusiastic and long-term interest in physics and by my conviction that any liberal arts physics course should emphasize the post-Newtonian concepts of our own time.

I hope that the Exercises and Thought Questions interspersed throughout the text give the book the flavor of a dialogue with the reader. (See To the Student for the nature and proper use of these questions.) There are Further Exercises and Further Thought Questions, along with a Checklist for review and a list of Recommended Readings, at the end of each chapter, with the exception of Chapter 1.

The book may be handled flexibly. Some historical, philosophical, and social topics may be assigned for reading only and mentioned just briefly or not at all in lecture. Nonscience students often excel in reading and can benefit from these enrichment topics even when they are omitted from the lecture. Alternatively, many of the social topics lend themselves to freewheeling class discussion, with divergent opinions freely expressed. Some of my most fruitful class periods have been devoted to this sort of debate.

The book is purposely a little long for a one-semester course. This allows the instructor the flexibility of assigning some topics for reading only or of omitting some topics entirely and thus tailoring the course to his or her taste. Any of the chapters and sections marked with asterisks in the Table of Contents can be omitted without affecting the remaining material. On the few occasions that the text refers to this starred material, the reference is not fundamental to the discussion and can be overlooked if the earlier section was omitted.

Some instructors may want to emphasize the philosophical and historical topics and delete most social issues. This would produce a fairly traditional physics course, developed along historical and philosophical lines and emphasizing modern physics. For a course of this sort, omit Chapters 10, 22, and 23, along with the starred sections of Chapters 11 and 13. Some other instructors may want to emphasize the social issues and delete many of the philosophical and historical topics. Such a physics-and-society course can be obtained by omitting Chapters 17* to 19, and 24. As yet another option, nuclear physics could be deleted by omitting Chapters 20 to 23.

Several themes recur throughout the book: energy as a unifying concept in physics and in society; the methods and validity of science; the multivalued social impacts of science; and the significance of post-Newtonian physics. Most of all, I hope that the book conveys the subtle web of *connections* among the individual scientist, the phenomena observed by scientists, physical principles, technology, the environment, and the attitudes and actions of people, including, in particular, the reader.

Art Hobson

*Although Chapter 17 is not one of the starred chapters, it may be skipped if Chapters 18 and 19 are also skipped.

The century would seek to dominate nature as it had never been dominated, would attack the idea of war, poverty and natural catastrophe as never before. The century would create death, devastation and pollution as never before. Yet the century was now attached to the idea that man must take his conception of life out to the stars. . . . A century devoted to the rationality of technique was also a century so irrational as to open in every mind the real possibility of global destruction. It was the first century in history which presented to sane and sober minds the fair chance that the century might not reach the end of its span. It was a world half convinced of the future death of our species yet half aroused by the apocalyptic notion that an exceptional future still lay before us. So it was a century which moved with the most magnificent display of power into directions it could not comprehend. The itch was to accelerate—the metaphysical direction unknown.

Norman Mailer
Of a Fire on the Moon

CONTENTS

PART 3   Transition to the New Physics

Light: an Electromagnetic Wave
The Decline of the Newtonian Universe

The Shape of the Universe
The Evolution of the Universe
Speculations

PART 1

A Science Sampler

1

PHYSICS AND YOU

We'll open with a few scenes designed to introduce you to the scope and flavor of this book. If you have difficulty with any of the following sketches, don't worry about it; the related science will be explained later.

24 May 1543. Churchmen lead a medieval procession through the cobblestone streets of Frauenberg, East Prussia. In the northwest tower of the fortified wall, an old man receives friends bearing the first printed copy of his life's work, On the Revolutions of the Heavenly Spheres. *He thumbs through the pages that will displace humans from their comfortable place at the center of God's universe and initiate an intellectual upheaval that continues to this day. The pages announce that ". . . the planetary theories of Ptolemy and most other astronomers, although consistent with the numerical data, seemed . . . to present no small difficulty. . . . A system of this sort seemed neither sufficiently absolute nor sufficiently pleasing to the mind. Having become aware of these defects, I often considered whether there could perhaps be found a more reasonable arrangement of circles. . . ."*

Copernicus could not have foreseen the discoveries in astronomy that would one day reveal the colossal scale of the heavens while consigning the human race to an apparently small role in the universe. This loyal and humble churchman could little know that, with the publication of this book, the sun was setting on the medieval world.

6 August 1945. 8 A.M., Japanese time. The coast is in sight now. The most important mission in the career of Colonel Paul Tibbets, U.S. Army Air Force, will soon be completed. It will, in fact, be the most important mission of the war, or of any war so far. An hour earlier, to warn of three approaching U.S. weather planes, the sirens of Hiroshima had sounded their too-familiar alarm. Citizens had for the most part gone on about their business, accustomed as they were to these routine flights. Unaccountably, no alarms are sounded now at the approach of the second group of three B-29s.

8:15 A.M. The bombay doors of the Enola Gay *swing wide over the city. Seconds later, the sky is lit with a fearsome torch as chain-reacting nuclei bring a small part of the earth's surface to a temperature approximating that of the center of the sun. Direct radiation, traveling at the speed of light, sears flesh in the silence before the arrival of the expanding shock wave. In an instant, 100,000 people are killed or doomed.*

Colonel Tibbets and the 135 flight-crew and 1700 ground-crew members of the 509th Composite Group of the 313th Wing of the 21st Bombing Command of the 20th Air Force had been training for these few moments since the fall of 1944. The blow delivered by the Enola Gay *had been preceded by 5 years and $2 billion worth of intensive research and development at Oak Ridge, Detroit, New York, Decatur, Milwaukee, Chicago, Hanford, Berkeley, Los Alamos, and Alamogordo. It was the* *largest coordinated development project in the previous history of the human race, a project known by its code name as Manhattan District.*

Captain Robert Lewis, copilot to Colonel Tibbets, maintained an informal log of the flight of the Enola Gay. *At 8:00 A.M., as his plane approached the mainland, Captain Lewis wrote, "There will be a short intermission [in the diary] while we bomb our target."*

The next entry read: "My God!"

The new device would come to be called an atomic bomb, although nuclear bomb would be more accurate. Within the next few days, humanity would become aware that a new age had dawned, that our species had, for the first time, the means for its own annihilation (Fig. 1.1).

Oak Ridge, Tennessee. In the depths of a reactor at Oak Ridge National Laboratory, the institution that separated the uranium 235 for the Hiroshima bomb, neutrons bombard the stable isotope cobalt 59 to produce the radioactive isotope cobalt 60. The Co-60 is sealed in lead and shipped to a medical center. Radiologists install the Co-60 in a lead housing that has a narrow "window" through which the high-energy gamma rays emitted by the decaying nuclei can be directed at cancer tumors. Gamma photons destroy the rapidly growing cancer cells more easily than the slowly growing normal cells. In this manner, the energy of the nucleus, released so catastrophically at Hiroshima, saves many lives every year (Fig. 1.1).

Zurich, Switzerland, 1895. An unorthodox 16-year-old dropout from a Munich high school applies for admission to the Federal Polytechnic University, but fails the entrance exams. To make up his deficiencies, he enrolls in high school in Aarau, Switzerland, where he is surprised to find a congenial and democratic learning atmosphere and where he begins dreaming the geometrical dreams that will develop into a cornerstone of twentieth-century physics.

Bern, Switzerland, 1905. Our hero, now an obscure young patent examiner, publishes an unusual paper entitled On the Electrodynamics of Moving Bodies. *The paper is accompanied by a brief note observing that ". . . we are led to the more general conclusion that the mass of a body is a measure of its energy-content; if the energy changes by E, the mass changes in the same sense by E/c^2. . . ." Thus the theory of relativity is born.*

In 1905, Albert Einstein, pacifist, could not have known that his theorizing would promote human knowledge of the atomic nucleus, nor could he have known the formidable consequences of this knowledge.

1900. Max Planck introduces the unprecedented concept that light is emitted in small bundles, or quanta, of energy. This unusual hypothesis, in seeming disagreement with a large body of previous knowledge, has only one known virtue: It leads to a correct quantitative prediction of the radiation emitted by heated solids, a task at which all other theories had failed. Thus is the quantum theory born. The new theory would lead to a rethinking of the concept of matter, and to a revolution in our perception of reality. The theory of the quantum would be crucial in furthering human understanding and control of the nucleus.

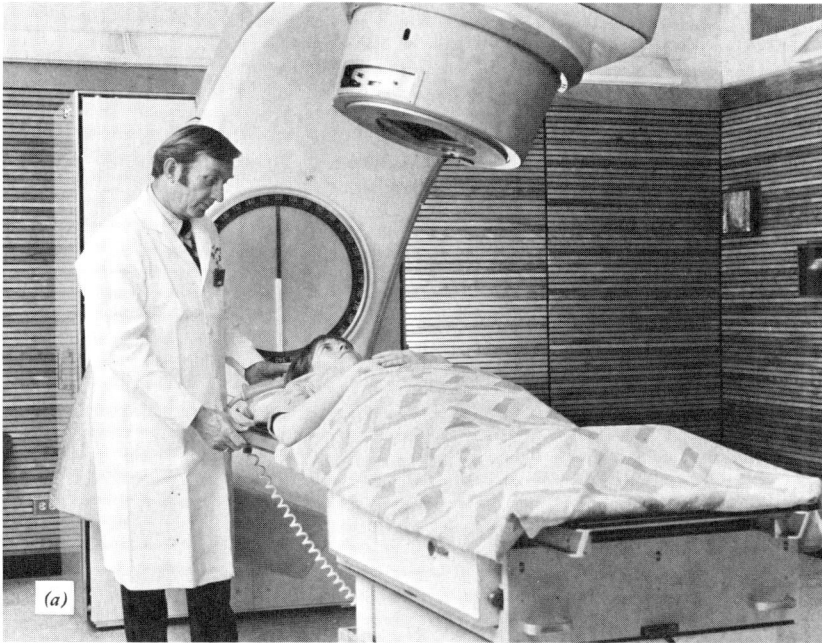

Figure 1.1 Two applications of nuclear physics. (a) Radiation therapy. (Courtesy of Central Arkansas Radiation Therapy Institute, Inc.) (b) Hiroshima after the bomb.

University of Chicago, 2 December 1942. Enrico Fermi, one of the many scientist-refugees from the Europe of Hitler and Mussolini, keeps his eyes on the quivering pen of the automatic recorder. A coworker stands before a 12-foot pile of graphite bricks interspersed with small lumps of natural uranium, his hands on the cadmium rods inserted in the pile to absorb neutrons and slow the chain reaction. At 3:20 P.M., Fermi calls for the final cadmium rod to be withdrawn an additional 6 inches. As the recorded graph climbs, excitement grows among the small crowd of participants in the strange events transpiring beneath the university stadium. The standing joke among them is: "If people could see what we're doing with a million and a half of their dollars, they'd think we are crazy. If they knew why we were doing it, they'd be sure we are."

The graph levels off.

Five minutes later, Fermi calls for the control rod to be pulled out another foot. "This is going to do it," he says. The recording pen climbs and continues to climb without leveling off. Fermi's face breaks into a broad smile. He orders the control rods replaced and the reactor shut down. Then he uncorks a bottle of Chianti. The assembled scientists drink to their success.

Today a plaque at the University of Chicago commemorates this event with the words:

> *MAN ACHIEVED HERE*
> *THE FIRST SELF-SUSTAINING CHAIN REACTION*
> *AND THEREBY INITIATED THE*
> *CONTROLLED RELEASE OF NUCLEAR ENERGY.*

1965. Total blackout on the east coast. Subways and elevators shut down without notice. Hospitals switch to emergency equipment. Not an energy shortage, really, just a temporary, local loss of power amplified by errors in the computerized east-coast electrical grid. The incident emphasizes how dependent we have become on our own creations.

1973. Embargo on Middle Eastern oil. U.S. energy consumption is reduced by 8 percent. Thermostats are lowered, travel decreases. The gross national product drops by $15 billion; new car sales are off by 35 percent; unemployment rises by 500,000; $700 million is lost in tax revenues. Five months later, the oil valves reopen.

1976. Heating fuel is in short supply. In the northeastern United States schools and businesses close, gas mains are empty and Americans relearn the virtues of warm clothes, thick blankets, and wood fires.

1978. U.S. purchases of foreign oil push the yearly trade deficit to record levels, triggering a decline in the foreign-exchange value of the dollar. The excess of imports over exports means that the American consumers are supporting more jobs abroad than foreigners support in the United States. The decline in the value of the dollar means that Americans pay more for imported goods and that domestic prices will rise. Thus our massive appetite for energy, and especially for liquid fuels, stimulates both unemployment and inflation.

1979. The price of oil jumps from $13 per barrel at the beginning of the year to over $30 at the end, further increasing the U.S. trade deficit. Political threats to the world's oil supply become visibly critical. Oil is in short supply. Gasoline lines develop across the country—the result of a nationwide 12 percent shortage of gasoline.

1981. Continued war between Iran and Iraq reduces the flow of Middle Eastern oil and contributes to the rising price of gasoline. The conflict threatens to spread beyond the boundaries of the two warring nations.

And yet, as I write these words, the national energy gluttony continues nearly undiminished. Automobiles, somewhat reduced in size, remain the preferred mode of transportation; land-use patterns continue to encourage an automobile-based society; heaters and air conditioners expend energy on leaky uninsulated buildings; enormous quantities of valuable energy-consuming products are thrown away daily.

1973. An artifact from Earth flashes past the giant planet Jupiter, cameras and spectroscopes silently recording the details of its gaseous hydrogen and helium envelope. Jupiter's gravity will swing the spacecraft in an arc around the planet and fling it on a journey beyond the solar system. Pioneer will be the first human-made object to escape our star. On board, a 6-inch-by-9-inch gold plate bearing news from Earth: the location of Earth and the date of launch written in an esoteric interstellar language. Strangest of all from the point of view of any extraterrestrial interceptor, the plate contains drawings of the weird two-legged creatures who launched the spacecraft. Traveling across the light-years at a mere 7 miles per second, the tiny vehicle will probably never enter the planetary system of any other star. Only a civilization capable of interstellar flight and able to execute interception and recovery operations in the vast regions between the stars will be able to recover the spacecraft. The chances seem small.

Powerful radio receivers at observatories in West Virginia, Puerto Rico, and Moscow tune in on the stars: Are we alone? If there is life throughout the universe, then perhaps humans are part of a great interstellar network of intelligent creatures; perhaps the human mind is a participant in a colossal drama—the drama of the universe striving to understand itself. And so, curious, lonesome creatures that we are, we search for recognizable signals from beyond our star, we send silent voyagers bearing our likeness across interstellar regions.

As the above paragraphs illustrate, science and technology are prime movers in today's world. The artist, the business executive, the political scientist, the educator, the writer, cannot hope to understand the present era without some understanding of science and, perhaps more importantly, of the processes of science. Depending on your professional speciality, you may or may not learn something from *Physics and Human Affairs* that will help you earn a living. I do hope, however, that you will learn some things that will help you live your life and that will extend your awareness of the beautiful and fearful world that surrounds you.

Thought Question 1*
Name several social/political or philosophical/religious issues related to science, other than the issues raised so far in this book.

*Most chapters contain open-ended Thought Questions and more specific Exercises. You should answer both types of questions during your reading. For further information, see To the Student in the front of the book.

To get you started on the thought questions, here are a few of the many issues related to science: What are the implications of biological research on DNA? On test-tube babies? Does science conflict with the Bible? How will scientific advances affect the communications media, and how will these changes affect our lives? Should science be controlled to ensure that it acts in the public interest?

Can you think of other science-related issues?

What Is Science . . .

What is science? Is it the search for knowledge? After all, the word stems from the Latin *scientia,* meaning "knowledge." But the Latin meaning is too broad for today's usage. The philosopher, the historian, and the poet are seekers of knowledge, yet we don't ordinarily classify them as scientists. Is it the invention of new devices and techniques: rockets, solar cells, heart transplants? We'll see below that such inventions are better classified as technology. Is it the use of computers and the manipulation of complicated mathematical tables and formulas? If so, then the occult art of astrology would have to be classified as a science.

Apparently *science* is one of those words, like *love* or *freedom,* that are difficult to define. Nevertheless (as you may have suspected!), I'm going to try to define it.

Science is the use of observation, logic, and a sense of beauty to devise general principles about the natural world.

Although this definition has a nice ring to it and would probably be accepted by most scientists, it is open to embarrassing questions. For instance, what does "natural world" mean? Well, it means the world "out there," the objective world, the opposite of the imagined or subjective world—obviously a slippery concept. I hope that in this book you will discover just how slippery, and fascinating, the natural world really is.

You may wonder what "a sense of beauty" has to do with science. I hope that later chapters, especially Chapter Two, will answer this question.

Our definition describes both the method (observation, logic, and a sense of beauty) and the object (general principles about the natural world) of science: how science studies and what it studies. Any science must fulfill both aspects of the definition. We will delve more deeply into this question later—the meaning of **science** is one of the recurrent themes of this book.

Technology can be distinguished from science. **Technology** is the application of science. For example, engineers and practicing physicians are technologists. They are more interested in constructing a particular bridge or removing a particular malignancy than in "general principles about the natural world."

We might ask whether the function of science in our society is to study the natural world or, instead, to serve technology. This is hard to answer. Some would say that the scientist must seek truth regardless of its effect on technology. In this view, science is really a branch of philosophy, and the scientist is a natural philosopher. Others would point out that pure, undirected science is rare these days, that without the drive for new technology the scientific enterprise would shrink drastically.

Thought Question 2

What do you think? Should we change our definition to read, "Science is the use of observation and logic to serve technology"?

These definitions and distinctions should be taken with a grain of salt. Definitions and distinctions help clarify our thinking, but they are necessarily a little fuzzy and arbitrary. Like the natural world, words are slippery.

... And What Is Physics?

Try giving a concise definition of these words: *geology, biology, chemistry, astronomy,* and, last but not least, *physics.*

(Pause, for thinking.)

Here, briefly, are my answers:

Geology is the study of the structure of planet Earth. (This definition has been complicated lately by the advent of lunar and Martian geology.)

Biology is the study of living organisms. (As an exercise, try to define *living*!)

Chemistry is the study of the ways atoms combine to form molecules.

Astronomy is the study of objects outside Earth.

Physics, though, is more difficult to define. Each of the preceding disciplines deals with a specific portion of the natural world (Earth, living organisms, etc.). Physics is in a different category. Rather than studying a specific portion of the natural world, physics seeks out those general principles which apply to *all* natural phenomena. As a simple example, geologists are interested in motions of Earth's crust and biologists are interested in the movement of animal groups, but physicists are interested in motion in and of itself. The geologists' principles of motion are likely to apply only to geological structures and the biologists' ideas will apply to animal movements, but the physicists' principles will apply to all objects: rocks, geese, antique Chevrolets, everything.

So, my definition of physics runs like this:

Physics is the study of those general principles underlying a broad range of natural phenomena.

It isn't enough to define physics as "the study of matter and energy," as is sometimes done. This definition is weak because the word *energy* is abstract and difficult to define and isn't usually defined until halfway through most texts (including this one), and, as we will see in Part 4 of this book, the word *matter* is even more elusive. Thus this definition isn't terribly meaningful.

Physics stems from the ancient Greek term *physis,* meaning "the essential or ultimate nature of reality." Today physics has a more restricted meaning than the study of *physis* had to the philosophical and mystical Greeks of the sixth century B.C. However, the term *physics* still carries with it the connotation of an idea-oriented, or philosophical, enterprise, which seeks the general principles (or essential nature) behind phenomena.

Thought Question 3

Which of the following fields are scientific disciplines, and why? Music, religion, history, political science, anthropology, psychology, medicine, mathematics.

Here are my answers. Take them with a grain of salt as there is a wide range of reasonable opinion on these matters.

Music and religion are not sciences because they are not primarily devoted to the study of natural phenomena (experimentally observable phenomena).

History, political science, anthropology, and psychology are difficult to classify. Some argue that these are sciences because they attempt to establish general principles, whereas others argue that these fields are not sufficiently based on experimentally observed phenomena and verifiable predictions to be classified as sciences. To indicate their partially scientific nature, they are called social sciences.

Medicine is not a science. It is partly a science-based technology, partly an art.

Mathematics is not a science, as it is not based on observation. In fact, mathematics is not concerned directly with the natural world at all. Mathematics deals with an "imagined world" of ideas invented by the human mind and manipulated by the rules of logic. Mathematical ideas such as numbers and geometrical figures are useful in dealing with the natural world. Scientists use these mathematical inventions in describing scientific ideas; that is, mathematics is used as a language in science.

Note that every one of these fields, including such obvious nonsciences as music and religion, uses at least some of the hardware, or methods, of science. Much contemporary music is based on electronic devices and an understanding of the physics of sound; much religious writing is tightly organized logic and is often concerned with natural phenomena; psychology and other social sciences make wide use of experiment and observation; and the mathematician's inventions are often suggested by natural phenomena. There is probably a little science in every field of human endeavor. On the other hand, even such so-called pure sciences as physics and chemistry employ many nonscientific methods and points of view. We will see throughout this book that intuition, aesthetics, and other humanistic elements play a large role in the development of science. Obviously, there is no clear boundary between the sciences and the nonsciences.

Thought Question 4
What is your opinion of my answers to Thought Question 3?

If your field is, for example, music, history, or religion, I hope that you aren't offended by the fact that in my opinion your field might not be "scientific." Don't succumb to the mistaken notion that every valid field of endeavor must ape the methods of science. Through such successes (?) as the A-bomb and orbiting satellites, science and technology have taken on so much prestige these days that there is a tendency for workers in every field to put on some of the trappings of science in order to validate their work. I hope it will be evident throughout this book that, although science has its advantages, it also has its limitations. There is no reason that every field of endeavor has to be scientific in order to be respectable. The nonscientific vantage points of religion or the arts, for example, are also valuable. It is, in fact, only through a serious consideration of all such vantage points that we can have any hope of arriving at some approximation of the truth.

2

THE ART OF SCIENCE

This chapter is about the **scientific method**—the methods scientists use. This topic is more significant than many people realize, especially in the interdisciplinary areas related to science. For instance, consider such science-and-society questions as the energy crisis, or such science-and-religion questions as creationism versus evolution. Any serious consideration of these topics must stem from an understanding of the limitations and validity of scientific knowledge, in other words, from an understanding of the methods of science.

The scientific method is sometimes defined as simply a certain sequence of activities, for example, observation, hypothesis, testing, theory, prediction, experimental checking. But any study of the actual history of science shows that science's methods are much more subtle and interesting than this. So I would like to introduce the scientific method by means of a historical example: the development of our picture of the place of Earth in the universe. This example is interesting and informative in its own right, and it provides illustrations of most aspects of the scientific method.

The scientific method is a central theme of this book. It is perhaps the most important lesson we can learn about science.

Ptolemy's Idea

Human beings have always been fascinated by the stars. They are beautiful and apparently eternal; they guide mariners in their travels; they used to help us determine the time to plant and to reap. We feel that by studying the heavens we can perhaps discover our own place in the scheme of things.

One of the earliest theories of **cosmology,** the study of the large-scale structure of the universe, comes to us from the early Greeks. Greek cosmology culminated in about 130 A.D. with the work of the great astronomer Claudius Ptolemy (Fig. 2.1).

The Greeks used common sense in setting up their cosmological scheme. The heavenly bodies (sun, stars, planets, moon) apparently go around Earth roughly once each day, and Earth is at rest in the middle. What could be more obvious? Surely the large and ponderous Earth on which we stand doesn't move! The intuitive notion of a fixed Earth and a moving sun is embedded in the very words and phrases we use: sunrise, sunset, the sun stood still over Jericho.

But the Greeks went far beyond these intuitive, **qualitative** (nonnumerical) notions. By Ptolemy's time, Greek cosmology had evolved into a beautiful and subtle **quantitative** (mathematical, numerical) theory, agreeing quite well with detailed measurements of the positions of the heavenly bodies and based on deep philosophical and aesthetic considerations.

The Greeks loved geometry. Plato and the esoteric band of philosopher-mystics known as the Pythagoreans (you might be familiar with their Pythagorean theorem)

FIGURE 2.1 Four giants of early astronomy. (a) Ptolemy. (The Bettmann Archive.) (Figures 2.1 b-d appear on following pages.)

regarded such geometrical objects as the square and the triangle as the purest examples of Platonic "ideal" objects. And among all possible geometric shapes the shape that stood out as the most nearly perfect was the circle. Unending, perfectly symmetric, the circle was, for them, the most ideal object in the universe of ideas.

The Greeks also loved the stars. Of all the objects in the observable universe, they felt that the stars, floating noble and eternal above this lowly Earth, had the greatest claim to perfection.

In Aristotle's theory of the physical universe, this imperfect world was made of the four base elements Earth, Air, Fire, and Water, and the heavens were composed of an entirely different kind of element, the Ether, or quintessence, (from the Latin term for "fifth element"), a pure and immutable substance befitting the stars. Each of the elements had its own particular type of natural motion and natural place: Earth was at the bottom or center of the universe, then, in ascending order, came Water, Air, and Fire, with Ether occupying the heavens. The natural motion of Ether, hence also of the heavenly bodies, was, of course, that most beautiful of all possible shapes, the circle.

Thus were the Greeks led, by common sense, aesthetics, and philosophy, to a cosmological theory in which heavenly bodies revolved in circles around a fixed Earth.

It should be mentioned that some Greek thinkers dissented from this view. Aristarchus maintained that a reasonable world system would result if one imagined

FIGURE 2.1 continued. (*b*) Copernicus. (AIP Neils Bohr Library.)

that the sun stood still at the center, circled by Earth and planets. But this sun-centered theory seemed unphilosophical and unpleasant to most Greeks, and it attracted few adherents.

Any attempt to set up a detailed, quantitative, Earth-centered theory ran into several complications. The most obvious problem concerned the five planets, or "wanderers," then known. The stars and the planets both circle Earth approximately once each day, but the planets do not keep precise step with the motion of the stars. Instead, the planets wander very slowly among the stars, occasionally even reversing the direction of their wandering.

Figure 2.2 illustrates this planetary wandering. In Figure 2.2 the dots represent the positions of a particular planet (Mars, for instance) as observed on the first of the month over a 7-month interval. The stars represent the background of "fixed" stars. The stars are said to be fixed because during the span of a human life their relative positions change very little. The general motion of the planet, as viewed against the background stars, is toward the right. However, during August and September, in this particular example, the planet moves toward the left, opposite the usual direction of motion. This backward motion is called ***retrograde motion.*** After about 2 months of retrograde motion the planet again reverses direction and resumes its previous course from left to right. Keep in mind that a planet's wandering is very slow and can only be observed over an interval of several nights. During a single night, all planets keep approximate step with the east-to-west motion of the stars.

FIGURE 2.1 continued. (c) Tycho Brahe. (The N.Y. Public Library Picture Collection.)

FIGURE 2.1 continued. (d) Kepler. (AIP Neils Bohr Library.)

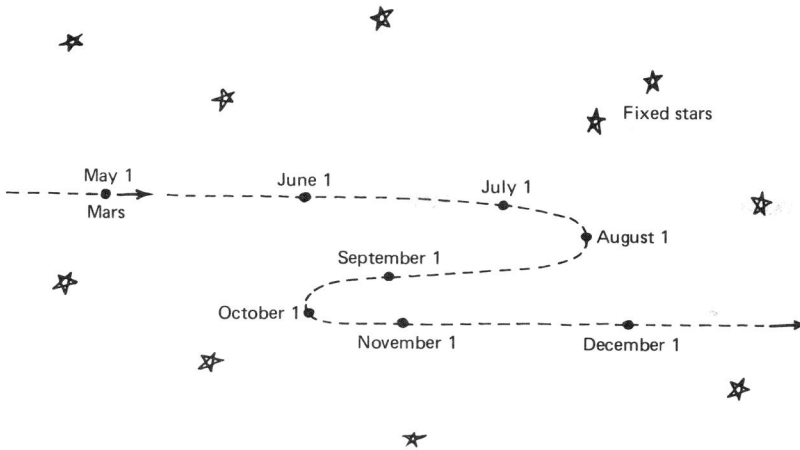

FIGURE 2.2 The wandering of Mars among the fixed stars.

An activity for some clear night would be to follow the stars as they move across a portion of the sky. Locate a particular star lying near the eastern horizon and another lying near the western horizon and then make observations of these two stars at 10-minute intervals. Note that the eastern star rises and crosses the sky toward the west, while the western star sets below the horizon. Note that all the heavenly bodies maintain fixed positions relative to one another during their nightly motion from east to west.

As a follow-up to the previous activity, locate a planet in the sky. Your physics or astronomy department or an astronomy magazine can tell you how to find one.* Observe this planet once each night over several weeks, noting that it appears in a slightly different position among the stars each night. Also note that the relative positions of the stars do not change.

Greek astronomers employed various subtle and beautiful ideas to develop a theory that explained the wandering of the planets, agreed quantitatively with the observations known at that time, and was based on the Greek vision of the heavenly bodies orbiting in eternal circles around a fixed Earth. This theory received its ultimate expression in the work of Ptolemy. The Ptolemaic scheme is a lovely example of a physical theory, and worthy of a somewhat detailed description despite the fact that it has been considered outmoded for the past several centuries.

Ptolemy assumed that all the heavenly bodies were part of a celestial sphere rotating around Earth once each day. Because the stars were assumed to be fixed to the inside of the surface of this sphere, they circled Earth once each day, in agreement with the observed facts. The sun, moon, and planets all participated in this motion of the celestial sphere, but since these bodies didn't quite keep step with the stars they were each assumed to move in separate, slow, circular motions that were in addition to the more rapid daily motion of the celestial sphere. A simple, unvarying circular motion of the planets with Earth as center wouldn't do the job though; for instance, retrograde motion cannot fit into such a simple scheme. So Ptolemy's system had

*Mars or Venus will work best, because they usually wander fastest. Of these two, Mars is preferable because Venus is usually not up for a long enough time in a dark sky.

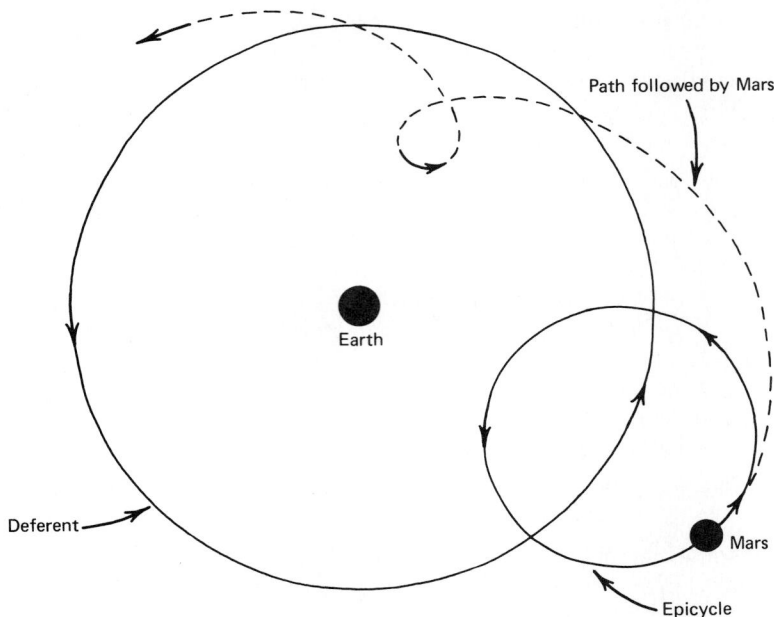

FIGURE 2.3 Ptolemaic system for a typical planet, such as Mars. Compare Ptolemy's final improved version of this system, Figure 2.5.

each planet moving in a combination of two circular motions. As shown in Figure 2.3, a typical planet, such as Mars, is attached to a small circle called the ***epicycle.*** The center of the epicycle moves along a second circle, called the ***deferent,*** having Earth at its center, while the epicycle itself (with Mars attached) rotates. Thus Mars moves in a circle around a point that moves in a circle centered on Earth. Figure 2.3 shows the loop-the-loop orbit that results from all of this. All the planets move around Earth in similar paths with their own epicycles and deferents, while the sun and moon move in ordinary circles (Fig. 2.4). With a suitable choice for the size and rate of rotation for each epicycle and each deferent, the Greeks found that they could obtain a theory that agreed almost precisely with measurements of the positions of the sun, moon, and the five then-known planets: Mercury, Venus, Mars, Jupiter, and Saturn.

Exercise 1*
Explain how one gets retrograde motion out of this scheme.

But wait! Theory and observation agreed almost, but not quite. There were small, but definite, disagreements, and the observations were carried out with sufficient care that the discrepancies could not be explained away as experimental error. Back to the drawing boards.

Ptolemy was not about to throw out the beautiful and nearly correct Earth-

*Most chapters contain specific Exercises and more open-ended Thought Questions. You should answer both types of questions during the reading. You'll find the answers to the exercises either in the back of the book or, when "Answer below" is stated in the exercise, in the text following the exercise. Be sure to try each exercise yourself before checking with my answer.

centered theory. To do so would have amounted to rejecting the philosophically satisfying and widely accepted foundations on which it was based. Instead, he did what most people do when faced with a discrepancy of this sort: he tinkered with the theory, making several small changes in order to obtain a theory that did agree with the observations while inflicting only minor damage on the original conception. Ptolemy made two changes in the previous theory (Fig. 2.5): (1) he moved the center C of each deferent a small distance away from Earth; and (2) he assumed the angular rate of rotation (the number of angular degrees covered per second) to be constant provided that the angle was measured not from the center of the circle but rather from a point called the **equant,** point Q in Figure 2.5.

With all these devices at his command, Ptolemy was free to play with the radii and rates of rotation of the epicycles and deferents and with the positions of the deferent centers and of the equants for each planet. Adjusting all of these quantities appropriately, Ptolemy obtained a scheme (or model, or theory) that fit all of the known observations to within the observation accuracy obtainable at that time.

Let me reemphasize: Ptolemy's theory was an impressive, accurate model that fit all known observations. It was scientifically correct. For the next 14 centuries, astronomers and astrologers and navigators based their calculations and predictions on Ptolemy's system and obtained results that agreed quite well with the observational

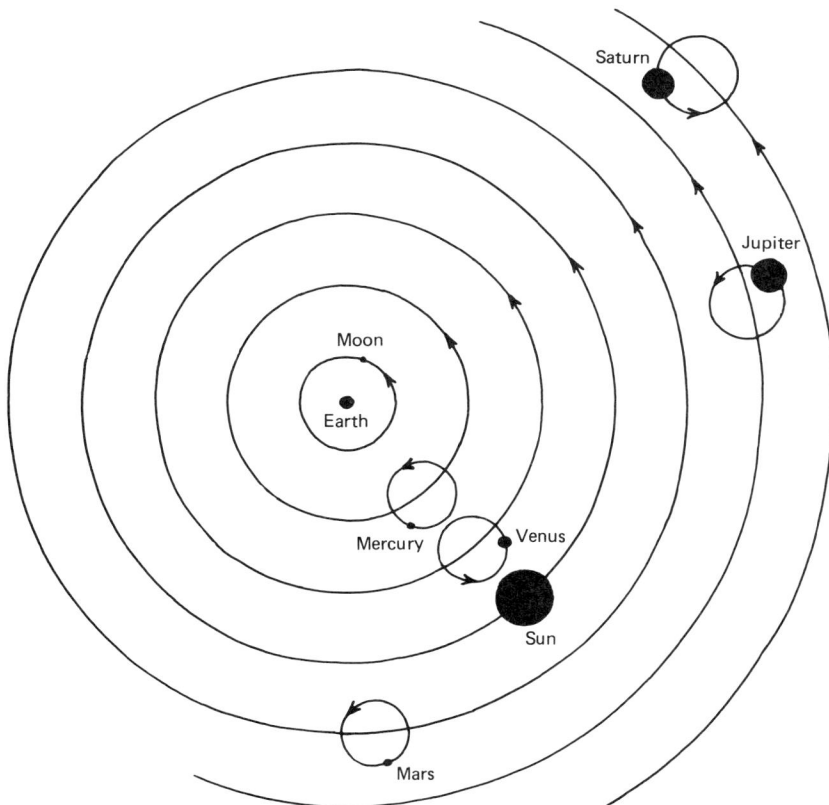

FIGURE 2.4 Schematic diagram of the Ptolemaic system.

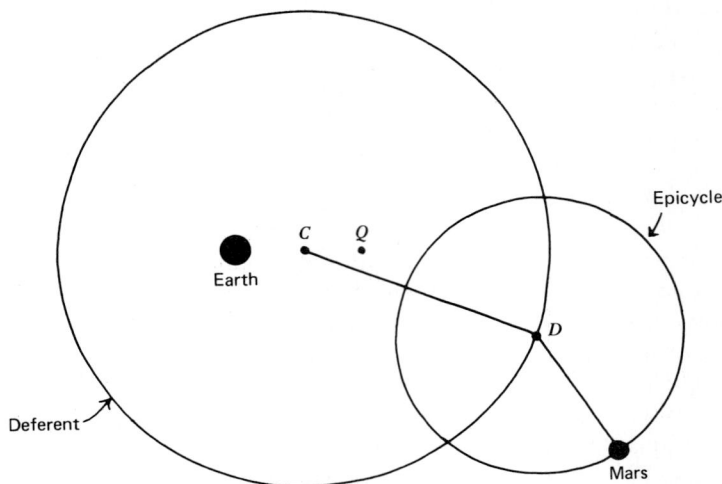

FIGURE 2.5 Ptolemy's improved version of the Earth-centered system for a typical planet, such as Mars. Earth is no longer at the center C of the deferent. Furthermore, the center D of the epicycle no longer rotates at a constant rate around C, but instead has a uniform rate around the equant, Q, which is some distance from C.

data. Ptolemy's system is thus one of the most successful scientific theories ever invented.

Thought Question 1

To what extent would you say the Ptolemaic theory is based on philosophical and aesthetic criteria, and to what extent on observational criteria?

Copernicus's Idea

During the thirteenth century A.D., Saint Thomas Aquinas linked Aristotelian philosophy with Christian theology. Ptolemy's theory, which rested on the aesthetics and philosophy of such ancient Greeks as Aristotle, was accorded a special place in church teaching. By 1500 A.D., Ptolemy's theory was in a strong position indeed.

But the fifteenth and sixteenth centuries were times of change in Europe. The intellectual and artistic flowering known as the Renaissance germinated in Italy in the fourteenth century, spreading to all of Europe within 200 years. Martin Luther led a frontal assault on church authority; Christopher Columbus took a memorable voyage.

Enter Nicolaus Copernicus (1473–1543), astronomer, mathematician, linguist, physician, lawyer, politician, economist, churchman. Copernicus didn't like Ptolemy's system. Not that Copernicus was a revolutionary. Quite the contrary. Copernicus objected to Ptolemy's theory on the grounds that, by using complicated circles on circles, by placing the centers of the different orbits at different points in space, and particularly by the introduction of the equant device (according to which the planets no longer revolved at angular speeds that were constant as measured from the centers of their orbits), Ptolemy had strayed too far from the original simpler and nobler conception of circular orbits having uniform rates around a common center.

Listen to Copernicus on his objections to Ptolemy's model:

> ...the planetary theories of Ptolemy and most other astronomers, although consistent with the data, seemed ... to present no small difficulty. For these theories were not adequate unless certain equants were also conceived; it then appeared that a planet moved with uniform velocity neither on its deferent nor about the center of its epicycle. Hence a system of this sort seemed neither sufficiently absolute nor sufficiently pleasing to the mind.
>
> Having become aware of these defects, I often considered whether there could perhaps be found a more reasonable arrangement of circles, from which every apparent inequality would be derived and in which everything would move uniformly about its proper center, as the rule of absolute motion requires.

Copernicus thought that any type of celestial motion other than uniform circular motion was obviously impossible. As he put it, ". . . the intellect recoils with horror" from any other suggestion; ". . . it would be unworthy to suppose such a thing in a Creation constituted in the best possible way."

Thought Question 2
Would you say that Copernicus's disagreement with Ptolemy was basically religious, or philosophical, or aesthetic, or scientific?

In other words, Copernicus just didn't like Ptolemy's scheme. So he tried another point of view. He imagined himself standing on the sun. How would the celestial motions appear from this vantage point? He proposed a model in which all the planets, including Earth, move in uniform circular motion around the sun as common center (Fig. 2.6),* and in which Earth spins on its own axis once each day. In Copernicus's theory, the daily rotation of the sun, moon, stars, and planets was caused by the daily spinning of Earth, while the much slower wandering of the planets, moon, and sun among the stars was caused by the circular motion of Earth and planets around the sun. As mentioned above, Ptolemy's theory had proven capable of predicting future planetary positions to within the observational accuracy obtainable at that time. By choosing appropriate sizes and rotation rates for the planetary orbits, Copernicus was able to obtain this same accuracy for his own sun-centered theory.

Note carefully: The schemes of Ptolemy and Copernicus *both* agreed with the data, to within experimental accuracy; there was no observational reason to prefer one over the other.

Thus were Ptolemy's displaced centers, equants, and nonuniform rotations swept away and replaced with the more primitive Greek conception of uniform rotations along simple circles. Very slick. But there was one slight problem: Copernicus's scheme required that Earth, this gross and sluggish body on which we stand and which is obviously in no way comparable to the stars and planets, *move* through the heavens. Worse yet, in order to obtain the apparent daily rotation of the heavens, Copernicus required Earth to spin about a north-south axis once each day.

Ptolemy's theory has a simple explanation for retrograde motion of the planets (see Exercise 1). How does Copernicus explain this phenomenon? The next time you

---

*I should point out for historical accuracy that, although Copernicus managed to get rid of the despised equants of Ptolemy's theory, he did find it necessary to mar the beauty of his own theory with a few small epicycles and slightly displaced centers in order to obtain agreement with experiment. Thus Figure 2.6 is simplified.

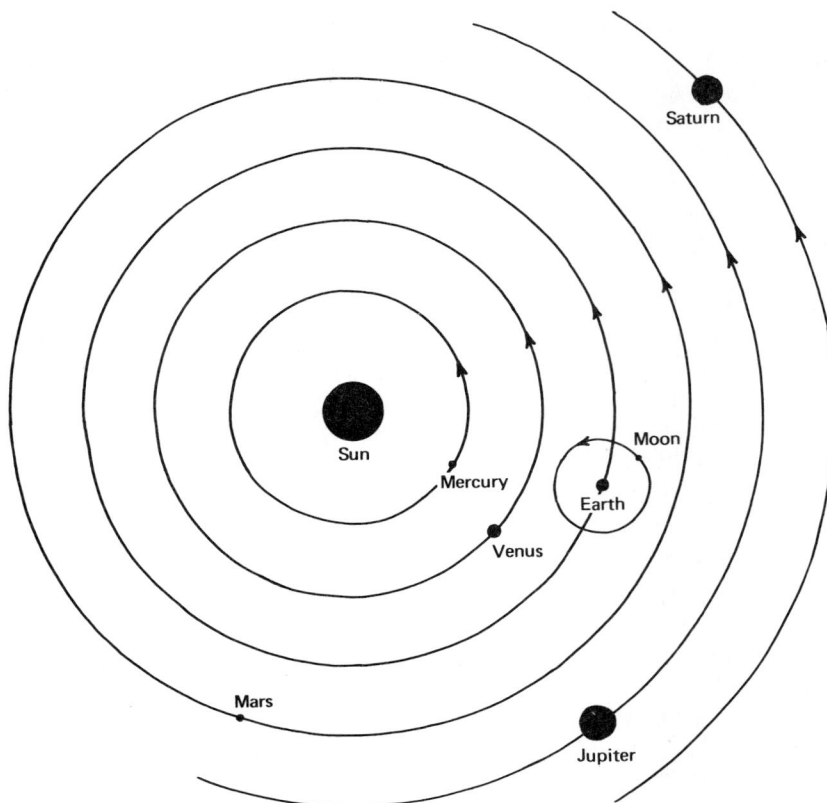

FIGURE 2.6 Schematic diagram of the Copernican system.

are a passenger in a car on the highway, notice the apparent motion of any car that your car passes, as seen against the background trees and houses. As you pass, the slower car appears for a few seconds to move backward, with respect to the background, simply because of the rotation of your line of sight. The same phenomenon occurs as Earth passes, say, Mars. As seen against the background stars (our only reference point for astronomical observations), Mars appears to move backward for a short time.

Copernicus's major work was finally published in 1543, the year of his death. In the world of the sixteenth century there were few reasons to support the new theory and many reasons to reject it. True, the Copernican scheme looked a little simpler on paper. However, its agreement with observations was no better than that of Ptolemy's theory (it was actually worse in some cases), and it upset the testimony of our senses that it is Earth which stands still while the heavens move. Above all, the new system undermined the elaborate philosophical basis of the Ptolemaic theory and hence struck at the Aristotelian underpinnings of church theology. For example, Aristotle had taught that the heavenly bodies were composed of a special substance, Ether, whose natural movement was eternal circular motion, while Earth was composed of other substances, whose natural motion was entirely different. Under Copernican cosmology this whole beautiful scheme was lost. Furthermore, Copernicus offered nothing to replace the lost Aristotelian physics, no underlying physical explanation as to why the planets moved around the sun and why Earth moved as it did. It would be more than a century before Isaac Newton would invent such an explanation.

Thought Question 3

In light of all these objections to the sun-centered theory, why in the world do you suppose Copernicus dreamed it up?

Thought Question 4

At this point in our story, who would you say was right, Ptolemy or Copernicus? Or neither? Or both? (Answer as if you were an objective observer living in the sixteenth century.)

Tycho's Observations

Tycho Brahe, born 3 years after Copernicus's death, loved to look at the night sky. Night after night for years he peered up at the stars, sighting along a device that was a glorified version of a long stick with a nail stuck upright at each end (the telescope hadn't been invented), patiently recording the angular positions of the planets with unheard-of precision (Fig. 2.7). His data often were precise to within a half-minute of arc (a **minute of arc** is 1/60 of an angular degree), an accuracy 20 times better than the 10 min of arc of the ancient Greeks or of Copernicus.

Tycho's improved data proved to be rather upsetting. Recall that both the Ptolemaic and the Copernican systems agreed with the data to within the accuracy of the observations made prior to Tycho's work. But Tycho's improved methods narrowed the experimental error considerably, and thus allowed less margin of error for the predictions of the Ptolemaic and Copernican theories. In fact, it was found that neither Ptolemy's nor Copernicus's predictions fit the new data to within Tycho's experimental accuracy.

Kepler's Idea

Enter Tycho's assistant, Johannes Kepler (1572–1630), mystic, philosopher, mathematician, astronomer—an intriguing individual who was at once a holdover from the medieval world and a forerunner of the modern world. Kepler loved geometry:

Why waste words? Geometry existed before the Creation, is coeternal with the mind of God, is God Himself (what exists in God that is not God Himself?); geometry provided God with a model for the Creation and was implanted into man, together with God's own likeness . . . and not merely conveyed to his mind through the eyes.

And he loved the sun:

The sun in the middle of the moving stars, himself at rest and yet the source of motion, carries the image of God the Father and Creator. . . . He distributes his

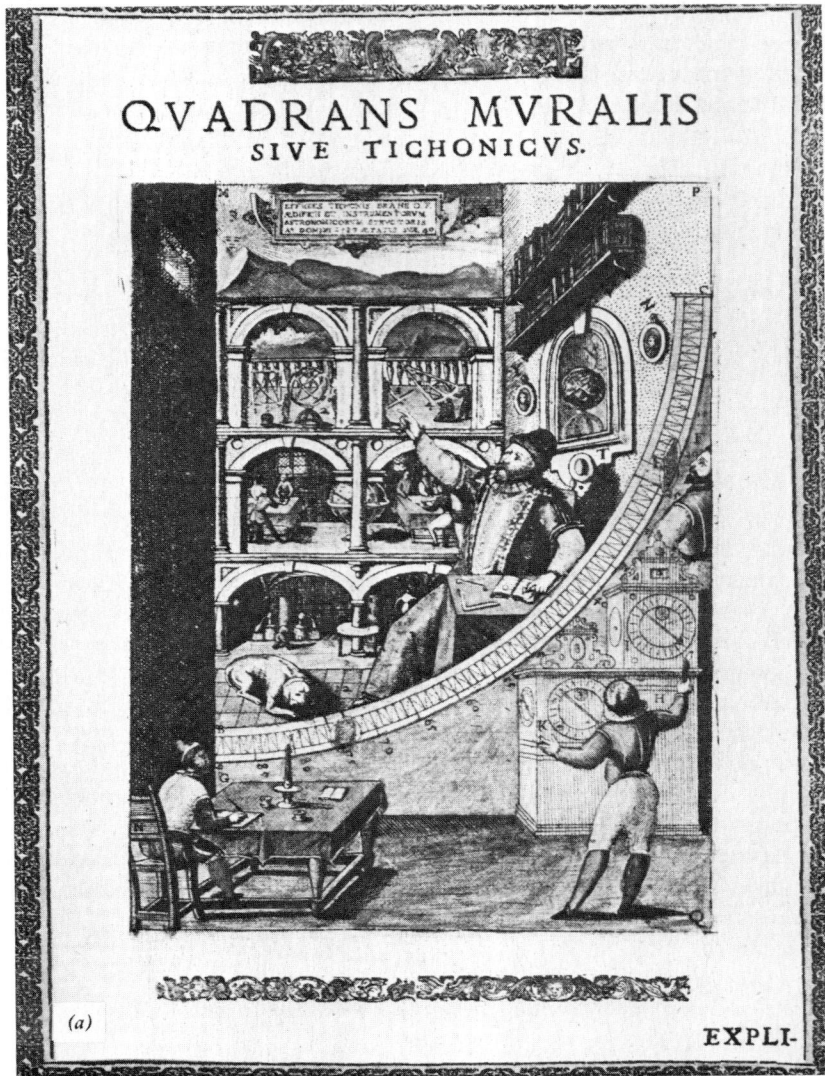

FIGURE 2.7 (a) Tycho Brahe studying the heavens from his Danish island observatory. (AIP Neils Bohr Library.)

motive force through a medium which contains the moving bodies even as the Father creates through the Holy Ghost. (Quoted in *The Watershed,* Arthur Koestler, Doubleday, 1960.)

It was natural, then, that Kepler's aim in life should be the perfection of the sun-centered theory, a theory whose beauty he contemplated with "incredible and ravishing delight."

Kepler realized that Copernicus's circular orbits didn't fit Tycho's data. But perhaps other circular orbits, differing slightly from the circles chosen by Copernicus, *would* fit the data. Kepler undertook the mathematical search for such orbits. After 4 years of theoretical calculations involving only the orbit of Mars, Kepler's project

ARMILLÆ AEQUATORIÆ MAXIMÆ
SESQUIALTERO CONSTANTES CIRCULO

FIGURE 2.7 continued. (b) Tycho's sextant for measuring the positions of the planets. (AIP Neils Bohr Library.)

ended in failure. The circular orbit coming closest to Tycho's data was still off by some 8 min of arc, an error many times greater than the experimental error in Tycho's data.

Kepler knew that Tycho's data was correct, so the Copernican theory had to be wrong. But Kepler felt that such a lovely theory, one agreeing so fully with his own philosophical presuppositions, couldn't be *far* wrong. Writing that "upon this eight minute discrepancy, I will yet build a theory of the universe," Kepler embarked on a 16-year search for those small improvements in the Copernican theory which would bring the theory into agreement with Tycho's data.

A lesson in the methods of science: Had Kepler been of a different philosophical bent, he could have spent most of his life trying to find small revisions in the Ptolemaic system in order to bring that theory into agreement with Tycho's data. Kepler could surely have found new equants, or extra circles within circles, or other devices, to obtain a revised Earth-centered theory that would fit the data. But for Kepler the way lay with Copernicus and the sun, not with Ptolemy and the earth. Even today, with telescopic data, laser ranging data, satellite fly-by data, and the like, yielding information exceeding Tycho's wildest dreams, we could find enough circles on circles, displaced centers, and so on, to fit a Ptolemaic-like Earth-centered system to the data. It is still possible to maintain a belief in a suitably revised Ptolemaic system of the universe!

Thought Question 5

To repeat the previous discussion question, which system is "true"? Why do we use the sun-centered system today? Reflect on the meaning of *scientific truth* in light of what you have read so far.

After many false starts, Kepler finally found a simple, harmonious, and quantitatively accurate sun-centered system. Briefly, he discovered that he could obtain agreement with all of Tycho's data, to within experimental error, if he assumed that each planet moved in an appropriate **ellipse** around the sun.

Now, an ellipse is a rather beautiful object. Let me describe it to you, by means of an activity you might enjoy carrying out. Stick two thumbtacks into a board and loop a piece of string around the tacks, as shown in Figure 2.8. Now, stretch the string tight with a pencil and draw a complete "orbit" around the tacks, as shown. You've just drawn an ellipse. The points where the thumbtacks are inserted are each called a **focus** of the ellipse.

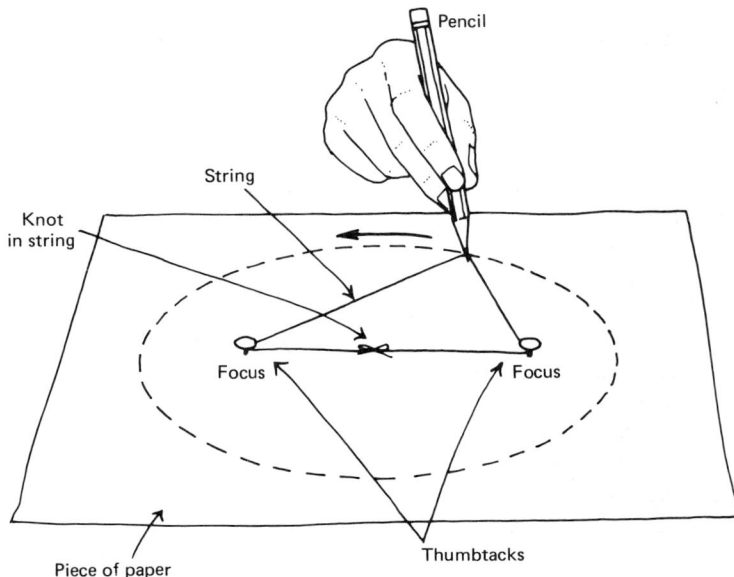

FIGURE 2.8 A method for constructing an ellipse.

Exercise 2

Is a circle a particular example of an ellipse? Explain how to get a circle out of the tack and string construction.

Kepler's elliptical orbits are very nearly circles, as indeed they must be if Copernicus's circular orbits were actually a good approximation to Tycho's data.

In addition to the notion of elliptical orbits, Kepler found several other satisfying relations. For example, the sun occupies one focus of each ellipse (there is nothing at the other focus). Kepler summarized his theory in three principles of planetary motion. The first of these just recapitulates what I've said above: Each planet moves in an elliptical orbit about the sun, with the sun at one focus. The second principle is a mathematical statement about the variation in speed of any planet as it moves along its elliptical orbit (it turns out that planets speed up as they come closer to the sun and slow down as they move farther away). Kepler's third principle relates the size of a planet's orbit to the time it takes for that planet to orbit the sun. These beautiful and quantitatively precise relations express that harmony of the spheres which the mystical soul of Kepler had always known must exist.

Thought Question 6

Many textbooks speak of the "discovery" of Kepler's principles of planetary motion, as though these principles exist in nature independently of the human mind and are "found" by scientists. Would you say that Kepler "discovered" such principles as the elliptical planetary orbits, or might it be more appropriate to say he "invented" them?

Kepler was elated:

What sixteen years ago I urged as a thing to be sought, that for which I joined Tycho Brahe . . . at last I have brought to light and recognize its truth beyond my fondest expectations. . . . The die is cast, the book is written, to be read either now or by posterity, I care not which. It may well wait a century for a reader, as God has waited six thousand years for an observer.*

The Copernican Revolution Today

Despite the enormous amount of new astronomical data obtained since Kepler's day, Kepler's scheme for the solar system still fits reasonably well. Minor adjustments are needed to bring Kepler's theory of planetary motion into line with the voluminous and accurate observational data now at hand, but these adjustments don't damage the basic structure, nor the simplicity, of Kepler's creation. Kepler had the aesthetic-prophetic-mathematical intuition to devise a scheme that was not only accurate and harmonious and simple at the time he invented it, but that remained so and that required only minor adjustments in the face of future findings.

Although Copernicus was no revolutionary by nature, the upheaval he initiated is justly called the Copernican revolution. Since Copernicus, we humans no longer stand at the center of the universe, surrounded by perfect and eternal heavenly objects. We stand now on the rather inconspicuous third planet in a series of nine planets orbiting the sun. The revolution has continued. According to contemporary astronomy, the sun is one very ordinary star in a collection of some 100 billion stars

*The observer referred to is Kepler's mentor, Tycho Brahe.

FIGURE 2.9 A typical galaxy, similar to our own Milky Way galaxy. Our sun is just one star among the billions of stars in our galaxy. (Palomar Observatory, California Institute of Technology.)

known as the Milky Way galaxy. There is more: The Milky Way galaxy is just one galaxy among a universe of at least 100 billion other galaxies, a universe that existed for billions of years before the earth and sun were born and that will survive our planet's death by more billions of years (Figs. 2.9 and 2.10).

The Copernican revolution has flourished in other, more general ways. Biologically, *Homo sapiens* appears as just one animal among countless species of animals. Biochemically, we appear to be merely well-organized collections of atoms and molecules, subject to all the laws that all other atoms and molecules obey. You see, post-Copernican science seems to have destroyed much of humankind's uniqueness. At least from the scientific point of view, we seem less special than we once did.

FIGURE 2.10 Our Milky Way galaxy is just one among billions of galaxies in the universe. Galaxies often occur in clusters, as in this photo of the Hercules cluster. How many galaxies can you count in this photo? (Palomar Observatory, California Institute of Technology.)

Indeed, this viewpoint has sometimes been carried to extremes in the twentieth century to picture the human race as a trivial, absurd sidelight in a universe that runs mindlessly onward like a cosmic computer, computing the destinies of all of us and the fates of all the stars.

Obviously, you are not obligated to accept the above mechanistic, impersonal view of reality. Other world views are possible. You might argue that science is just one of several viewpoints toward reality. For example, the human condition might be described more accurately by such artistic works as Shakespeare's *Hamlet* or Beethoven's *Ninth Symphony* than by biochemical theories of biological structure. Or you could argue that the importance of the human race in the overall scheme of things is not affected by the position Earth happens to occupy in the physical universe. Or that science has no legitimate role in making value judgments such as those stated in the preceding paragraph.

The grim vision of humanity displaced from the center and of the world as machine is sometimes called the Newtonian world-machine, stemming as it does from the *mechanistic* ("like a machine") *physics* of Isaac Newton. The title is appropriate, despite the fact that Newton himself would not have subscribed to such a world view. The Newtonian age, the period dominated by the mechanistic physics of Newton, was initiated by the Copernican revolution. Figure 2.11 captures the essence of the transition to the Newtonian age.

One of the themes of this book is that, from the scientific point of view, the Newtonian age is over. You will discover in later chapters that the conceptual underpinnings of the mechanistic world view were destroyed by new findings in physics around the turn of this century. The new physics of relativity and quantum theory suggests more open and humane world views.

On Scientific Truth

You are, I hope, convinced by now that scientific research is not simply a matter of finding "the facts." The processes and results of science are more subtle and more interesting than this. The major lesson of this chapter is that there is no simple scientific method. Science is a very human activity. Like all human activities, any neatly formulated description of it will be an oversimplification. Nevertheless, guided by the

FIGURE 2.11 The pre-Copernican and the post-Copernican world. (The Bettman Archive.)

history described in the preceding pages, let's try to isolate some of the features of science.

One of the most characteristic features of science is the scientific theory or scientific principle. A **scientific theory** is an idea, or collection of interconnected ideas, that is fundamental or basic in the sense that it "explains" a fairly broad range of observed phenomena. Note that a theory is an idea, a mental construct invented by human beings. The theories of science are in a different category from the facts (the specific observations) of science. A theory is not, and can never be, a fact, regardless of how widely the theory is accepted. Several words are used interchangeably with *theory*—*law, principle, model*—they all mean about the same thing. I prefer *theory* or *principle* to the word *law*, as *law* sounds too absolute. Because the ideas of science are always tentative and often of only limited validity, *law* seems inappropriate. In this book, we'll use the words *theory, model* and *principle* in preference to *law*. However, there will be one exception: Newton's principles of motion were so firmly entrenched for so long that everyone speaks of them as laws, even despite the fact that they are now known to have limited validity.

Kepler's three principles of planetary motion are a good example of a scientific theory. These principles explain all of Tycho's observations of the planets, not to mention an incredible number of other observations made before and after Tycho, in the following sense. If we assume that the planets always move according to Kepler's laws, then it follows that those planets will be found in the places where they actually were observed by Tycho. That is, we can deduce, or predict, the specific observations of Tycho, if we assume that Kepler's theory is true.

Thought Question 7

Since Tycho's data actually did agree with the predictions of Kepler's theory, did Tycho's data prove that Kepler's theory was true?

This thought question has a definite answer: Tycho's data does not prove that Kepler's theory is true, since there is no way we can observe all the planets, for all time past and future, in order to ascertain that Kepler's predictions are always valid. For instance, perhaps Mars moved completely out of its Keplerian orbit last night at midnight, did a quick loop-the-loop around Jupiter, and jumped back into its standard orbit, only nobody was looking so nobody noticed that Kepler's theory was violated. More plausibly, perhaps Mars will deviate from its Keplerian orbit because of the temporary influence of a passing asteroid (a large rock orbiting the sun). Do you see? There is no way to prove the truth of a scientific theory. No amount of observation will do the job. Scientific theories are always tentative and can never be established as fact.

Here's another argument, also showing that scientific theories are unprovable. The Ptolemaic and Copernican theories were both capable of explaining the experimental data known prior to Tycho. This is the general state of affairs in science. It is always possible to devise several different theories to explain any set of observed data. No matter how much data we collect, we will never be able to use that data to establish the validity of one theory and the invalidity of all possible competing theories.

Although observations cannot prove a theory, observations that agree with the predictions of a theory obviously lend support to the theory. We have much more faith in a theory after a large number of its predictions have been found to agree with observations. For example, the Greeks and medieval Europeans made numerous observations of the planets, and they invariably found that these observations enhanced scientists' confidence in Ptolemy's theory.

Thought Question 8
Can we disprove a scientific theory?

The answer: Yes, we can disprove a theory. A single, clear-cut observation that contradicts the predictions of the theory can suffice to disprove the theory. For instance, the theory of Ptolemy and also the theory of Copernicus were disproved by Tycho's data. Scientific theories lead a precarious life, never capable of being proved but always in danger of that single observation that will disprove them.

Theories are usually not disproved and discarded easily. In practice, when a widely accepted and previously successful theory is confronted with new observations contradicting the theory, many scientists try to preserve the theory by either disputing or ignoring the observations. Some of the scientists of Kepler's day, for example, were unwilling to look at the heavens through the newly invented telescope, because the telescopic evidence contradicted the Ptolemaic theory. They ignored the telescopic evidence or disputed the validity of the new instrument.

But if the observations cannot be ignored or dismissed in this manner, then we expect that the old theory will be rejected, right? Wrong. In practice, scientists don't junk the old theory, especially if it is well loved and of venerable age and good reputation. Rather than consign the old theory to the graveyard, scientists try to patch it up with minor surgery to cure it of its disagreements with observation. Thus Ptolemy performed successful minor surgery on the Greek Earth-centered theory when he introduced the equant device to clear up several small discrepancies between theory and observation. Kepler performed minor surgery (although Copernicus would have considered it major, since it gave up the precisely circular orbits that seemed to him so essential to the perfection of the heavenly bodies) when he replaced the Copernican circles with nearly circular ellipses. Only after the failure of all such attempts do scientists finally relinquish the old theory and replace it with something really new. The final abandonment of a previously well-established theory is usually done reluctantly and over a long period of time. Often the new theory is not completely accepted until all the old-timers who advocate the old theory have died.

The example of Ptolemy and Copernicus shows that scientific theories are partly a matter of point of view. Copernicus and Kepler took a different point of view from Ptolemy. This view turned out to be more fruitful, rather than more true, than Ptolemy's theory. The important scientific insights, the inspired breakthroughs, are usually sparked by a new point of view, a different way of seeing, an original perspective into which the details fit more comfortably. Kepler, commenting on his discovery of the theory he had sought, wrote that "it was as if I awoke from a deep sleep and saw a new light." Kepler's experience was much the same as that of the painter who, after much adjustment and readjustment, finds that particular arrangement of shapes on the canvas and breathes, "Ah . . . now I have it."

It should be obvious by now that the ideas, or theories, of science can in no sense be called facts. The closest things to facts in science are the specific data resulting from particular observations or experiments. But even these so-called facts are tentative, because experimental uncertainties are always present: observations are often made at the extreme limits of the abilities of the equipment, the observer may be influenced by a strong bias, and, in any case, the observations are only made with a certain limited precision.

Hypothesis is another useful word. A hypothesis is simply a guess at a theory or part of a theory. In other words, a hypothesis is a highly tentative principle that hasn't yet been subjected to many observational checks. When Kepler found that Tycho's data disproved the original Copernican theory, he tried several hypotheses as to how to repair that theory. Most of these hypotheses turned out to disagree with the data, so Kepler rejected them. But the hypothesis of elliptical orbits worked. This hypothesis, coupled with Kepler's two other principles of planetary motion, soon received enough observational support to merit the title **theory.**

The symbiotic (mutually supporting) relationship between theory and observation is undoubtedly the most characteristic feature of science. Without this relationship, a field of study should not be called a science. A key feature of this relationship is that a theory or a hypothesis must make specific, checkable assertions about the world— predictions that can actually be checked by observations. These specific observations must be carried out, and must agree with the theoretical predictions, before an idea can justifiably be called a scientific theory.

Finally, let's return to our definition of science (Chapter One). **Science is the use of observation, logic, and a sense of beauty, to devise general principles about the natural world.** Please read this definition slowly, and ponder it in light of what you have learned in this chapter.

Checklist

Ptolemy	ellipse
qualitative, quantitative	Copernican revolution
Ptolemaic theory	Newtonian age
retrograde motion	theory
Copernicus	principle
Copernican theory	law
Tycho Brahe	model
Kepler	hypothesis
Kepler's theory	science

Further Thought Questions

9. A standard objection to Darwin's theory of the origin of the species is that it is only a theory, rather than a fact, and is thus suspect. How does this attitude indicate an incorrect understanding of the scientific use of the word *theory*?

10. What do we mean when we say that a certain scientific principle is "true"? Is it true that the earth goes around the sun? Was this Earth-centered theory true in Ptolemy's day? Was it true in Ptolemy's day that the sun went around the earth?

Further Exercises*

3. Suppose that Ptolemy's theory predicts that Mars will be found at an angle of 25 deg and 20 min of arc above the horizon at midnight next Saturday (recall that 60 min of arc is 1 deg). Copernicus's theory predicts 25 deg and 27 min of arc. An actual

*No answers are provided for the Further Exercises.

observation, accurate to within 10 min of arc, performed at midnight on Saturday yields a result of 25 deg and 19 min of arc. Does Ptolemy's theory agree with this observation? Does Copernicus's theory agree?

4. Suppose that the observation in the preceding exercise had been five times as accurate, so that the error was estimated to be at most 2 min of arc. Would Ptolemy's theory then agree with the observed angle of 25 deg and 19 min of arc? Would Copernicus's theory agree?

5. Suppose that one of the tabloid newspapers sold at grocery store checkout counters announced, "SCIENTISTS PREDICT THAT THE UNIVERSE AND EVERYTHING IN IT WILL DOUBLE IN SIZE NEXT NEW YEAR'S EVE AT MIDNIGHT." Could this conceivably be a scientific statement? Why? (*Hint:* Are there any experiments that could test this prediction?)

6. Use the string and thumbtack construction to draw an ellipse that is almost a circle. Draw another that is highly elongated (flattened).

7. Venus often rises (in the east) just before the sun, or sets just after the sun. Why? Why isn't Mars so likely to play this role? (*Hint:* See Figure 2.6.)

Recommended Reading

1. Angus Armitage, *The World of Copernicus,* EP Publishing Ltd., Wakefield, England, 1972. The life and times of Copernicus.
2. Leon N. Cooper, *An Introduction to the Meaning and Structure of Physics.* Harper & Row, New York, 1968. A physics textbook with a cultural-historical emphasis. See Chapters 5, 7, and 8 for material on early theories of the solar system and on the methods of science.
3. Gerald Holton, and Stephen G. Brush, *Introduction to Concepts and Theories in Physical Science,* 2d ed., Addison-Wesley, Reading, Mass., 1973. An excellent, nontechnical physical science textbook, emphasizing cultural and historical aspects of science. Chapters 1 through 5 discuss early theories of the solar system.
4. Arthur Koestler, *The Sleepwalkers,* Grosset and Dunlap, New York, 1963. A beautiful study of humankind's changing vision of the universe from the early Greeks to Galileo and Newton, by a well-known novelist and essayist. The chapter on Kepler, entitled "The Watershed," has also been published as a separate book.

3

ATOMS

Scientists believe that all matter is made of tiny particles, or atoms. This idea explains an extraordinary range of observed phenomena and thus provides a fine example of a scientific theory. This atomic idea underlies much of the physics in this book.

The Atomic Idea

Suppose that God appeared on the television news tonight and made the following announcement:

> Due to the foul-smelling mess that you humans have made of my Earth, I have decided to replace *Homo sapiens* with a new, improved model. At tomorrow noon, eastern standard time, I shall destroy the human race and every trace of human civilization with the following single exception: for each field of human knowledge, I will allow one sentence of 25 words or less to be preserved for the benefit of the future inheritors of Earth.

Besides upsetting everyone's evening meal, this announcement would present an interesting problem for the world's physicists: What single sentence should physics bequeath to the future? One good choice would be: Every material object is composed of tiny particles; these particles are much too small to be seen with the naked eye.

Exercise 1
The physicists' statement is an example of (*select one*) (a) a scientific fact; (b) an experimental observation; (c) a scientific theory; or (d) a scientific hypothesis. (Answer below.)

The physicists' statement is a good candidate for the selected sentence because it is useful in explaining an extraordinary range of apparently unrelated phenomena and because it is not at all obvious.

We will call this idea the **atomic theory.** It has certainly been experimentally tested often enough to be called a principle or theory rather than a hypothesis. It is, however, incorrect to call this idea a fact or an experimental observation (see Exercise 1), because it is a general statement referring to *every* material object and so cannot be proved by observation, just as Kepler's principles of planetary motion couldn't be proved (Chapter Two).

The tiny particles referred to above are called **atoms.** For reasons to be explained, nobody has actually seen an atom and nobody ever will. Nevertheless, our faith in atoms is strong because we have observed a wide range of phenomena that can be explained in terms of atoms and that would be difficult to explain in any other way. The atomic theory is useful because it explains, and hence unifies, a wide range of phenomena. This is what we mean, and this is all we mean, when we say it is true. This is all we *ever* mean when we say that any scientific theory is true.

For now you can think of atoms as small objects, like fantastically tiny peas. Although this "small pea" picture of an atom is useful for some purposes, it is a gross oversimplification of the contemporary view of the atom. An object like a pea is the way it is because it is composed of an unimaginably large number of atoms. Its properties result from the interactions among these atoms. Thus it is inconsistent to think of a single atom as like a tiny pea. As we shall see in Part 4, a single isolated atom is an elusive ghostlike object, quite a different creature from a pea, even a very small pea!

Atoms and Molecules

There are a limited number of different kinds of atoms. At last count this number stood at 105. We number them from 1 to 105 and we give them names: hydrogen (number 1, abbreviated H), helium (number 2, abbreviated He), and so on. Figure 3.1 gives the complete list. Only the first 92 of these occur naturally on Earth;* the others have been made in laboratories. Any substance composed of just one type of atom is called an **element.**

It isn't difficult to identify hundreds of different substances in your immediate surroundings: paper, ink, fingernails, leather watch bands, and more. How can we get all these different substances out of just 92 naturally occurring elements? The answer is that the smallest building blocks of most substances are not atoms, but are collections of atoms called **molecules.** For example, the smallest particle of water that still has the characteristic properties of water (freezes at 0 degrees Celsius, quenches your thirst, etc.) is not a single atom. It is one oxygen atom stuck to two hydrogen atoms: H_2O in the scientist's notation. If you break the H_2O molecule down further, into two separate H atoms and one O atom, you won't have water anymore. You'll have hydrogen and oxygen instead, two substances that have properties radically different from those of water.

A molecule can be a fairly complicated creature. The simple sugar known as glucose, a typical **organic molecule** (related to the chemistry of living organisms), has the chemical formula $C_6H_{12}O_6$.

Exercise 2
How many atoms does a glucose molecule contain?

A typical protein molecule may contain a million atoms. Having 92 elements to play with, nature can come up with an incredible variety of molecules.

Some substances are composed of more than one type of molecule. Air, for example, is a mixture of about 80 percent N_2 (the nitrogen molecule) and 20 percent O_2 (the oxygen molecule), with a smattering of Ar (argon), CO_2 (carbon dioxide), H_2 (hydrogen molecule), and gaseous H_2O (water vapor).

---

*More precisely, only 89 of the first 92 elements occur naturally on Earth. Numbers 43, 61, and 85 do not occur naturally.

Number	Element	Symbol	Number	Element	Symbol
1.	hydrogen	H	54.	xenon	Xe
2.	helium	He	55.	cesium	Cs
3.	lithium	Li	56.	barium	Ba
4.	beryllium	Be	57.	lanthanum	La
5.	boron	B	58.	cerium	Ce
6.	carbon	C	59.	praseodymium	Pr
7.	nitrogen	N	60.	neodymium	Nd
8.	oxygen	O	61.	prometeum	Pm
9.	fluorine	F	62.	samarium	Sa
10.	neon	Ne	63.	europium	Eu
11.	sodium	Na	64.	gadolinium	Gd
12.	magnesium	Mg	65.	terbium	Tb
13.	aluminum	Al	66.	dysprosium	Dy
14.	silicon	Si	67.	holmium	Ho
15.	phosphorus	P	68.	erbium	Er
16.	sulfur	S	69.	thulium	Tm
17.	chlorine	Cl	70.	ytterbium	Yb
18.	argon	Ar	71.	lutecium	Lu
19.	potassium	K	72.	hafnium	Hf
20.	calcium	Ca	73.	tantalum	Ta
21.	scandium	Sc	74.	tungsten	W
22.	titanium	Ti	75.	rhenium	Re
23.	vanadium	V	76.	osmium	Os
24.	chromium	Cr	77.	iridium	Ir
25.	manganese	Mn	78.	plantinum	Pt
26.	iron	Fe	79.	gold	Au
27.	cobalt	Co	80.	mercury	Hg
28.	nickel	Ni	81.	thallium	Tl
29.	copper	Cu	82.	lead	Pb
30.	zinc	Zn	83.	bismuth	Bi
31.	gallium	Ga	84.	polonium	Po
32.	germanium	Ge	85.	astatine	At
33.	arsenic	As	86.	radon	Rn
34.	selenium	Se	87.	francium	Fr
35.	bromine	Br	88.	radium	Ra
36.	krypton	Kr	89.	actinium	Ac
37.	rubidium	Rb	90.	thorium	Th
38.	strontium	Sr	91.	protractinium	Pa
39.	yttrium	Y	92.	uranium	U
40.	zirconium	Zr	93.	neptunium	Np
41.	niobium	Nb	94.	plutonium	Pu
42.	molybdenum	Mo	95.	americium	Am
43.	technetium	Tc	96.	curium	Cm
44.	ruthenium	Ru	97.	berkelium	Bk
45.	rhodium	Rh	98.	californium	Cf
46.	palladium	Pd	99.	einsteinium	Es
47.	silver	Ag	100.	fermium	Fm
48.	cadmium	Cd	101.	medelevium	Md
49.	indium	In	102.	nobelium	No
50.	tin	Sn	103.	lawrencium	Lw
51.	antimony	Sb	104.	rutherfordium	Rf
52.	tellurium	Te	105.	hahnium	Ha
53.	iodine	I			

FIGURE 3.1 The elements.

The States of Matter

Substances come in three different "states" on Earth: solid, liquid, and gas. Solids are distinguished by the fact that they retain their shape, liquids by the fact that they spread out over the bottom of a container, and gases by the fact that they fill up a closed container.

According to the atomic theory, the differences among the three states of matter is a consequence of the different arrangements of molecules in the three states.

A typical *gas,* such as oxygen, consists of molecules that are separated widely and moving rapidly. The distance between molecules is large compared to the size of one molecule. The molecules move in straight lines, at constant speed, except when they occasionally strike either another molecule or a container wall.

A *liquid* consists of molecules packed fairly tightly together but not so tightly as to be locked into place. The molecules of a liquid are able to slide past each other and move throughout the liquid.

A typical *solid,* such as coal, consists of atoms, carbon atoms in the case of coal, locked into position next to each other in a regular pattern.* Although the atoms in a solid are not free to move around within the body of the solid, they do vibrate.

The Odor of Violets and Other Wonders

The atomic theory explains many things.

For instance, did you ever wonder how you can smell a bowl of violets from across the room? No? Well, then, stop and wonder for a few moments.

(Please pause and wonder.)

Ready? Here's how it works. A horribly complicated type of organic molecule, let's call it an OV molecule, for "odor of violets", is occasionally shaken loose from a violet. Suspended in the atmosphere, these OV molecules are jostled by the rapidly moving air molecules, and this jostling gradually disperses the OV molecules throughout the air. Once in a while one of the molecules winds up inside your nose. This OV molecule then reacts chemically with a smell receiver, or olfactory receptor, in your nose, which stimulates an olfactory nerve fiber to send an electrical signal to your brain. And you smell violets. Fantastic.

Suppose you sip a glass of brandy after an evening meal and leave the glass half-filled on the table. The following morning the glass is nearly empty, although you haven't touched it since last night. You taste it. It is bitter. It has lost most of its alcoholic content. Let's explain.

Why is the glass nearly empty? Apparently, the brandy evaporated. Let's take a closer look. An 80-proof brandy is a mixture of about 60 percent water (H_2O molecules), 40 percent ethanol, or grain alcohol, (C_2H_5OH molecules), plus a smattering of other molecular types (organic molecules from grapes) that give the brandy its characteristic flavor. In the liquid state all these molecules are in close contact, sliding past one another in continual motion throughout the container (Fig. 3.2). Their speeds vary widely, some moving slowly, some rapidly. These molecules have a natural attraction for each other, which is why liquids hold together as, for example, raindrops do. Despite this attraction of each molecule for its fellow molecules,

*Exception: Amorphous solids have no regular atomic pattern. Glass and tar are examples.

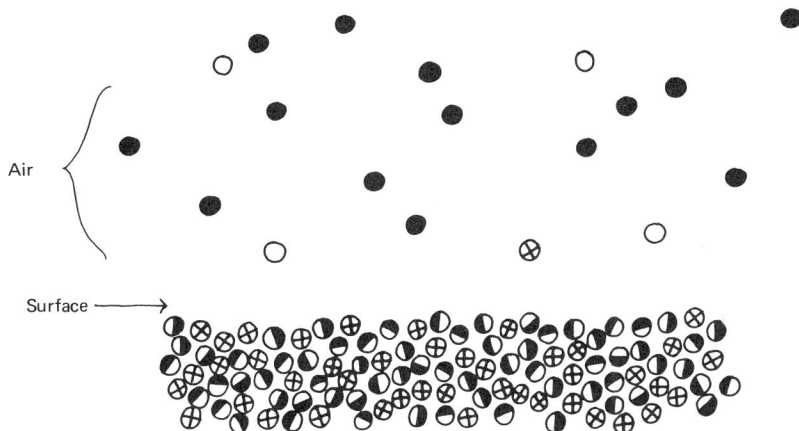

FIGURE 3.2 A small portion of the surface of a glass of brandy. $\bigcirc$ = oxygen; $\bullet$ = nitrogen; $\oplus$ = C_2H_5OH; $\ominus$ = H_2O. An alcohol molecule has just evaporated from the surface.

an occasional H_2O or C_2H_5OH molecule is moving fast enough, is sufficiently near the surface, and is moving in the proper direction, that it can break through the surface and enter the air above. This gradual change of liquid to gas is called *evaporation.*

Why is the remaining brandy bitter? In the brandy, molecules have combined, or reacted chemically, with oxygen from the air, and the resulting combination tastes bad. This is why liquors, and especially wines, are carefully sealed.* Corked wine should always be stored on its side, to keep the cork wet and therefore expanded. Over a period of time, sizable amounts of air can filter into the bottle through a dry cork.

The alcohol content of the brandy has decreased relative to the water content. Explanation: The C_2H_5OH molecule evaporates more easily than the H_2O molecule (Fig. 3.2), because the C_2H_5OH molecule is not attracted as strongly to its neighbors as the H_2O molecules is to its neighbors.

Let's turn now to another example of the explanatory power of the atomic theory. In a famous experiment conducted in 1654, Otto von Guerieke, burgomaster of Magdeburg, placed two copper hemispheres together to form a sphere whose diameter was somewhat less than 2 ft. A ring of leather soaked in oil was placed between the two hemispheres as an airtight joint, and most of the air was evacuated from inside the sphere. Two teams of eight horses each, attached to opposite hemispheres, were unable to separate them.

The hemispheres were held together, even against the best efforts of 16 horses, by the impacts of enormous numbers of air molecules striking the outside of the hemispheres (Fig. 3.3). Because the inside of the hemispheres was partially evacuated, there were few impacts on the inside to balance the impacts on the outside. Thus the hemispheres were pushed tightly together by the air on the outside.

Any surface exposed to air is exposed to the impact of air molecules. Even

---

*However, a limited amount of oxidation (combination with oxygen) is actually desirable. Thus wines should be uncorked and allowed to "breathe" for 30 minutes before consumption.

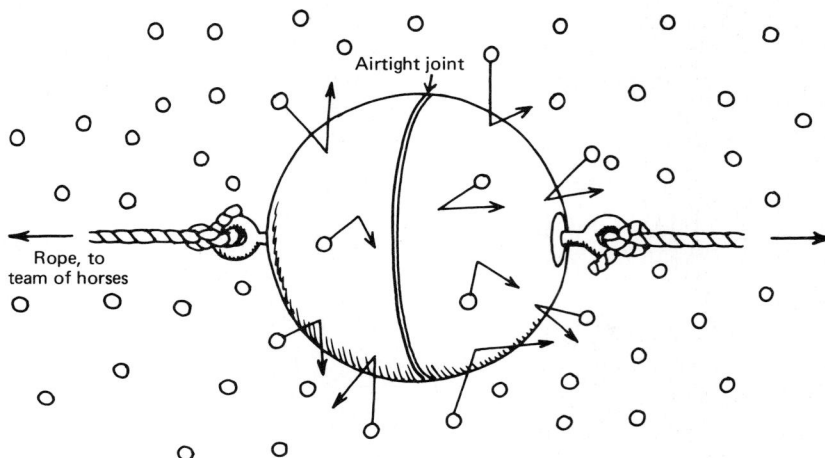

FIGURE 3.3 A few molecules bouncing off the Magdeburg hemispheres.

though each impact is weak, there are so many impacts per second that the overall effect is great. The air exerts a push of about 15 lb on every square inch of surface exposed to air. This push is called *air pressure.*

When you pump air into a tire you are increasing the number of air molecules inside the tire, which naturally increases the number of tiny impacts per second on the inside of the tire, which means that the pressure increases. For example when you fill a high-pressure bicycle tire to 90 lb/sq in., the air inside the tire exerts a push of 90 lb on every square inch of the inside of the tire.

Air pressure is greater than we usually imagine. It was great enough to hold the burgomaster's 2-ft hemispheres together. It's also big enough to exert a push of several tons on the surface of your body. Let's take a moment to figure out roughly how many tons. Fifteen pounds are exerted on every square inch of your body. To roughly estimate the number of square inches of surface area on your body, make the unaesthetic but simplifying assumption that your body is just a solid block of dimensions 1 ft × 1 ft × your height. The surface area of this block is roughly four times the area of one of the long sides (we neglect the small top and bottom of the block). If you are 5 ft tall, the area of one of these sides is 1 ft × 5 ft = 5 sq ft, so the total surface area of your body is about 4 × 5 = 20 sq ft. But in every square foot there are 12 × 12 = 144 sq in., so the surface area of your body is about 20 × 144 = 2880 sq in. Finally, the total push exerted on you by the air is 15 × 2880 = 43,200 lb, or about 22 tons (there are 2000 lb to a ton). Think about that when you get out of bed tomorrow morning!

Exercise 3
With all this pressure, why don't you collapse?

Exercise 4
A deep-sea diver descends 300 ft below the ocean surface, where the pressure is 10 atm, that is, ten times atmospheric pressure. How big is the total push on the surface of her body due to the surrounding water? Why doesn't her body collapse?

Air pressure accounts for all sorts of things. For example, air pressure is responsible for the suction produced in a vacuum cleaner tube. The vacuum cleaner motor removes air from the tube. Atmospheric pressure, always pushing from outside, is then opposed by less pressure pushing back from inside, so dust is pushed (not really pulled or sucked) into the tube by the outside air.

Exercise 5
Could you drink soda through a straw if you were standing on the surface of the moon?

Exercise 6
Why is it so difficult to remove the lid from a vacuum-sealed jar?

Q. Why is it so easy to compress a gas? **A.** Because its atoms are far apart, it is easy to crowd them closer together.

Q. Why is it so easy to deform a liquid, that is, why does it flow so easily? **A.** Because its atoms are not rigidly attached to each other.

Q. Why is it difficult to compress a solid or a liquid? **A.** Because its atoms are already about as close together as they can be.

Q. Why are solid objects so hard to deform? **A.** Because their atoms are rigidly attached to each other.

Q. Why does a feather float slowly to the ground instead of plummeting like a rock? **A.** Both the feather and rock bump large numbers of air molecules on the way down. The feather is slowed by this action, but the heavier rock is not. This retarding effect of air is called *air resistance.*

Exercise 7
An astronaut drops a feather and a rock onto the moon's surface. Compare their motions.

Perfume of violets, evaporation of brandy, air pressure, floating feathers . . . the list goes on and on. An incredible variety of apparently unrelated phenomena are explained by the atomic theory. These phenomena are joined together, unified, by the atomic theory. And this is the primary thrust of the scientific enterprise: to unify an ever-widening range of phenomena by means of a few underlying principles.

As mentioned before, nobody has seen an atom. We accept the atomic theory solely because of its unifying value in explaining a wide range of phenomena and not because it has been directly observed to be true. If you begin by assuming the atomic theory you can deduce (explain, predict) the perfume of violets, evaporation, and all sorts of amazing phenomena. But you can never be certain of the truth of your starting point (the atomic theory, in this case). Science is always tentative.

Atoms and Empty Space

The atomic idea isn't new. It was first formulated 2400 years ago by the Greek philosophers Leucippas, Democritus, and Epicurus, who used the atomic theory to give an acceptable atomistic explanation of many properties of solids, liquids, and gases.

Democritus stated the atomic theory in a stark and far-reaching form: "Nothing exists except atoms and empty space. All else is mere conjecture."

Democritus speculated that beneath the infinite variety of appearances lay a single, unifying reality: atoms moving silently in the void. "By convention bitter is bitter," he wrote, "by convention hot is hot, by convention cold is cold, by convention color is color. But in reality there are atoms and the void. That is, the objects of sense are supposed to be real, and it is customary to regard them as such, but in truth they are not. Only the atoms and the void are real."

If we accept the philosophy of Democritus it follows that there is no bitter, no hot, no cold, no color. There are only atoms and empty space. Bitter and hot and cold and color are not real, they are only our sense impressions of the truly real objects, the atoms. Let's carry this thinking a little further. If sense impressions such as color are not real, then it is difficult to see how a person's perception of, say, a painting, could be real. So it seems that a concept such as beauty is not real either. In fact, this philosophy seems to imply that there is no beauty, no love, no freedom, no dignity, for each person experiences these as mere sense impressions. According to this view, we can entirely account for bitter, hot, cold, color, beauty, love, freedom, dignity, by describing the motion of tiny particles moving through empty space. We may talk about cold and color and freedom and dignity but if we truly desire to penetrate the reality beneath mere appearance then Democritus tells us that we must understand these concepts in terms of atoms. For, according to this thinking, atoms are indestructible (the Greek name *atom* means "indivisible"), and hence they are the most fundamental objects in the universe. All else—the table on which my writing paper rests, this book, your body—is destructible and ephemeral. The atoms alone are real.

This is a form of the philosophy known as **materialism.** It holds that only matter (i.e., atoms) is real and that all else can be fully explained in terms of the workings of matter.

Thought Question 1
Do you believe this? Why or why not?

In one form or another the materialistic concept that a wide range of phenomena can be reduced to atoms (i.e., can be explained in terms of atoms) runs as a theme through the history of science. As we will see, Newtonian physics is quite compatible with this philosophy (although Newton himself was far from being a materialist).

Personally, I don't agree with the materialist position. Perhaps you disagree with it, too. Most people don't think of themselves or their surroundings as mere collections of unthinking atoms. And yet the materialistic philosophy can be difficult to refute. For example, you might argue that a symphony concert, or the taste of a grape, or feelings of love, cannot be reduced to atoms. But recall our earlier explanation of the odor of violets. It does seem that the experience of the odor of violets can be reduced to atoms. Similar atomic explanations apply to sound, to taste, and, once the biochemistry of the human brain is more fully understood, probably even to human emotions such as anger or love. In my opinion, our culture, or at least our science, is so accustomed to **reductionist** explanations of this sort (in which experiences are *reduced* to atoms) that other ways of viewing the universe have become difficult to accept or even to conceive.

If you happen not to buy this materialistic-reductionist view of things, do not despair. There are respectable arguments on the other side.

1. All things can be looked at from many points of view. Science is just one such point of view. We can view the odor of violets scientifically, or aesthetically, or religiously—all viewpoints have their own validity. To say that we have explained the odor of violets in terms of atoms is not to say that we have explained it away. The odor still really exists even though we can explain it in terms of moving molecules. Reality has many facets, and the scientific or atomistic facet is just one of them. Other nonscientific points of view can also be useful, even crucial.

2. As we will see later in this book, twentieth-century physics has nearly destroyed the scientific underpinnings of the materialistic philosophy. Relativity theory and quantum theory have introduced a subjective element into the scientific viewpoint itself by emphasizing the importance of the human observer and the consciousness of that observer.

For elegant statements of the antimaterialist view, see References 1 and 3.

Thought Question 2

What is your opinion of these two arguments? Try to think of additional arguments. How might a materialist respond to these points?

Distance and Powers of Ten

The next section deals with the size of atoms, so it's necessary at this point to discuss the system used for distance measurements. The standard international system of measurement is the **metric system.** In the metric system, distances are usually measured in **meters.** One meter is about 39 in., a little longer than 3 ft.

An important advantage of the metric system is that different units are related by factors of ten, which makes it easy to convert from one unit to another. For instance the **kilometer** is 1000 meters, or three factors of ten ($10 \times 10 \times 10$) larger than 1 meter. The **centimeter** is 1/100 meter, or two factors of ten (10×10) smaller than the meter.

The kilometer, meter, and centimeter are abbreviated km, m, cm.

Exercise 8

Convert 5 km to meters and then to centimeters. Now convert 5 miles to feet, and then to inches (there are 5280 ft in a mile). Which system is most convenient for conversions?

Another nice feature of the metric system is that a standard set of prefixes is used to show the relationship between units. Here are the most popular prefixes:

Micro means one-millionth (1 divided by 1 million).

Milli means one-thousandth.

Centi means one-hundredth.

Kilo means one thousand.

Mega means one million.

Exercise 9
What is a microsecond? A millimeter? A millisecond? How many watts are there in a kilowatt? In a megawatt?

In the United States the most popular system of units is the **English system,** with distances measured in inches, feet, miles, and so on. There are at least two good reasons for preferring metric units over English. (1) The rest of the world uses the metric system; in fact, even the English no longer use the English system! (2) The metric system is more straightforward than the English system (see Exercise 8). The United States is in the midst of a (slow) transition to the metric system. We'll use metric units in this book.

To develop a feel for metric distances it's worth remembering that 1 cm is about 0.4 in. (less than half an inch), 1 m is a little more than a yard, and 1 km is about ⅝ miles (a little more than half a mile).

Exercise 10
Using metric units, estimate the length of your thumbnail, your thumb, your foot, your arm, your body, a football field, a marathon (27 miles), distance around Earth (25,000 miles), distance to the sun (93 million miles).

Thought Question 3
What problems are likely to be created by our switch to the metric system? What problems might arise if we don't switch? Should we do it?

It is convenient to express very large numbers, such as the larger distances in Exercise 10, in an abbreviated form called **powers of ten.** The 40,000 km distance around Earth can be expressed as four times 10,000, or 4 multiplied by 10 four times, or $4 \times 10 \times 10 \times 10 \times 10$. This is abbreviated as 4×10^4, or "four times ten to the fourth power." The 150,000,000 km distance to the sun is abbreviated 1.5 $\times 10^8$, or "1.5 times ten to the eighth power."

Very small numbers may also be abbreviated as powers of ten. For instance, 5 mm is .005 m, which may be obtained from 5 m by dividing by 10 three times. This is abbreviated 5×10^{-3}, or "five times ten to the negative three power."

If you are given a number in powers of ten notation, you can find the number in ordinary decimal notation by moving the decimal place forward (for positive powers) or backward (for negative powers) the appropriate number of places. For instance $5 \times 10^4 = 50,000$ (start from 5.0 and move the decimal four places forward), and 7×10^{-6} is .000007 (start from 7.0 and move the decimal six places backward).

George Washington's Dying Breath

Did you know that there are probably about five molecules from George Washington's dying breath in your lungs right now? Think about it. George Washington exhaled so many molecules as he expired that every cubic meter of Earth's atmosphere contains a hundred or so of them. Not that old George was abnormally full of air. We all breathe out that many molecules with each breath.

That little example gives you a feeling for how many molecules there are in a single breath of air and for how small each molecule and atom must be. A single atom is about 1×10^{-8} cm (.00000001 cm) in diameter. (Henceforth, we'll leave off the 1 in numbers like 1×10^{-8} and just write 10^{-8}, because multiplying by 1 doesn't change anything.)

Another example: If *every* atom in a thimbleful of water were blown up to the size of a pea, our planet would be flooded with peas to a depth of several miles.

Figure 3.4 will give you an appreciation of the range of sizes in the universe. The atomic nucleus, at 10^{-15} m, is among the smallest objects known. Galaxies, at about 10^{21} m, are among the largest objects known.

And *Homo sapiens*, measuring about 2 m, stands midway between the atoms and the stars.

No human being has ever or will ever see an atom. The reason is that visible light, the instrument for all our seeing, has a wavelength about 5000 times the size of an atom. So trying to see an atom with visible light is like trying to find a cork afloat on an ocean by looking for the disturbance the cork produces in large ocean waves. The wavelength is too large for the waves to be sensitive to such a small object. However, we can detect (but not see directly) atoms by using nonvisible waves having a suitably short wavelength. For example, under appropriate conditions the so-called matter waves produced in an electron microscope have the right wavelength. Figure 3.5 was made with an electron microscope.

The Simplified Atom

Physicists often use simplified descriptions to gain the insight needed to move on to a more sophisticated description. Strictly speaking, every description in physics is simplified. It is sometimes said, for example, that a baseball in flight travels along

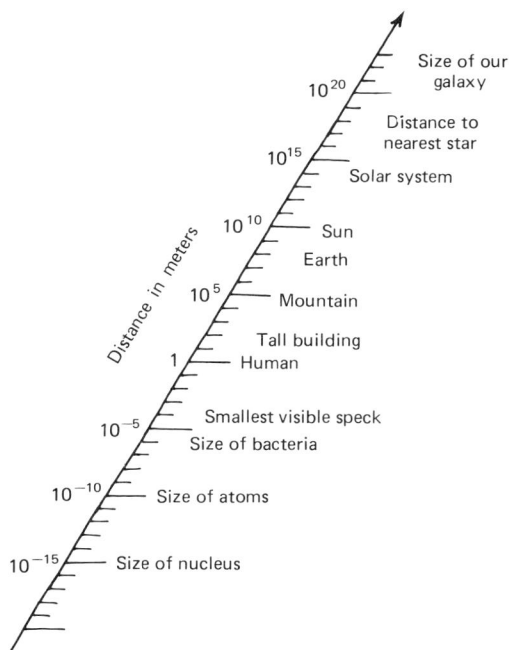

FIGURE 3.4 The range of sizes in the universe.

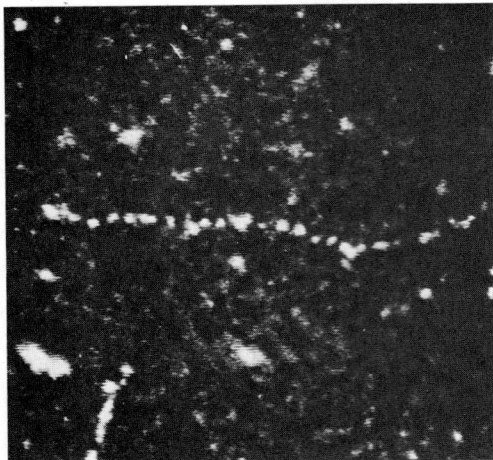

FIGURE 3.5 The strings of dots are thorium atoms, detected with an electron microscope. They wouldn't have been detected by a microscope using visible light: we cannot see atoms.

an elliptical path. But the actual path of a baseball is not really an exact ellipse. The elliptical path is a simplified description that omits the effects of air resistance, the random jiggling of the ball as it is deflected by individual air molecules, the thickness of the ball, and so on, and so on. What our original statement actually meant was that, for many purposes, we can treat the motion of a baseball as if the ball were moving along an elliptical path.

This section presents a simplified description of the atom. For many purposes, this description is good enough.

Figure 3.6 shows "pictures" of the carbon and helium atoms. The center, or *nucleus,* contains two types of particles: *protons* and *neutrons.* A third type of atomic particle, the *electron,* orbits the nucleus. The orbits of the electrons happen to be ellipses, just like the orbits of the planets around the sun. Protons and neutrons are relatively large and heavy, whereas electrons are much smaller and lighter, so that some 99.9 percent of the weight of an atom resides in its nucleus. Ordinarily the number of electrons equals the number of protons.

Protons and electrons are electrically charged particles, which means that they are capable of exerting electric forces on other electrically charged objects (more about electric forces in Chapter 12). The electrons are held to the atom by the electric force of attraction between protons and electrons. The nucleus, on the other hand, is held together by nonelectric nuclear forces. The proton carries a positive charge and the electron carries an equally large negative charge. Because the number of electrons equals the number of protons, the atom is electrically neutral (no *net* electric charge) overall. A neutron, true to its name ("neutral one"), carries no charge, so it does not feel electric forces.

The theory of the atom described in the preceding two paragraphs is known as the *planetary model of the atom* because it is a picture, or model, that resembles the sun-centered theory of the solar system.

Exercise 11
List some similarities between the solar system (i.e., the sun and the planets and other objects that orbit the sun) and an atom. Now try listing some differences.

The planetary model of the atom is the second atomic model presented in this book. The first was the small-pea model presented earlier in this chapter, in which the atom was pictured as simply a very small object with no particular internal structure. We have already seen that even the extremely simplified small-pea model is able to explain a wide variety of phenomena. Our more sophisticated planetary model is capable of explaining an even wider variety of phenomena, and so it is regarded as a more accurate or more complete model than the small-pea model. However, for some purposes, even the planetary model of the atom isn't good enough. Thus during the 1920s, physicists invented an even more sophisticated theory, the **quantum theory of the atom.** This theory (the word **model** seems inappropriate because the quantum theory of the atom cannot really be pictured) explains an even wider variety of phenomena than the small-pea model or the planetary model. It is regarded today as our most accurate theory of the atom. We'll be coming back to it later.

With the help of the planetary model of the atom we can understand the differences of the 105 different types of atoms discussed earlier. They differ in that they have different numbers of protons in the nucleus and thus they have different numbers of orbiting electrons. It turns out that the chemical properties of atoms are strongly dependent on the number of electrons orbiting the nucleus. Therefore, atoms that differ in their number of protons will have quite different properties, and this is in fact why we speak of them as different types or as members of different elements.

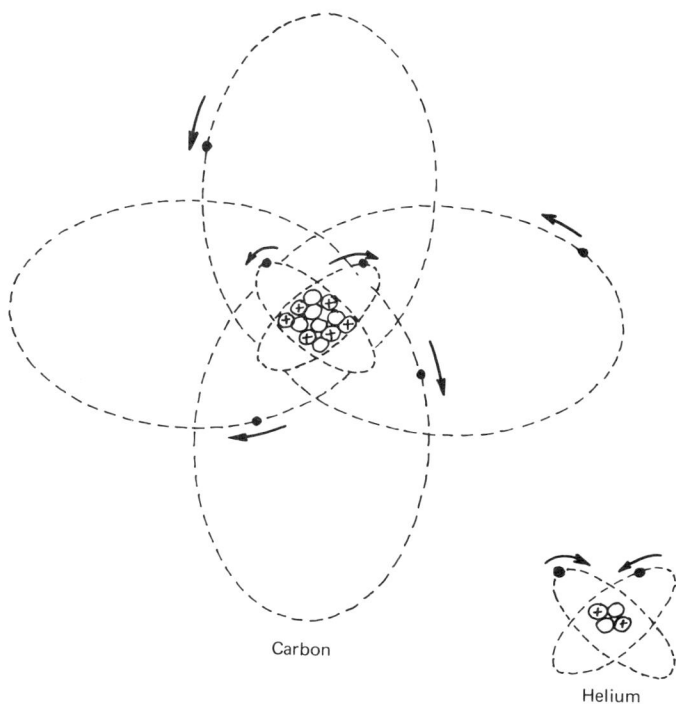

Carbon

Helium

FIGURE 3.6 Simplified pictures of the structures of the helium and carbon atoms. ⊕ = proton; ○ = neutron; ● = electron; dashed lines are the paths followed by the electrons.

The different types of atoms are named and numbered according to the number of protons in the nucleus. This number is called the **atomic number** of the atom. Atom number 1, called hydrogen, has one proton in its nucleus; atom number 2, helium, has two protons, and so on.

Atoms are really spaced out. Imagine one atom enlarged so that the nucleus is the size of a basketball. On this scale, a typical orbiting electron would be a pea some 20 miles away. Thus if you fire a small particle, such as a neutron, into a batch of atoms, such as a sheet of iron, your neutron bullet will probably penetrate deeply into the iron before striking a nucleus. It is precisely this sort of experiment that first suggested the atomic model being described here. Such experiments, and a host of other experiments, helped confirm (but cannot prove) this model of the atom.

Chemical Reactions and You

Recall that a molecule is the smallest part of a substance that still has the characteristic properties of that substance, and that molecules are composed of atoms that are attached to each other. Under certain circumstances the molecular structure of a substance might change. Any such reorganization of the molecular structure of a substance is called a **chemical reaction.**

Example: When oxygen and carbon are mixed under certain conditions (the right conditions happen to be high temperature), they react to form carbon dioxide while producing heat. The heat from the reaction keeps the reaction going because it keeps the remaining oxygen and carbon hot enough to continue reacting. We symbolize this reaction as:

$$O_2 + C \rightarrow CO_2 + \text{heat}$$

The symbols tell us that individual oxygen molecules (O_2) combine with individual carbon atoms (C) to produce carbon dioxide molecules (CO_2) and that heat is produced during this combination. The name for this particular reaction is burning, or **combustion.** When a log burns, carbon from the organic molecules in the log combines with oxygen from the air to produce carbon dioxide. Under certain conditions the deadly gas carbon monoxide, CO, is produced, rather than the nonpoisonous CO_2; the combustion is then said to be incomplete. As a result of incomplete combustion, the automobile engine contributes some 75 percent of the carbon monoxide pollution so prevalent in our cities.

Question: How do animals, humans, for instance, obtain their energy,* that is, their ability to move around? Answer: By eating and breathing. For a more detailed explanation, let's reduce eating and breathing to atoms and molecules. The energy is actually acquired via a chemical reaction known as **respiration.** In symbols, this reaction is:

$$O_2 + C_6H_{12}O_6 \rightarrow CO_2 + H_2O + \text{energy}$$

What happens is that the O_2 that you breath combines with carbon from glucose ($C_6H_{12}O_6$) to produce CO_2 and H_2O plus energy. This reaction of oxygen with carbon is nearly identical to combustion. We animals literally burn all the time! The primary difference between combustion and respiration is that about half of the liberated

---

* I'll use the word *energy* loosely for now. In animals, *energy* means, roughly, "ability to move around." You'll find a detailed discussion of energy later.

energy of the respiration reaction goes, not into heat, but into synthesizing a high-energy molecule known as ATP (adenosine triphosphate), the energy-bearing molecule in animals.

This reaction keeps all of us going. So let's look at more of the details. Take a deep breath and concentrate on the air being "pulled" into your lungs.

Exercise 12

Recall that suction effects are always caused by air pressure pushing from behind. How does the air actually get into your lungs?

This air you inhale is 20 percent O_2. Your lungs are lined with small, thin-walled blood vessels known as capillaries. Some of the O_2 from the air in your lungs passes through the walls of these capillaries and is captured by hemoglobin molecules in your blood. The flowing blood transports this hemoglobin with its attached oxygen to a cell in, for example, your thumb, where the O_2 dissociates from the hemoglobin and passes across another thin capillary wall and into the cell. In the cell the respiration reaction occurs. This reaction consumes glucose and oxygen, which is why you must eat and breathe in order to live, and it produces carbon dioxide, water, and excess energy. About 50 percent of this excess energy goes into manufacturing ATP molecules in the cell. For every molecule of O_2 consumed, some 40 molecules of ATP are manufactured. The energy of the ATP molecule is then used for muscle contraction when you move your thumb. The CO_2 produced as a by-product of the respiration reaction passes back into your bloodstream, attaches itself to a hemoglobin molecule, flows back to your lungs, and is exhaled. The water produced by respiration is excreted as sweat, urine, and as water vapor in exhaled air.

Thus all animals, except for a few non–oxygen-using bacteria, consume oxygen and manufacture carbon dioxide.

Now, what about the energy mechanism in plants? Whereas animals get their energy from ATP by way of the reaction of glucose with oxygen, plants manage to get their energy directly from the sun. The reaction which does this is:

$$CO_2 + H_2O + \text{solar energy} \rightarrow C_6H_{12}O_6 + O_2$$

Carbon dioxide and water vapor, both from the atmosphere, react together with the help of sunlight to synthesize the high-energy molecule glucose, leaving oxygen as a by-product. The process is called, appropriately, **photosynthesis** ("putting together by light"). Unlike the respiration reaction, the photosynthesis reaction consumes outside energy. The energy comes from the sun and goes into manufacturing the organic molecule glucose. The sun is thus the ultimate source of the energy used to produce the glucose that animals use in respiration because animals eat plants, or they eat other animals that ate plants, or they eat animals that ate animals that ate plants, and so on. Any way you look at it, it all comes back to the sun.

The animal kingdom depends on the plant kingdom in more ways than one. Besides producing our food, the photosynthesis reaction produces O_2. This is the source of almost all our atmospheric oxygen. If green plants were suddenly wiped from the face of the earth the animal kingdom, besides starving, would eventually run out of oxygen. The relationship between plants and animals is mutually beneficial, or symbiotic: animal respiration produces CO_2 for photosynthesis in plants, and plants reciprocate by producing O_2 for animal respiration. It really is one world!

Checklist

atomic theory	metric system
atom	milli-, kilo-, mega-
element	powers of ten
molecule	planetary model of atom
organic molecule	proton, neutron, electron
the three states of matter	atomic number
evaporation	chemical reaction
air pressure	combustion
Democritus	respiration
materialism	photosynthesis
reductionism	

Further Thought Questions

4. Do you think that all facets of human consciousness (love, for example) will eventually be explained in terms of atoms and empty space? Why or why not?

5. What motivations (practical, aesthetic, religious, etc.) do you suppose scientists have for trying to explain nearly everything in terms of atoms and empty space?

Further Exercises

13. The airship *Hindenburg* was filled with hydrogen before it disastrously burned in 1937. From this fact, and from your knowledge of the chemical composition of air, make a guess as to the chemical reaction that hydrogen undergoes when it contacts air.

14. For safety, lighter-than-air balloons are usually filled with helium. From this fact, how do you suppose helium acts in the atmosphere?

15. If a gas is compressed (confined to a smaller volume) the pressure on the inside walls of the container increases. Explain in terms of atoms.

16. If you could add a proton to the nucleus of every atom of a sample of germanium, what would you get? (Use Figure 3.1). Would you want the resulting substance to get into your stomach?

17. Gasoline, heating oil, and natural gas are hydrocarbon fuels composed entirely of hydrogen and carbon. What chemical reaction do you suppose produces the energy that runs an automobile?

18. Two pollutants produced by the burning of hydrocarbons in car engines are nitrogen oxide and nitrogen dioxide. Which elements must combine to form these molecules?

19. Continuing the preceding exercise: In view of the fact that gasoline contains *neither* nitrogen nor oxygen, where do you suppose these oxides of nitrogen come from?

20. Put these in order from lightest to heaviest: water molecule, oxygen atom, hydrogen atom, glucose molecule, electron, proton.

1. Theodore Roszak, *The Making of a Counter Culture,* Doubleday, Garden City, N.Y., 1969. An elegant statement of the antimaterialist and antiscience position. See especially Chapter 7, "The Myth of Objective Consciousness," by which Roszak means (in part) the myth of materialism.
2. Bertrand Russell, *Mysticism and Logic,* Doubleday, Garden City, N.Y., 1957. Essays by a foremost twentieth-century philosopher and logician that set forth a materialist and proscience position. See especially the essays entitled "A Free Man's Worship" and "The Ultimate Constituents of Matter." Perhaps Russell's book could be contrasted with the Roszak or Wheelis books, also listed here.
3. Allen Wheelis, *The End of the Modern Age,* Harper Torchbooks, New York, 1971. A moving and scientifically literate account of the crumbling foundations of the materialist-reductionist position.
4. Louise B. Young, ed., *The Mystery of Matter,* Oxford University Press, New York, 1965. An anthology of scientific essays. See especially Part 2, "Is Matter Infinitely Divisible?" and Parts 6, 7, and 9, which deal with living matter.

4

LIFE BEYOND EARTH?

Let's use our knowledge of the methods of science and of stars and atoms to discuss an interesting and possibly important question: Does life exist beyond Earth?

The Search

Are we alone? A few decades ago very few scientists studied this question seriously. The topic was considered a little weird, a little eccentric. Today several governments are putting money into the search for extraterrestrial life. It is no longer a crazy question.

Thought Question 1
What are the criteria for an idea to be crazy? Is astrology crazy? If a friend soberly explained to you that he had seen a 50-ft gorilla walking the streets of your town last night, would that be crazy? Keep in mind that crazy ideas are sometimes correct. Prior to 1800 most people were incredulous of the idea that rocks fell to Earth from outer space. When two Yale professors reported on several such rocks that fell in Connecticut, politician and naturalist president Thomas Jefferson reflected prevailing opinion by remarking "It is easier to believe that Yankee professors would lie, than that stones would fall from Heaven." Yet today meteorites are a commonplace idea.

The search has begun. Radio-astronomy search projects are under way in the USSR, the United States, and Canada; NASA satellites look for signs of intelligent communication via ultraviolet waves; Pioneer and Voyager spacecraft travel beyond our solar system bearing messages from Earth; Viking landers have searched Martian soil for signs of microscopic life.

Figure 4.1 is a reproduction of the message from Earth borne by the Pioneer spacecraft. Pioneer was launched in 1972 with the primary mission of observing the planet Jupiter and the asteroids lying between the orbits of Jupiter and Mars. As it passed Jupiter, the spacecraft was accelerated by the giant planet's gravity to become the first human artifact to eventually leave the solar system.

It seemed appropriate that the spacecraft should carry a plaque carrying a message from its makers to any intelligent civilization that might happen upon this space derelict some millions or billions of years in the future. The message, designed to be easily decipherable, is written in the language most likely to be comprehended by all members of the galactic technological community: science. It describes the location of our star (the sun) in the Milky Way galaxy, the nine-planet structure of our solar system, the planet from which the spacecraft was launched, and the year (relative to certain specific events in our galaxy) of launch. From the point of view of any extraterrestrial eyes that may see this plaque, the most mysterious objects will be the unusual creatures on the right-hand side.

Pioneer will probably never be intercepted. Moving at a mere 7 miles per second, 80,000 years would be required just to reach the vicinity of the nearest star, even assuming the spacecraft were pointed toward the nearest star. But the craft is not moving toward the nearest star, nor, as far as is known, directly toward any star. It could well be an interstellar wanderer forever. *The Cosmic Connection* (Ref. 2) by Carl Sagan, designer of the plaque, is a fascinating account of the search for, and prospects of, extraterrestrial life. Sagan's ideas are the inspiration for much of this chapter.

The message on the two Voyager spacecraft, launched in 1977, is more elaborate than that of Pioneer's plaque. The Voyagers each carry a 12-in. copper phonograph record, complete with cartridge, turntable, and pictorial instructions for use. The records contain such sounds of Earth as whales, rain, automobile gears, rocket lift-off, greetings in all major languages, and a musical variety show of 27 selections featuring everything from Bach to Navaho chants to Japanese flutes to Louis Armstrong. A major portion of the disk is devoted to 110 pictures encoded into sound. Any

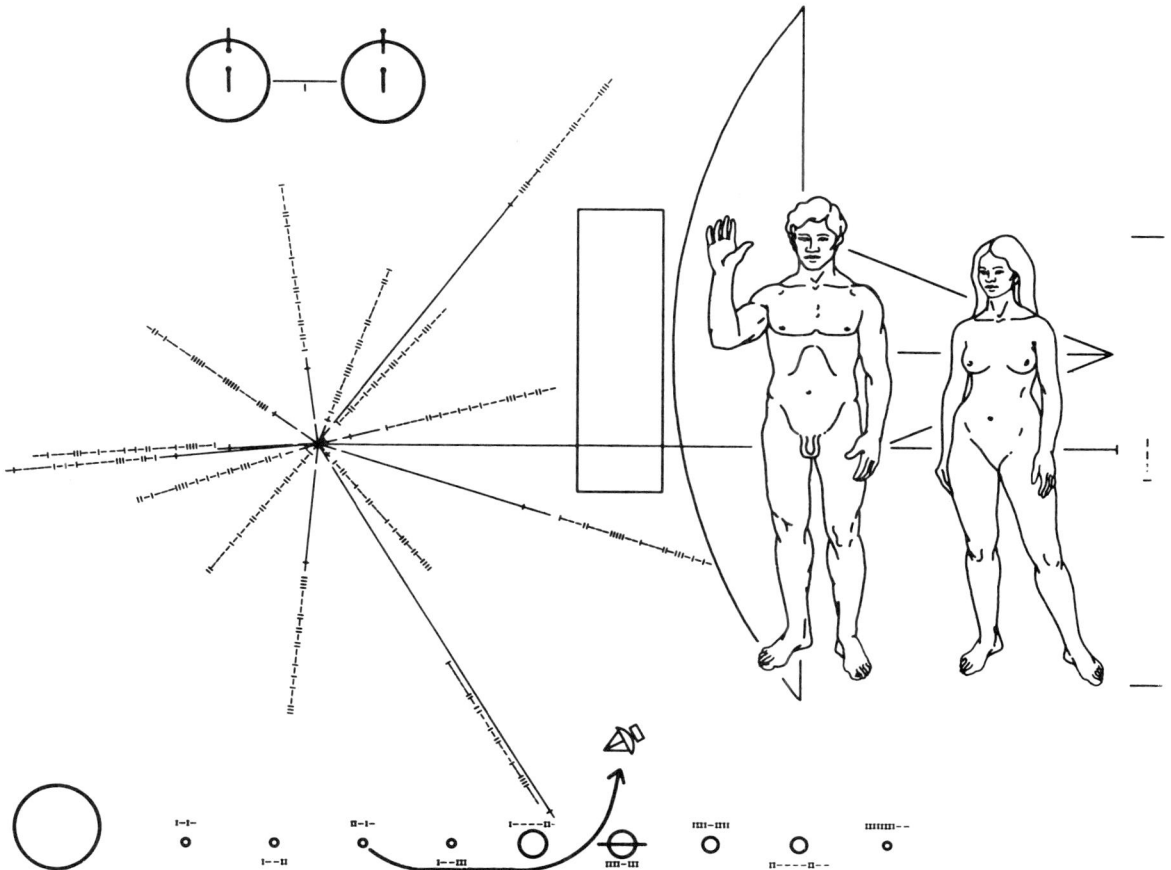

FIGURE 4.1 The plaque aboard the Pioneer spacecraft. (From *The Cosmic Connection* by Carl Sagan, Dell Publishing Company, Inc., New York.)

civilization with the know-how to have recovered the spacecraft should be able to decode these sounds. If they do they'll discover a gallery that includes Mars, DNA, birth, the Snake River, a leaf, a crocodile, a mountain climber, a Chinese dinner, Boston rush-hour traffic, and the Toronto airport.

Thought Question 2
What in the world do you think they'll make of all this?

In August 1975, the United States launched two Viking satellites toward Mars. Each satellite orbited Mars and sent a smaller lander down to the surface. A primary mission of the two landers was the search for life. Because of extreme climatic conditions, few people expected to find large-scale animals or plants on Mars. There was some hope, however, that microscopic life forms, such as those that thrive even in arctic and desert conditions on Earth, might be found in the soil.

Viking performed several experiments designed to detect life. For example, the labeled-release experiment was designed to detect the presence of microscopic animallike life or LGPs ("little green people").* The lander scooped up a cupful of Martian soil and combined it with a "soup" concocted of various nutrients. The idea was that any LGPs in the soil would eat the soup. The soup, like all food on Earth, consisted mainly of various carbon-containing molecules. Some of these molecules had been made radioactive, so that they could be detected by radiation-measuring devices, by incorporating the radioactive form of carbon known as carbon-14 rather than the normal, nonradioactive forms of carbon. The radioactivity was low enough that it shouldn't harm the LGPs but high enough to be detectable. It was hoped that the LGPs would find this free lunch palatable and thus incorporate some of the carbon-14 into their bodies.

If LGPs are the least bit similar to Earth animals they should perform respiration (Chapter Three), producing CO_2 gas, with the carbon coming from their cellular structure. Thus the captive LGPs should eventually exhale radioactive CO_2 gas, and a radiation detector above the sample should detect the presence of radioactive gas. To summarize, the purpose of this experiment was to search for signs of animal respiration.

The experimental results: As soon as the nutrient solution was added to the soil, radioactive gas appeared. In a control experiment, that is, an experiment to partially check the validity of the original experiment, a Martian soil sample was sterilized by heating, unfortunately murdering any LGPs present, and then the soup was added. As expected, no radioactive gas appeared. Whatever mechanism had produced the gas in the original experiment was shut off by the sterilization process.

Not only the labeled-release experiment, but also two other entirely independent experiments, gave strong positive indications of life on Mars. Things were looking good for the hypothesis that life does exist on Mars.

Now for a lesson in the scientific method. In addition to the three experiments already mentioned, scientists conducted a fourth experiment, a direct chemical search for organic molecules in Martian soil. If life exists in Martian soil, then this soil should contain organic molecules. To date no such molecules have been found, despite a chemical analysis that should have revealed them if they were there.

---

*The traditional term is LGMs ("little green men").

So, three experiments support the hypothesis of life on Mars, while one contradicts it. Three-to-one in favor, so the hypothesis is true, right? Wrong. The hypothesis is false. The 3-to-1 score in favor of the hypothesis isn't sufficient to sustain it. Neither would a 1000-to-1 score suffice. Scientific theories and hypotheses always dangle on the edge of disaster. A single piece of contradictory evidence can disprove it (see Chapter Two on this). So most scientists now feel that life does not exist on Mars and that some other explanation must be found for the strong positive results of the first three biology experiments. (Disclaimer: By the time this book reaches your hands, scientific opinion might have swung around in favor of life on Mars. The negative results of the chemical search for organic molecules are not absolutely conclusive. Isn't science fascinating?)

Is Anybody Out There?

There is a wide range of opinion on this question, because we have so little data to go on. Some people feel that our planet is special, that Earth was especially created as a home for humans, the sole intelligent species in the universe. The other side of this argument is that our planet doesn't seem to be special in any other way. Earth circles a very ordinary star in a very ordinary location in a very ordinary galaxy. There are billions of other stars in our galaxy and billions of other galaxies in the universe. What's so unusual about Earth, anyway, that intelligent life should arise only here?

Many scientists who have given much thought to the matter hypothesize that life very likely exists at other places in the universe. But there is great uncertainty as to whether technological life exists out there, that is, life forms capable of sending out radio waves or of communicating with us in some other way.

One argument for the existence of life beyond Earth is that there are several places in our solar system alone where life might conceivably have evolved: Earth (obviously), Venus, Mars, Titan (a moon of Saturn), perhaps the gas clouds of Jupiter. Although these places certainly don't have conditions that could support Earth-like life, they could support (or might have supported in the past) certain types of organic molecules and perhaps living organisms. So, the argument goes, if around our star alone there are several places that could possibly support life, then among the 100 billion stars in our galaxy and the 100 billion galaxies in the universe there may be a fantastic number of other places that could support life. It is very likely that life has arisen in many of these places.

The following quantitative argument will give us some idea of the number of places where life might have arisen, and a feel for the kinds of scientific problems that relate to this question. This argument is speculative, so don't take it too seriously. Here goes.

In our Milky Way galaxy there are about 100 billion stars. How many of these stars have planets orbiting them? About 50 percent of the stars in the galaxy are actually not single stars at all, but instead are small systems of two or more stars in orbit around each other. Any planets orbiting such a multiple-star system would probably move in such an erratic orbit that life would be impossible. It is difficult to estimate what fraction of the remaining 50 percent have planets, since stars are too far away for us to be able to detect planetary systems by direct observation. It is thought that most single stars evolved in a manner similar to that of our own sun, so it may be reasonable to assume that nearly all of the single stars have planets.

Next question: How many planetary systems have planets or moons with conditions that could conceivably support some form of life? We have seen that our solar system alone contains several such places. So perhaps 10 to 100 percent of all planetary systems have at least one planet or more with life-supporting conditions. Let's be conservative and say 10 percent.

But did life actually evolve in all these places? Or on some fraction of them? Or only on Earth? Many biochemists hypothesize that life will arise, through a gradual chemical process, under a wide variety of conditions. One of the remarkable properties of the amino acids and other organic chemicals is the ease with which they form spontaneously in the laboratory. Once a large number have formed, they tend to cling together in isolated droplets of highly concentrated organic material. Such droplets then compete with each other to absorb further organic molecules. Very highly organized droplets may eventually evolve through such chemical competition. In this manner, perhaps, did the first, simple, single-celled organisms evolve on Earth.

Let us, for the sake of argument, estimate that 10 percent of those planets with life-supporting conditions actually developed life.

Exercise 1
Based on these estimates, calculate the number of places in our galaxy at which life evolved. (Answer below.)

If you figured correctly, you figured that 10 percent of 10 percent of 50 percent of 100 billion is 500 million. And that's in our galaxy alone. Now recall (Chapter Two) that there are some 100 billion other galaxies out there. It seems that our universe may teem with life!

Since we're interested in the possibility of communicating with whatever LGPs are out there, we would like to know how many of these 500 million places in our galaxy support intelligent, technological life. First, consider intelligence alone. Although cynics might consider the point debatable, it does seem that intelligent life evolved at least once, namely on Earth, in the form of *Homo sapiens*. Furthermore, nature has independently evolved another, distinctly nonhuman, form of intelligent life on Earth.

Exercise 2
Any ideas as to what this second, independent form of intelligent life on Earth might be? (Answer below.)

Answer: dolphins.* The dolphins and their relatives the whales have large and highly complex brains. Perhaps dolphins would have developed a technology if they had hands instead of flippers. Since intelligent life developed in two entirely independent ways on Earth, intelligent life might be reasonably likely to develop wherever life exists. As a conservative estimate, let's say that intelligence develops in some 10 percent of the places where life has developed.

Next, how about technology? The dolphins have never developed a technology, but humans did. This gives us very little to go on in our speculations about the

*The great apes are, of course, also highly intelligent, but they are closely related to humans so they don't represent an independent evolutionary path.

likelihood of the development of technological life. Just for fun, let's assume that an intelligent life form has a 10 percent chance of developing a technology.

Exercise 3
At how many places did technology develop? (Answer below).

Answer: 10 percent of 10 percent of 500 million (Exercise 1) yields 5 million places in our galaxy alone where technology developed!

But how do we know that technology still exists in all these places? Perhaps most of these technologies started up in the distant past and were destroyed long ago by nuclear war or throw-away beer cans or traffic jams. Some may have been destroyed by natural catastrophies such as the death of their planet's sun. Certainly a nuclear war could destroy our civilization in short order. On the other hand, it is conceivable that the human race will emerge into a long era of peace and stability. Thus it is extremely difficult to estimate how long technological civilizations survive. We won't try to estimate this figure.

Our galaxy has been here for some 10 billion years. It seems reasonable to suppose that technological civilizations have been cropping up at all times during the history of our galaxy. There's no reason to suppose that they all started roughly when ours did.

Exercise 4
In this case, how often does a new technological civilization start up in the galaxy? (Answer below.)

If we disperse our 5 million instances of technology (Exercise 3) evenly among the 10 billion years of galactic history, we conclude that a new technological civilization started up every 10 billion/5 million = 2000 years. Thus if your typical technological civilization survives for only a few centuries then, at any given time, there is only one, or (more likely) none, in our galaxy. In that case, we're it for the present.

Thought Question 3
How long do you expect we will survive as a high technology civilization?

So the current number of technological civilizations in our galaxy may be just one (us) or it may be as many as 5 million (Exercise 3), depending on how many of them survive the possibly harmful consequences of their own technology.

Exercise 5
There are about 100 billion stars in our galaxy. On the assumption that there are 5 million technological civilizations in our galaxy today, roughly how many stars would astronomers have to scan to have a reasonable chance of finding one star with technological life? (Answer below.)

Answer: 100 billion/5 million = 20,000. Now, scanning 20,000 stars for signs of life is no mean task. Even if there are as many as 5 million technological civilizations in our galaxy, it won't be easy to locate them.

Listening for Life

Dialogue with an extraterrestrial civilization (ETC) will probably be impossible. Radio waves travel at the speed of light, and at this speed a radio wave from Earth takes about 100 years to travel to even the relatively nearby stars. Not your snappiest conversation! But we can send and receive one-way messages. Since any message we transmit must wait a century or more for a reply, most of us are more interested in receiving than sending. As mentioned earlier, listening projects are already in operation in the USSR, the United States, and Canada. But given the large number of stars that must be surveyed to find a single ETC, if any even exist, and given the uncertainties as to what frequency an ETC might use to communicate, if they are sending messages at all and if they are using electromagnetic waves as the communication medium, our present small-scale efforts are not likely to be successful anytime soon.

But what if we do succeed? Many scientists feel that the receipt of a message from another civilization would be the most important event in the history of our planet. Any such message would almost certainly come from an ETC far in our technological future. After all, at a mere 50 years into the technological age we are in our technological infancy. Most technological ETCs would be far more mature. A large portion of the wisdom of their civilization could be gotten into a few hours of radio signals. There is no telling what they would want to communicate to us. My guess is that, among other things, they would try to tell us how to survive the consequences of our own technology.

Will They Visit Us? ... Have They Already?

Good arguments based on presently known science can be made for the proposition that no ETC has visited Earth in the past and for the proposition that there is some possibility that we have been visited.

Astronomer Carl Sagan, in *The Cosmic Connection,* makes a good case for the first proposition. Here is Sagan's argument. If there are ETCs out there, how would they know we are here? Radio and television and other electromagnetic signals are limited to the speed of light, so our signals from Earth will not begin reaching them for another century. The only way they could be aware of life on Earth, then, is by making a random fly-by or random visit here. Now let's make the very optimistic assumption that a large number of ETCs, say 50,000, is continuously engaged in sending out spacecraft to randomly search for life on other worlds. How many spacecraft per year would each one of them have to send out in order for there to be a reasonable probability, say 1 percent, that one of these spacecraft will investigate our solar system? The answer comes out to be several thousand launches per year. It seems unreasonable that any ETC would be interested in putting this kind of effort into the search for further ETCs; the energy and material requirements would be enormous.

Thought Question 4
Can you think of any rebuttals to this argument?

A good case for the contrary proposition, that a past visitation is possible (not that it has necessarily actually occurred), has been made by jet-propulsion scientist T. B. H. Kuiper and astronomer M. Morris in their article "Searching for Extraterrestrial Civilizations" (*Science*, May 6, 1977). Rather than assume that an earth visitation would have to result from a random fly-by of Earth by an ETC based around some single star, they consider the possibility that at least a few ETCs have engaged in colonization of planets around stars other than their original home star. Using such known energy sources as nuclear fusion, and assuming reasonable speeds for interstellar travel, Kuiper and Morris argue that such colonization is possible. Any ETC with an instinct for colonization would then go on spreading from one star to another until a large number of stars, spread over a large region of our galaxy, were inhabited. In this case, it is plausible that a star near our sun has already been colonized and that, starting from this nearby star as base, a random fly-by of Earth has occurred.

Thought Question 5
Can you think of any rebuttals to this argument?

So, good arguments can be made on both sides. Note that the argument is between the proposition that we have not been visited and the proposition that it is *possible* that we have been visited. To the best of my knowledge there is no solid evidence supporting the hypothesis that we actually have been visited, despite the frequent sensationalist UFO reports displayed so prominently and so tastelessly in tabloid newspapers (no artifacts from ETCs, no unambiguous photographs, etc).

Because there seems to be no theoretical evidence in the form of a convincing argument that we probably have been visited and no observational evidence in the form of artifacts or photographs, my opinion is that we have never been visited.

Thought Question 6
But I might be wrong. What's *your* opinion?

Thought Question 7
How would you feel if a representative from an ETC came to Earth and, for instance, addressed the United Nations General Assembly on live television tomorrow morning? Would such a visitation by a being from a civilization far in advance of ours have any effect on your self-image?

I would like to close this chapter with a personal response to the above question. Many centuries ago, Copernicus displaced humankind from the center of creation. Despite the fact that we often act as though this planet were ours to exploit as we see fit, I accept the idea that the universe was not made specifically for *Homo sapiens*. So it won't bother me, and I think it won't bother many others, to find that there are other intelligent creatures out there. On the contrary, I find appealing the notion that the universe is filled with life, filled with intelligence. Life and intelligence take on a larger significance if *Homo sapiens* is part of a living universe. Human civilization on this remote planet then becomes part of a vast natural experiment. What we do here truly matters, for it contributes to the evolution of an increasingly self-aware universe. And so it becomes cosmically important that our earthly experiment with life be successful.

Checklist

labeled-release experiment

life in the universe

intelligent life in the universe

technology in the universe

communication from extraterrestrial life

visits by extraterrestrial life

Further Thought Questions

8. Why might ETCs want to communicate with us? What would they communicate?

9. What effects might the receipt of a message from an ETC have? What effect might such a message have on religion, world peace, science, the space program, the arts?

10. Do we have anything to fear from ETCs?

Further Exercises

6. Could the labeled-release experiment described in this chapter have detected the presence of microscopic plant life in the Martian soil? Why? *Hint:* Recall that this experiment detects radioactive CO_2 gas in the space above the soil sample.

7. Suppose that Martian soil contained organisms that perform photosynthesis. What type of organic molecule would you expect to find in the soil?

8. Recalling that photosynthesis produces O_2, would the presence of O_2 in the atmosphere above a Martian soil sample necessarily mean that the soil contained plantlike life? Explain.

9. Suppose that you injected a Martian soil sample with water containing radioactive oxygen atoms. Describe how you might then detect the presence of Martian plant life.

Recommended Reading

1. William K. Hartmann, *Astronomy: The Cosmic Journey,* Wadsworth Publishing Company, Belmont, Calif., 1978. Chapter 25 of this text for nonscientists is devoted to a discussion of extraterrestrial life.

2. Carl Sagan, *The Cosmic Connection,* Dell Publishing Co., New York, 1973. A fascinating and authoritative perspective on extraterrestrial life. Excellent science writing for nonscientists.

PART 2

The Newtonian Universe

5

MOTION: HOW THINGS MOVE

Let's pick up the historical thread of physics and astronomy where we left it in Chapter Two. By the mid-seventeenth century Aristotelian physics and Ptolemaic cosmology were fading, whereas Copernican cosmology was coming into bloom. Enter Isaac Newton (1642–1727). Newton invented a few simple but profound ideas that tied together the ideas of Copernicus, Kepler, and others, giving a comprehensible physical foundation for the new cosmology. As Alexander Pope expressed it at the time:

> Nature and Nature's laws lay hid in night.
> God said, Let Newton be! and all was Light.

Newton's physics dominated science in the eighteenth and nineteenth centuries but was finally superseded at the beginning of the twentieth century by quantum mechanics and Einstein's theory of relativity. So, you may be wondering, why on earth should we study Isaac Newton's old ideas when newer and presumably better ideas are available? Here's why.

For one thing, nearly every mechanical device can be best understood in terms of Newton's physics rather than in terms of the modern theories of quantum mechanics and relativity. Automobiles, skyscrapers, steam generating plants, bridges, airplanes, electric motors, bicycles, you name it—Newton's physics deals with these things best. The reason is that, for these devices, Newtonian physics gives essentially the same results as the modern theories, and Newtonian physics is much easier to use and to comprehend. It would be fair to say that at least 90 percent of all twentieth-century technology is based on pre-twentieth-century physics.

For another thing, it is true today, as it was true in Newton's day, that the structure of our solar system is best understood in Newtonian terms.* In addition, many other aspects of astronomy (structure of galaxies, star clusters, multiple stars) are basically Newtonian.

But the most important reason for studying Newton's physics today is its enormous hold on our view of ourselves and our environment. The Newtonian world view, the philosophical implications of Newtonian physics, has permeated our culture so deeply and so subtly that most of us are unaware ot its presence. Even though modern science is based on post-Newtonian theories, most of us still live philosophically in the Newtonian age.

Newton invented two theories: a theory about the general relationship between forces and motion, and a theory of gravity. The first theory, known as Newtonian mechanics, is summarized in Newton's three laws of mechanics, to be presented in this chapter and the next. This chapter will concentrate on motion; Chapter Six will emphasize forces. Together, they add up to mechanics: the study of forces and

---

*This statement needs some qualification as a few solar system phenomena require the theory of relativity for their explanation.

motion. Chapter Seven will deal with Newton's theory of gravity, and Chapter Eight will present a few weird and wonderful applications of Newton's physics.

Aristotle on Natural Motion

Thought Question 1

Imagine dropping a book, throwing a rock, rolling a ball. In light of these examples, what type of motion seems to be the natural motion of a solid object? That is, if an object (a rock, for instance) is allowed to just do whatever it is naturally inclined to do, without any help from the outside, how will it behave?

Aristotle asked himself Thought Question 1. So have many others. It is perhaps the most basic question one can put to nature.

The answer to this question seems fairly obvious. When you drop a book, it falls to the floor. When you throw a rock, it moves through the air for awhile and eventually falls to the ground. When you roll a ball, it eventually comes to rest. Obviously it is natural for such objects to be at rest on the ground. Falling, in an effort to reach this natural state, is then the natural motion of solid objects.

This is how most of us look at this question. It is also how Aristotle looked at it. In addition to its intuitive appeal, this answer receives further support from Aristotle's theory of the physical world. This theory states that all earthly objects are mixtures of the four elements Earth, Air, Fire, and Water, and that each of these pure elements has its own natural place of repose: Earth at the bottom (the center of the universe), next Water, then Air, with Fire at the top. The natural motion of an object is determined by the element most evidently present. For example, a rock, composed primarily of Earth, should fall, whereas steam, containing Fire, should rise until it loses its Fire, whereupon it reverts to Water and falls to Water's natural place, above Earth but below Air. But according to Aristotle's theory, the four earthly elements exist only in the sublunar domain, below the moon. Beyond the moon, the universe is constructed of the heavenly and perfect element Ether, or quintessence, (from the Latin for "fifth element"), a substance whose natural motion is that most symmetric and eternal of all conceivable motions, endless circles.

Thus Aristotle's physics and Ptolemy's astronomy compliment one another nicely. A rather neat and sensible system, all things considered. A system that satisfies our criteria for a good theory (Chapter Two): a few related ideas that explain a broad range of observed phenomena and that lead to definite, checkable predictions about the natural world. Furthermore, Aristotle's and Ptolemy's theories agree with the obvious testimony of our senses, that solid bodies fall to the ground and that heavenly bodies move in circles around Earth.

Exercise 1

When an air-filled balloon is heated, it expands. Use Aristotle's physics to explain this phenomenon. The modern atomic explanation is that heating causes the air molecules inside to move faster, so they hit the sides of the balloon harder, which increases the pressure on the inside of the balloon, which makes the balloon expand.

If our air-filled balloon is heated sufficiently, it will rise spontaneously. Explain, using Aristotle's physics. The modern explanation is that the expansion of the balloon makes it less dense than it was before. That is, a cubic centimeter of the balloon weighs less because of the expansion. Thus the balloon floats upward in air, just as a piece of wood floats to the surface in water.

Moral: Even "out-of-date" theories can be useful. In these exercises we have another instance of observed phenomena that are explainable by either of two theories (Aristotelian physics and atomic physics). Today we regard atomic physics as the better theory because it can explain a much broader range of phenomena than can Aristotelian physics.

Aristotle concluded that the natural motion of objects like books, rocks, and balls is downward, because they are composed primarily of the element Earth, and that the natural state of such objects is the state of rest on the ground. Any other motion, such as sliding along a tabletop, or moving upward, or moving through space in a straight line, would then have to be unnatural, violent, or forced motion.

Galileo on Natural Motion

Galileo Galilei (1564–1642), inventor, astronomer, physicist, perhaps the first true experimentalist, father of the modern scientific method, contemporary of Kepler and predecessor of Newton, took a different point of view. In fact, Galileo had an original point of view about almost everything. His iconoclastic ways of thinking and his outspoken manner kept him in trouble most of his life. Maybe you can see some of these qualities in his portrait, Figure 5.1.

Galileo offended his scientific colleagues by disputing Aristotelian physics and offended church authorities by advocating the Copernican cosmology. Using his newly invented telescope, he discovered mountains on the moon, spots on the sun, and moons circling Jupiter, discoveries that contradicted the Aristotelian-Ptolemaic views that the heavenly bodies are perfect and that all such bodies circle Earth. Unable to shut him up, church authorities finally locked him up instead. Galileo was placed under house arrest for the last 8 years of his life, but he nevertheless managed to write *Two New Sciences*, his fundamental work on mechanics.

In his investigation of natural motion, Galileo studied the motion of solid objects as they moved horizontally, rather than as they fell vertically. He observed that when a smooth ball rolled along a smooth table, the ball always slowed down if the table was inclined slightly upward, sped up if the table was inclined slightly downward, and kept moving almost indefinitely if the table was horizontal. In his mind's eye, Galileo extrapolated to the ideal case: A perfectly hard, smooth ball on a perfectly flat, level table of infinite extent would roll forever once it was started. Even though Galileo couldn't carry out this idealized thought experiment because of the lack of an ideal ball and ideal table, he was of the opinion that this was the really basic result, that the ideal ball rolling forever along the ideal table was truly natural motion. He concluded that a naturally moving object, that is, a body subject to no forces (pushes or pulls), must move in a straight line at constant speed.

FIGURE 5.1 (a) Aristotle. (The New York Public Library-
Picture Collection.)

The distinction between Aristotle's view and Galileo's turned out to be quite basic. Natural motion is one of the foundation blocks in the house of science. Nearly every other concept depends, at least indirectly, on what kind of motion is natural (unforced) and what kind is thus unnatural (forced). Remove this foundation block and the entire structure falls. In removing the Aristotelian block labeled *natural motion* and replacing it with an invention of his own, Galileo laid the foundation on which Isaac Newton would build a new physics.

Galileo's principle was eventually incorporated into Newton's physics as the first law of mechanics,* so it is appropriate at this point to state:

Newton's First Law of Mechanics (Galileo's principle of natural motion): Any body that is subject to no outside influences persists in its state of rest or of straight-line motion at constant speed.*

Exercise 3
A typical rock, or meteoroid, in space moves through our solar system for billions of years. What keeps such a rock moving? Is anything keeping it moving? (Answer below.)

---

*We'll speak of Newton's *laws*, because everyone else uses this term. Newton's *principles* would be a more appropriate term. Don't get the impression that these laws are absolute!

*Historical note: For Galileo, this principle applied only to motion on a horizontal surface. Newton extended Galileo's idea to motion in any direction.

FIGURE 5.1 continued. (*b*) Galileo. (The Bettmann
Archive.)

The answer: Nothing is keeping it moving; it keeps moving because there is no outside influence to stop it. Fine, you may say, we don't need to explain the meteoroid's continued motion once it has been started. But what started it moving in the first place? Unfortunately, astronomers don't know the answer to this one. Meteoroids are bound to our solar system (i.e., they can't escape our solar system), so their origin is probably tied up with the obscure origins of the sun and the planets.

Exercise 4
I hold a rock above the ground and release it from rest. Once released, and while falling through the air, is there an outside influence on the rock? (Answer below.)

According to Newton's first law, the answer is: Yes there must be an influence on the rock, because it does not persist in the state of rest that it had at the instant of release. Or, again, its speed is not constant during the fall (it speeds up). The outside influence operating here came to be called **gravity.**

Exercise 5
Imagine a book sliding along a tabletop. It slows down and stops. Would Aristotle have said that an outside influence acted on the book? Would Galileo have said so? What name would they have given to this outside influence?

Unnatural Motion

Obviously there are many kinds of motion other than straight-line motion at constant speed. These unnatural motions are best described in terms of the concepts of *speed, velocity,* and *acceleration.*

The **speed** of a moving object is just the distance it travels divided by the time required to travel that distance:

$$\text{Speed} = \frac{\text{distance traveled}}{\text{time of travel}}$$

Exercise 6
A bicycle racer cycles 150 km in 5 hours. What is the racer's speed?

Note that speed, as defined here, really means average speed. The cyclist in the exercise was surely not maintaining exactly 30 km/hr (kilometers per hour) for the entire race. The next exercise illustrates this.

Exercise 7
You desire to travel from Dallas to Los Angeles. You take a cab from your downtown Dallas hotel to the Dallas–Fort Worth Airport, 40 km out in the country. Because of the early evening rush, this trip takes 1 hour. You immediately board your waiting jet and fly 2000 km to the downtown Los Angeles International Airport in 4 hours. Unfortunately, your suburban home is 60 km out in the country, so you take a bus from the airport. It seems that you have managed to hit the late evening rush in Los Angeles; furthermore, you must change buses once; furthermore, there are detours and delays caused by highway construction. So, this final leg of your journey takes 3 hours. (a) How on earth did you manage to hit rush-hour traffic in both Dallas and L.A., when the flight took 4 hours? (b) What was your speed during each of the three legs of your trip? (c) What was your average speed for the entire trip (total distance divided by total time)?

In most people's vocabularies, *speed* and *velocity* mean the same thing. Not so in physics. In physics it turns out to be useful to define the **velocity** of a moving object as its speed together with its direction of travel.

Exercise 8
One jogger is headed north at 12 km/hr, another is headed south at 12 km/hr. Do they have the same speeds? Do they have the same velocities?

Now this distinction between speed and velocity might look artificial to you, but it turns out to be useful. These kinds of seemingly trivial distinctions often turn out to be crucial in getting a clear picture of the physical world. We could have defined velocity in any way we wanted. For instance, we could have defined it to mean simply speed. Physicists have chosen to adopt the above definition because it is useful, as we will see. We have here another instance of the creative, inventive element in science. The very words that scientists use, words like *speed, velocity, acceleration,* are given their meanings for reasons of economy of ideas rather than because nature ordains that these words must be given these particular meanings.

Exercise 9

A bicyclist moves at a steady 20 km/hr throughout a 90-deg curve. Is the speed constant? Is the velocity constant?

Exercise 10

Space satellites in low circular orbits around Earth all move at about 18,000 mph or 30,000 km/hr. It is about 40,000 km around Earth. How long does one complete orbit take? Is the satellite's speed constant? Is its velocity constant?

You can see from these exercises that circular motion at constant speed is not at constant velocity, since the direction of the motion is always changing. Keep this in mind; it will be important later.

Velocity is useful for describing moving objects, that is, objects whose *position* is changing. We've also seen examples of objects whose *velocity* is changing, and it is useful to have a concept to describe this idea as well. Physicists call this concept **acceleration.** This word is used in a very precise way to refer to changes in velocity, that is, *an object is accelerated whenever its velocity is changing.* Quantitatively, the acceleration of an object is the change in the object's velocity divided by the time required to make that change:

$$\text{Acceleration} = \frac{\text{change in velocity}}{\text{time to make change}}$$

People ordinarily use the word *acceleration* to mean speeding up, and *de-acceleration* to mean slowing down. The official physicist's definition of acceleration might seem artificial to you, since it is only loosely related to the ordinary usage. According to the physicist's meaning of the word, any object whose velocity is changing in any manner whatsoever is accelerated. An object is accelerating not only when it is speeding up, but also when it is slowing down. Moreover, it is accelerated when it is simply turning a corner, even at constant speed, since if the object is changing its direction of motion it is changing its velocity and so (by our definition) it is accelerated. We define acceleration in this strange manner because it turns out, later, to be useful.

Exercise 11

For each of the following examples, fill in the corresponding blanks in the table below with *yes* or *no*.

(a) A car moving down a straight, level road at a steady 100 km/hr.
(b) A car moving uphill on a straight incline at a steady 30 km/hr.
(c) A car just starting up from rest along a level road.
(d) A car rounding a curve at a steady 50 km/hr.
(e) A car slowing down from 40 km/hr to 30 km/hr along a straight road.
(f) A car stopping (coming to rest) along a straight road.
(g) A rock that has just been dropped and is falling through midair.
(h) A NASA satellite orbiting Earth at a steady 30,000 km/hr.
(i) Earth orbiting the sun at a steady 100,000 km/hr.
(j) The sun orbiting the center of the Milky Way galaxy at a steady 800,000 km/hr.

	a	b	c	d	e	f	g	h	i	j
Is the speed increasing?										
Is the speed decreasing?										
Is the direction changing?										
Is the object accelerated?										

Compare your answers to Exercise 11 with my answers in the back of the book. If your answers are correct, you probably have a good grasp of acceleration. Note that in each part, if your answer to any one of the first three questions is *yes,* then your answer to the fourth question should also be *yes.*

To further sharpen your thinking on this, let's work out a simple numerical example. Suppose a NASA rocket carrying a satellite into Earth orbit accelerates from 0 to 500 m/sec (1120 mph—about 6 percent of the final orbital speed) in 10 sec. Let's calculate the acceleration. Since the motion is along a straight line, the acceleration is due to the changing speed rather than to any change in the direction of motion. This change in speed is 500 m/sec, so that

$$\text{Acceleration} = \frac{\text{change in velocity}}{\text{time to make change}} = \frac{500 \text{ m/sec}}{10 \text{ sec}} = 50 \frac{\text{m/sec}}{\text{sec}}$$

You can read the units as "meters per second per second," sometimes abbreviated as m/sec², or meters per second squared. The answer says that during each second, the speed *changes* by 50 m/sec. So the acceleration is 50 meters per second *per second.* Get it?

Falling

Pick up a light object, such as a penny, and a heavier object, such as a rock or a book. Hold one in each hand. Does it seem to you that, if dropped at the same time, one will fall faster than the other? Now drop them. Which falls faster?

Many people expect that the heavier object will fall faster than the lighter object. But you probably discovered when you did the experiment that the two objects fell at the same speed.*

Galileo is reputed to have performed this experiment by dropping various weights from the top of a tall tower, such as the Leaning Tower of Pisa, thus dramatically demonstrating that light objects fall as fast as heavy objects. Whether it is historically accurate or not, this story jibes well with Galileo's forceful personality and with his flair for discrediting the traditional ideas of Aristotle and Ptolemy.

It had seemed natural to Aristotle that heavier objects, containing more of the element Earth, should strive harder toward the center of the Earth and so should fall faster than lighter objects. This point plays only a minor role in Aristotle's theory; he never even stated it very clearly. But Galileo, believing that Aristotelian physics and Ptolemaic astronomy were fundamentally erroneous, seized on this obscure and ambiguous point and interpreted it in the worst possible light so that it seemed to say

*Provided that air resistance on each object is negligible. This point is discussed in more detail below.

that bodies fall with a speed that is proportional to their weight, that is, a 2-lb ball should fall twice as fast as a 1-lb ball. After actually reading Aristotle's original statement, it is not clear that Galileo's interpretation bears much resemblance to it. Once he had set Aristotelian physics up for the kill, it was not difficult for Galileo to drop various objects (cannon balls and musket balls, reportedly) from the tower to prove Aristotle's fallacies to the learned world.

Thought Question 2

Was this a dishonest tactic? Remembering that the new physics of Galileo and Newton proved to be more successful than Aristotelian physics and that it was difficult to persuade the seventeenth-century world to open their minds to the new theories, should Galileo have employed this tactic?

Try dropping a light coin and a heavy coin simultaneously. Listen for the clicks as they hit the floor. Do they actually fall at the same speed? If you did this experiment carefully you heard the two objects strike at the same time. Well, nearly the same time. The two clicks, if you heard two rather than one, should have occurred so close together that the difference between them could be ascribed to such experimental errors as releasing the coins at slightly different times or holding them at slightly different distances above the floor. This is always the way it goes with scientific experimentation. The results are never completely clear-cut, and the experimenter must decide which results are central to the experiment and which are merely caused by experimental error.

After performing this experiment with a large number of falling objects and finding that they all fall at the same speed, we finally reach the conclusion that some sort of general principle is operating here. Let us call it:

Galileo's Principle of Falling: Any two objects dropped simultaneously from the same height will hit the ground simultaneously.

Thought Question 3

Is this statement best classified as a fact or as a theory?

Exercise 12

(a) Suppose you performed Galileo's experiment with a large lead ball and a large Styrofoam ball. Would the two hit at roughly the same time? (b) What if you used a lead ball versus an air-filled balloon? (c) What if you tried it with a helium-filled balloon? (Answer below.)

Apparently there are exceptions to Galileo's principle of falling. This principle is violated in all three examples in Exercise 12. In fact, there are some objects, like the helium-filled balloon, that don't even fall when you drop them. The list of violations of Galileo's principle could be extended *ad nauseum*: a piece of paper falls at a different rate from a stone, a strong wind might affect the rate of fall, and so on, and so on.

These exceptions to Galileo's principle illustrate an important point: Nearly every scientific theory has only a limited range of validity. The limitations on a theory do not make the theory wrong, or useless . . . they simply make it limited. Often the

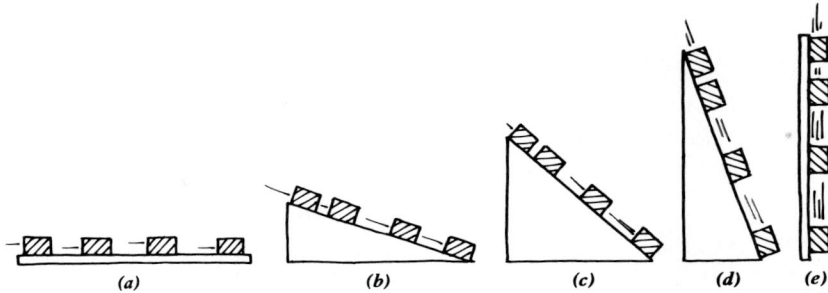

FIGURE 5.2 An assortment of inclines.

limitations on a theory are not clearly stated and must be read into the theory. For example, Galileo's principle of falling is limited to situations in which air resistance is negligible and in which no strong winds are blowing.

Have you ever been to a science museum? If you wander into a museum that has do-it-yourself physics experiments, you might find one on objects falling in vacuum, that is, in the absence of air. A large jar (vacuum chamber) contains a rock and a feather. With air in the jar the rock falls like a rock and the feather floats down like a feather. The viewer can push a button to activate the vacuum pump. With the jar half-evacuated, the feather falls much faster; with the jar completely evacuated, the feather falls just like the rock. It's a good demonstration of the effect of air resistance on falling.

More Falling

In Galileo's study of motion, he investigated the behavior of objects started from rest and allowed to move down an assortment of inclines. Figure 5.2 pictures a smooth block sliding down several smooth inclines, such as a piece of ice sliding down an icy slope.

Exercise 13

In which of the cases shown in Figure 5.2 is the block accelerated? In which cases is it unaccelerated?

In Figure 5.2e the motion of the block is the same as the motion of a falling object. Galileo was interested in falling objects, but he didn't have a clock accurate enough to time their motion. So he used inclines, as in Figure 5.2b through 5.2d, to slow down the falling motion. He then extrapolated (extended in a reasonable manner) his results to the case of falling, Figure 5.2e.

Galileo found that the velocity of a block sliding down an incline increases uniformly with time. In other words, the block picks up the same amount of velocity in each successive time interval. On the incline of Figure 5.3, for example, the block picks up 2 m/sec with each successive second.

Exercise 14

Remembering that acceleration means change in velocity divided by the time to make

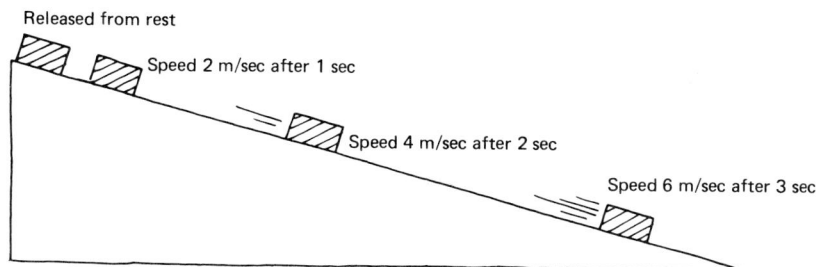

Released from rest

Speed 2 m/sec after 1 sec

Speed 4 m/sec after 2 sec

Speed 6 m/sec after 3 sec

FIGURE 5.3 Another incline.

that change, what is the numerical value of the acceleration of the block in Figure 5.3? Include the proper units.

Since the speed of the block in Figure 5.3 is known to increase uniformly with time, we can figure out how fast it would be moving after 4 sec, after 5 sec, and so on, assuming that it doesn't come to the end of the incline in this time. For instance, after 4 sec it would be moving at 8 m/sec because it picks up an additional 2 m/sec during *every* second (this is the meaning of the 2 m/sec^2 acceleration found in Exercise 14).

Exercise 15
How fast is the block moving after 5 sec? After 10 sec? After 2.5 sec?

Moving to steeper and steeper inclines, we will find in each case that the velocity increases uniformly with time but that this increase is larger for steeper inclines, that is, the *acceleration* is larger for steeper inclines. For the case in Figure 5.2e, where the block is simply falling, we naturally get the largest value for this acceleration. Experimentally, it turns out to be roughly 10 m/sec^2.* This number is called the *acceleration due to gravity.*

The acceleration due to gravity can be found experimentally by measuring the velocity at various times after release. When we make these measurements, we get roughly the following results (assuming, as we have all along, that air resistance can be neglected):

Time Elapsed Since Release	Velocity Acquired Since Release
0 sec	0 m/sec
1 sec	10 m/sec
2 sec	20 m/sec
3 sec	30 m/sec
4 sec	40 m/sec
5 sec	50 m/sec

You can see from this table that the block gains an additional 10 m/sec during each successive second, so the acceleration is 10 m/sec^2.

*The measured value is closer to 9.8 m/sec^2. We'll use 10 m/sec^2 because it's simpler to work with and it introduces an error of only 2 percent.

Galileo was also interested in the distance fallen by objects in various amounts of time. Because the velocity of a falling object keeps getting larger as the object falls, the distance covered during each successive second must keep increasing. For instance, the object moves farther during the fourth second than during the third second. Unlike the velocity, which increases uniformly (i.e., by equal amounts in equal times), the total distance fallen increases nonuniformly. An experimental measurement of distance fallen for various times of fall bears this out. The experimental results are:

Time Elapsed Since Release	Total Distance Fallen
0 sec	0 m
1 sec	5 m
2 sec	20 m
3 sec	45 m
4 sec	80 m
5 sec	125 m

Exercise 16

How far does a rock fall *during* the first second? *During* the second second (i.e., during the time 1 sec to 2 sec in the table)? During the third second?

Exercise 16 shows that the distance does not increase by equal amounts in equal times, but instead increases by increasing amounts in equal times.

A close look at these experimental results revealed to Galileo that the velocity is proportional to the time elapsed since release, whereas the distance is proportional to the square of the time elapsed since release. That sounds like a mouthful. Let's slow down and see what these proportionalities mean. The downward velocity is 10 m/sec after 1 sec, twice that after 2 sec, three times that after 3 sec, and so on. Because doubling the time doubles the velocity, and tripling the time triples the velocity, we say that velocity is *proportional* to the time elapsed. But the distance fallen is 5 m after 1 sec, four times that after 2 sec, nine times that after 3 sec, and so on. Because doubling the time quadruples the distance, and tripling the time gives us nine times the distance, we say that distance fallen is *proportional to the square of the time elapsed*.

These quantitative relationships might seem unimportant, but they are just the sort of relationships that interest people like Galileo. The numbers given in the distance table look fairly meaningless at first glance, but once we search out the "distance is proportional to the square of the time" relation hidden in the table, these numbers take on a new significance. We have found a regularity in nature that had not been apparent beforehand. In a similar fashion, Kepler studied Tycho's data and found an elliptical orbit for Mars. Our scientific understanding of nature comes from just this sort of search for a pattern or shape or regularity.

Once such a regularity is discovered, it seems reasonable to try to explain it. Galileo was able to partially explain the regularities in our two tables. Using a more quantitative version of the reasoning that we used to argue that an object should fall farther during each successive second, Galileo showed that the first regularity (velocity proportional to time) implies the second (distance proportional to square of time).* In other words, if the experimental data show that velocity is proportional to the time, then it necessarily follows that distance is proportional to the *square* of the time. But Galileo still couldn't explain why nature exhibited the first regularity, that is, why objects fall with an unchanging acceleration. Later, Newton would answer this question.

Checklist

Aristotle	average speed
Aristotle's five elements	velocity
Aristotle's theory of natural motion	acceleration
Galileo	Galileo's principle of falling
Galileo's theory of natural motion	acceleration due to gravity
Newton's first law	velocity proportional to time
gravity	distance proportional to square of time
speed	

Further Thought Questions

4. Who would you say was correct in his theory of natural motion, Aristotle or Galileo? Or perhaps both? Or perhaps neither? Discuss.
5. We have seen that many scientific theories have limitations. Does this line of thinking extend to science itself? Are there overall limitations on science? If so, list a few.

Further Exercises

17. A ball is thrown straight up. By how much do you suppose the speed will *decrease* in every second? (Answer: It slows down by 10 m/sec during each second.)
18. A ball is thrown up at an initial speed (just after release) of 20 m/sec. What is its speed and direction of motion at the end of 1 second? 2 seconds? 3 seconds? 4 seconds? (Partial answer: After 1 second, the ball is moving upward at 10 m/sec.)
19. If you leaned out over the top of the Eiffel Tower and threw an ice cream cone downward with an initial speed of 6 m/sec, how fast would it be moving after 1 second? 2 seconds? (Be careful: The cone doesn't start from rest.) (Partial answer: After 1 second, the cone is moving at 16 m/sec.)
20. After 1 second, a rock has fallen 5 m and is moving at 10 m/sec. Using the two proportionality relationships discussed above, find the speed and the distance fallen after 10 seconds, and also after 50 seconds. In English units, the speed after 50 seconds turns out to be 1100 miles per hour. But that's ridiculous. Falling objects on Earth don't move this fast. Think, for example, of falling raindrops or of occasional cases of people surviving falls from airplanes. Where have we gone wrong? Moral: The blind application of scientific principles can lead to absurd results.
21. On the moon the acceleration due to gravity is much less than it is on Earth. Objects fall onto the moon with an acceleration of only 1.7 m/sec². So after 1 second,

---

*In fact, it is possible to show that Distance = ½ × acceleration × square of time, which agrees with the distance table. In symbols, the velocity table tells us $v = at$, while the distance table tells us $d = \frac{1}{2} at^2$. Physicists find equations like these helpful.

a rock falling onto the moon is moving at 1.7 m/sec. How fast is it moving at the end of 2 seconds? 3 seconds? 10 seconds? 20 seconds? Will it *really* be moving this fast after 20 seconds, or will it be moving much slower (compare Exercise 20, remembering that the moon is airless)?

22. The acceleration due to gravity for objects falling onto the giant planet Jupiter is 25 m/sec². What speed does an object falling onto Jupiter attain in 1 second? in 10 seconds? Jupiter's atmosphere and surface are rather unusual (to Earthlings): the gaseous atmosphere gradually thickens as you move downward toward the surface, until the atmosphere finally turns into a liquid. Jupiter is like a gigantic slushy snowball with ill-defined edges and no definite surface. Under these conditions do you suppose that an object dropped in Jupiter's atmosphere would attain anything like the above predicted speed after 10 seconds?

Recommended Reading

1. C. B. Boyer, "Aristotle's Physics," *Scientific American*, pp. 48–51 (May 1950). A brief presentation of Aristotle's theory of the physical world.
2. B. Brecht, *The Life of Galileo*. A play based only loosely on the historical facts, which treats Galileo as a symbol of the modern scientific age. The play has been produced as a film entitled *Galileo*.
3. E. A. Burtt, *The Metaphysical Foundations of Modern Science*, Doubleday, Garden City, N.Y., 1955. An influential study of the shift from the medieval to the modern view of the universe. Chapter 3 is entitled "Galileo."

6

FORCE: WHY THINGS ACCELERATE

We have seen that objects subject to no outside influence move with constant velocity. In this chapter we will study the effect of outside influences. These outside influences are called *forces.*

Force

A force means a push or pull. You exert a force whenever you push or pull something. An object has a force acting on it if it is being pushed or pulled. Examples of forces are: your hand pushing upward on a rock in order to hold it up or to lift it; air resistance pushing backward on a moving car; friction pushing backward on a book as the book slides across a table; your hands pulling on a rope in a tug-of-war; Earth's gravity pulling down on one of Galileo's cannonballs dropped from the Leaning Tower of Pisa. Note that one object can exert a force on another object even if the two are not touching. For instance, Earth exerts a downward gravitational force on a falling rock even though Earth and the rock are not in contact. As another example, a magnet can exert a force on a nail even though the magnet and nail are not touching.

In the United States, pushes and pulls are often measured in pounds, and the *pound* is in fact the official unit of force in the English system. In the metric system the standard unit of force is an unfamiliar quantity called the *newton,* abbreviated N. One newton equals about one-quarter of a pound, so 4 newtons equals about a pound. To give you a feel for the newton, think of it as the weight of a quarter-pound stick of margarine. Despite its unfamiliarity, we'll use the newton rather than the pound in this book, because of its connection with the metric system.

Consider a meteoroid hurtling through space, far from any planets or stars. The meteoroid is surely isolated from outside influences, so no forces, no pushes or pulls, act on it. Thus Newton's first law (Chapter Five) implies that the meteoroid moves at constant velocity, that is, at zero acceleration. That is, Newton's first law implies that objects on which no forces are exerted move with no acceleration.

What about objects that do have forces acting on them? Since objects on which no forces are exerted move with no acceleration, Newton concluded that objects on which forces are exerted do have an acceleration. Furthermore, he gave a quantitative relation between the amount of force and the amount of acceleration. Note carefully that Newton is saying that forces cause acceleration. Aristotle would have said that forces cause velocity, so that a push or pull would be needed to explain continued motion. Newton maintained that continued motion at constant speed in a straight line requires no push or pull, but that any change in the motion of an object, any slowing down or speeding up or change in direction, does require a push or pull.

Mass

We have seen that objects respond to forces by accelerating. But some objects respond more strongly than others. For example, suppose you swing a golf club at a golf ball and then at a large ball made of lead. The golf ball will be knocked some distance, but the lead ball won't do much at all beyond ruining your golf club and perhaps your hand as well. If your swing was the same in both cases, however, each ball felt the same force from the club. Thus two objects subjected to identical forces don't necessarily exhibit the same response. The golf ball picked up a large speed, so it must have experienced a large acceleration, whereas the lead ball experienced hardly any acceleration. To describe this difference in the response of the golf ball and the lead ball, physicists use the concept of *mass*. That is, we say that the balls' responses are different because their masses are different.

The *mass* of an object means roughly the quantity of matter or amount of material in it. When we say that one object is more massive than another, we mean that it has more material, or more matter than the other.

Figure 6.1 shows Dick and Jane on a seesaw that is pivoted at the center. If Dick overbalances Jane, then we feel that in some sense Dick's body has more matter in it than Jane's. This balance-beam idea is actually the basis of the official definition of the amount of mass in an object.

> The **mass** of an object is the number you get when you use a balance scale to compare the object with the standard kilogram mass.

You may have seen a standard kilogram mass in a science lab. On Earth, a kilogram weighs a little more than 2 pounds. All kilograms balance each other on a balance scale, and they are all said to have a mass of one kilogram. To find the mass of any other object, your body for instance, you should climb into one pan of a balance scale and see how many kilograms must be placed in the other pan to balance you (Fig. 6.2). This is your mass in kilograms.

The *kilogram,* abbreviated kg, is the standard metric unit for mass. The corresponding English unit of mass, the *slug*, is hardly ever used and will not be used in this book.

Another useful concept is similar to mass, and, unfortunately, is often confused with mass. This concept is *weight.*

> The **weight** of an object is the force of gravity on it.

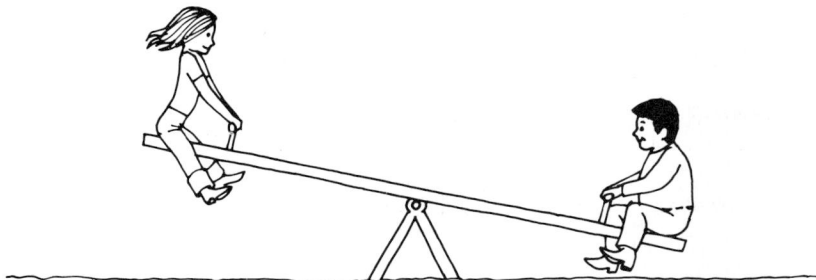

FIGURE 6.1 Dick and Jane on a seesaw. By definition, Dick is more massive than Jane if he overbalances her.

FIGURE 6.2 One way to determine your mass.

For example, when we say that a book weighs 4N (about 1 lb) we simply mean that Earth's gravity is pulling downward on the book with a force of 4N. Weight, because it is a force, is measured in newtons or pounds, whereas mass is measured in kilograms.

Suppose a particular rock, on Earth, balances a standard kilogram mass. If you take the rock, along with the balance scale and the standard kilogram, to the moon, the rock would surely still balance the standard kilogram (even though the pull of gravity on both objects is now reduced). So the rock's mass is 1 kg no matter where it is, on Earth or on the moon. This agrees with our idea that mass is a measure of the amount of matter in an object, since the amount of matter in the rock is surely not changed by simply transporting it to the moon. On the other hand, the rock *weighs* more on Earth than on the moon, because the pull of gravity is about six times as strong on objects near Earth as it is on objects near the moon.

Here are a few more examples.

Exercise 1
A dog, Rover, has a mass of 20 kg and a weight of 200 N on Earth. Find Rover's weight in pounds. Rover is rocketed into outer space, far from any planet or star, where the force of gravity is essentially zero. What are Rover's new mass and weight?

Exercise 2
Is it easier to lift a Volkswagen on the Earth or on the moon?

Exercise 3
Which has greater mass, a ton of feathers or a ton of iron (1 ton = 2000 lb)? Which has greater weight? Which is largest?

Exercise 4

Would you rather your hunk of gold weighed 1 N on the earth or on the moon? Does it make any difference?

Newton's Main Idea

Now we're ready for Newton's idea relating force and acceleration. Newton's relation is about the simplest you can imagine. He said that force and acceleration are just proportional to each other. In other words, if you double (or triple) the force on an object, the object will respond by doubling (or tripling) its acceleration.

Exercise 5

The weight of an object on the planet Mars happens to be only 40 percent of its weight on Earth, that is, the force of gravity is 40 percent as large. (a) How great is the acceleration of an object falling onto the surface of Mars? (b) If a rock is dropped above the surface of Mars, how fast will it be moving after 2 sec? (c) How fast would a rock dropped above the surface of Earth be moving after 2 sec?

Exercise 6

Imagine that you deliver a horizontal blow with a sledgehammer to a brick and then to a heavy iron anvil, as shown in Figure 6.3. As this is an imaginary experiment anyway, let's suppose it is conducted on a frictionless surface, such as a smoothly frozen pond. (a) Assuming that the two blows are delivered with the same force, which object will be moving faster after the blow (i.e., when hammer and object are no longer in contact)? (b) Which object had the larger acceleration during the blow (while the hammer and object were in contact)? (c) How big is the acceleration of each object after the blow?

In Exercise 6, the two objects had the same force exerted on them during the blow, yet they had different accelerations. The less massive object had the largest acceleration during the blow. On the basis of examples such as this, Newton came to the conclusion that acceleration and mass are *inversely proportional*. For instance,

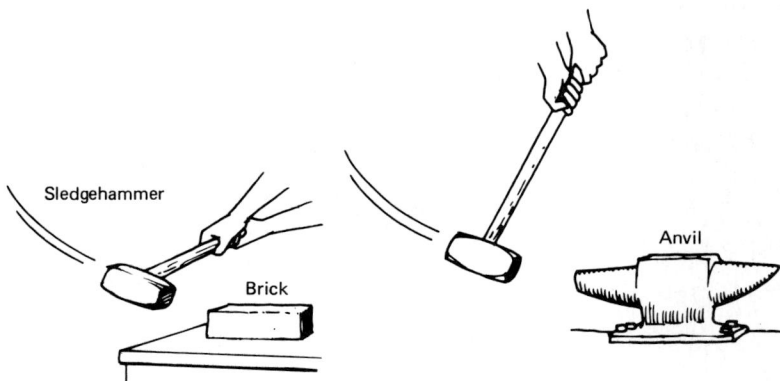

FIGURE 6.3 Which will be moving fastest afterward?

if the anvil has twice the mass of the brick, then the anvil will have one-half the acceleration of the brick: doubling the mass gives one-half the acceleration; tripling the mass would give one-third the acceleration, and so on. This relationship is what is meant by inversely proportional.

Exercise 7

The father-and-son circus team of Alphonse and Alphonse have an act in which they are both shot from cannons. Alphonse the elder is twice as massive as Alphonse the younger. The explosive charge, and hence the force, behind each is the same. If Alphonse the elder is accelerated at 3 g's (three times the acceleration of gravity) while being sped up inside the barrel, what is the acceleration of Alphonse the younger?

Now we're ready to wrap this all up in the form of:

Newton's Second Law of Mechanics. The acceleration of an object is proportional to the net (or total) force on the object by its environment, and inversely proportional to the object's mass. This acceleration is in the direction of the net force.

We can state the essentials more briefly:

$$\text{Acceleration} \propto \frac{\text{net force on object}}{\text{mass of object}}$$

where $\propto$ means "is proportional to."* We'll call this principle Newton 2 for short.

Because more massive objects have smaller accelerations for a given amount of force, the mass of an object is a measure of its reluctance to accelerate. More massive objects are more sluggish and more difficult to accelerate: it's more difficult to stop them, more difficult to start them, and more difficult to make them turn corners. In other words, more massive objects have a greater tendency to maintain their velocity. An object's tendency to keep on doing what it is doing is called **inertia.** We say that more massive objects have greater inertia.

For an interesting demonstration of inertia, get hold of a heavy metal anvil and a sledgehammer, talk your instructor into lying down, and gently place the anvil on his or her stomach. (Don't drop it—the moving anvil's large inertia will carry it a goodly distance into your instructor!) Now whack the anvil from above with the sledgehammer (Fig. 6.4). Don't miss. You might lose your course credit if the instructor expires. However, your intrepid instructor should barely feel a thing. Do you see why? The anvil is so massive, has so much inertia, that even a very large force accelerates the anvil only a small amount, carrying it only a short distance into your instructor's stomach. If the anvil had been made of some light material like aluminum, this experiment might have been uncomfortable indeed for your instructor.*

As a simpler do-it-yourself version of this experiment, put an unabridged dictionary on your stomach or your head and ask a friend (!) to hit the top of the dictionary with his or her fist.

*In symbols, $a \propto F/m$. If we measure a in m/sec^2, F in newtons, and m in kilograms, the proportionality actually becomes an equality: $a = F/m$. In other words, $F = ma$. This last equation is the physicist's usual way of writing Newton 2.

*Don't try this experiment on your own as you could get hurt.

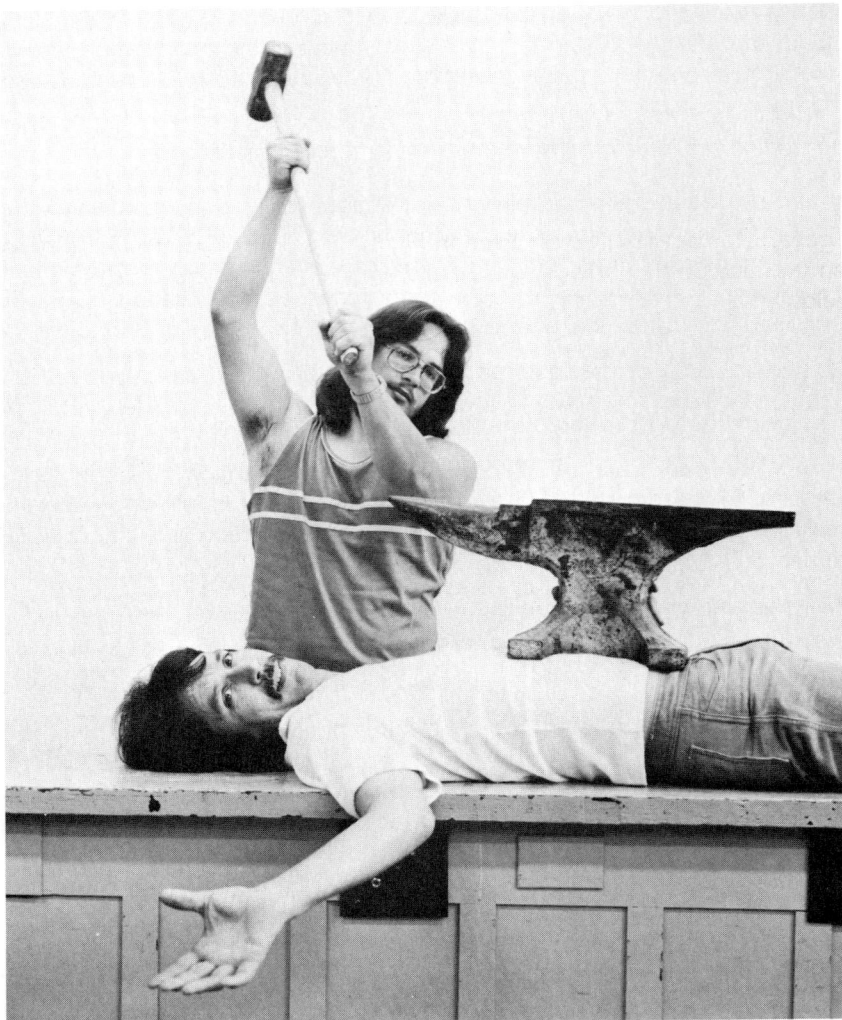

FIGURE 6.4 Saved by Newton 2!

Let's return to Galileo's Leaning Tower experiment (Chapter Five). Recall that Galileo found that light and heavy objects fall at the same rate, but he couldn't explain why. Now we can explain why. Imagine a 1-kg rock and a 2-kg rock, both falling from the tower. The 2-kg rock has twice the weight, and also twice the mass, of the 1-kg rock. Thus the 2-kg rock has twice as much force on it as does the 1-kg rock, but the 2-kg rock also has twice as much inertia. The acceleration, proportional to the ratio of force to mass, is therefore the same for the two rocks: they fall with the same acceleration and are even with each other all the way down.

Thought Question 1

We have explained, or deduced, Galileo's principle from Newton's laws. Does this prove that Galileo's principle is true? Can we now assert with complete confidence that in the absence of air resistance, wind and so on, Galileo's principle will hold in every experiment?

The next exercise illustrates the final sentence of Newton 2, which states that the acceleration is in the direction of the force.

Exercise 8

Study Figure 6.5. In the situation depicted in Figure 6.5a, which way will the spool move when you pull on the thread? Which way will the spool move in Figure 6.5b? (Answer below.)

Most people predict that the spool pictured in Figure 6.5a will move to the right. This answer is correct and is in agreement with Newton 2, because the force (the pull on the thread) is toward the right and hence the spool must start up (accelerate) toward the right. In the situation pictured in Figure 6.5b, many people predict that the spool will move to the left. This answer is incorrect. The force on the spool is still toward the right, because the thread is being pulled toward the right, thus Newton 2 demands that the spool accelerate toward the right. If you didn't get the right answer, it may be because of an intuitive feeling developed while playing with yo-yos as a child. But this experiment differs from experiences with yo-yos because yo-yo strings pull vertically, whereas the experiment of Figure 6.5 uses a horizontal string.

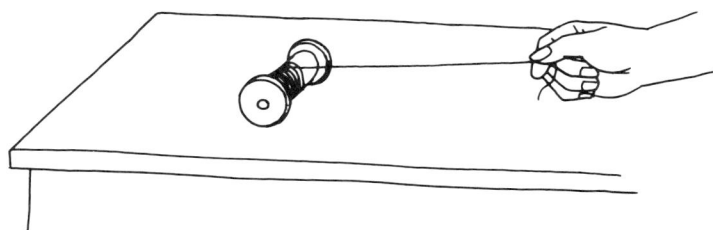

(a)

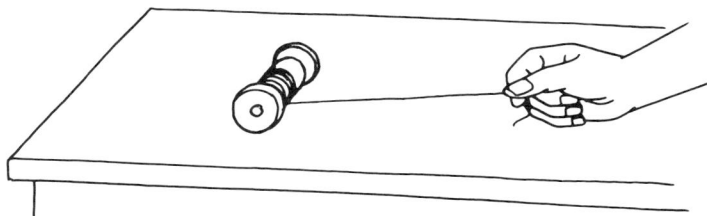

(b)

FIGURE 6.5 Two experiments with spool and thread.

If you don't believe the predicted result, try this experiment at home with a spool of sewing thread. You'll find that in the situation of Figure 6.5b, the spool actually rolls to the right, rolling up in the process.

There is a lesson here: Beware of your intuition or common sense when thinking about physics. Your common sense is sometimes right, sometimes wrong, but Newton 2 is always right.*

Another little catch in Newton 2 is the word *net*. The law states that an object's acceleration is proportional to the net force on it. I think you can see what the word *net* means by considering the following exercises.

Exercise 9
During lift-off, a 10,000 N rocket develops a 50,000 N thrust, that is, an upward force of 50,000 N acts on the rocket because of the action of the engines (Fig. 6.6). How big and in what direction is the net force on the rocket?

Exercise 10
What would happen to the rocket in Exercise 9 if the engines could develop only a 10,000 N thrust? How big is the net force in this case?

Newton's Third Law

Slap on a table. Hard. It hurts your hand, doesn't it? In other words, the table slaps back. Stand on a smooth floor in stocking feet (to reduce friction) and push against a wall. The wall pushes back. Still in your stockings, find some immovable object to grab hold of and pull against it. You are pulled toward the object, right? The object pulls back.

Newton figured that a general rule was operating here. We'll call it:

Newton's Third Law of Mechanics. Forces always come in pairs, in the following sense: Whenever one object exerts a force on a second object, the second object exerts a force in the opposite direction on the first object. Furthermore, these two forces are equal in magnitude.

So, you can't touch without being touched. Touch your friend's face with your hand, and your friend's face touches your hand. And you can't hit without being hit. Pound a wall with your fist, and the wall hits back—a good way to break your fingers. Forces must come in pairs, sometimes called **action-reaction pairs.**

Newton's third law asserts that the two forces are equal in magnitude. This seems reasonable. If you slap a table lightly, it slaps back lightly, whereas if you slap it hard, it slaps back hard.

You might have heard somewhere that "for every action there is a reaction." That's the idea behind Newton 3. But the action-reaction terminology can be confusing, because the question of which force is the action and which the reaction is fairly arbitrary. For instance, if two stags butt against each other, it makes little sense to assert that one force (for example, by the left-hand deer against the right-hand deer) is the action while the other (by the right-hand deer against the left-hand deer) is the reaction. The essence of Newton 3 is simply that forces come in pairs.

*Well, nearly always. As we'll see later, there are limitations to the validity of Newton 2.

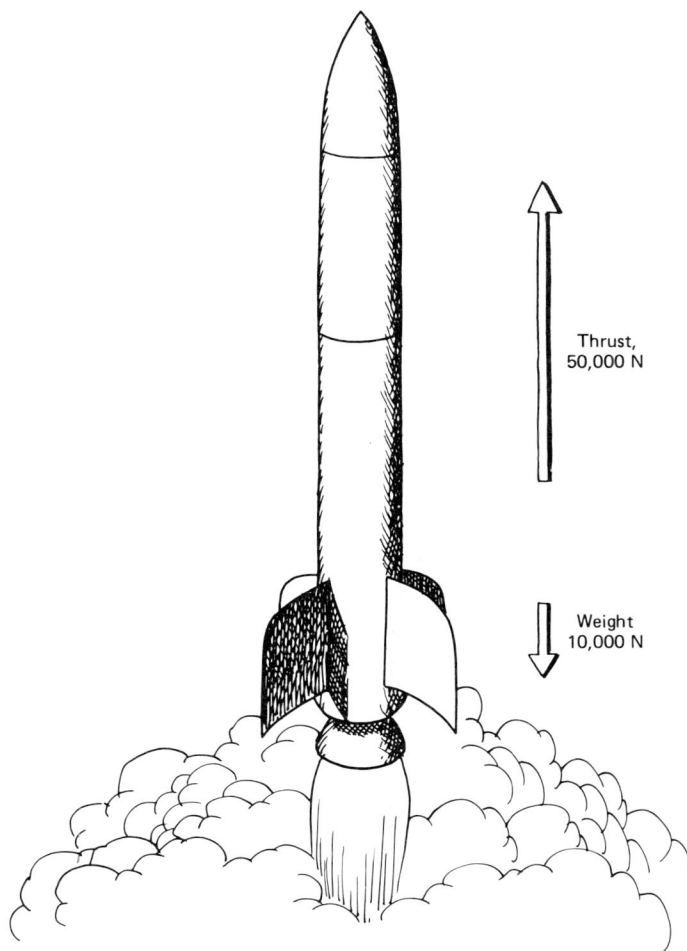

FIGURE 6.6 Rocket lift-off. How big is the net force?

Consider an apple hanging from a tree (Fig. 6.7). Suppose the apple weighs 1 N. The apple is acted on by two different forces: the 1 N weight of the apple acting downward, and the force of the branch pulling upward. This second force must also be 1 N, since the net force on the motionless apple must be zero. However, the weight of the apple and the pull of the branch cannot form an action-reaction pair, because both these forces are on the same object (the apple). In fact, each of these forces is part of a different action-reaction pair. The two action-reaction pairs are: (1) The branch exerts 1 N upward on the apple (this holds the apple up), and the apple exerts 1 N downward on the branch (Newton 3 tells us that this force must be present); and (2) Earth exerts a 1 N gravitational force downward on the apple, and the apple exerts a 1 N gravitational force upward on Earth. The last of these forces, the upward pull by the apple on Earth, might sound farfetched, but Newton 3 tells us it must be there. Note carefully that the 1 N upward pull by the branch on the

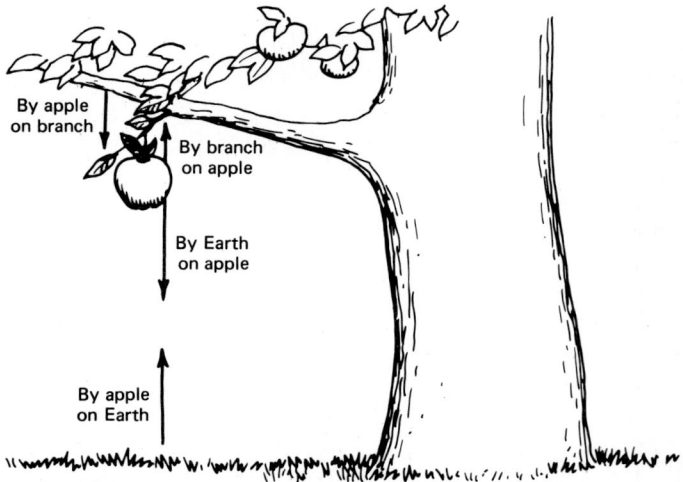

FIGURE 6.7 Two action-reaction pairs.

apple and the 1 N downward pull by Earth on the apple do not form an action-reaction pair; if you don't understand this, you'd better go to the beginning of this section and start over.

Various Drive Forces

Suppose you are at rest and you want to start moving. Newton 2 says that in order to get started (i.e., to accelerate) you must have a net force exerted on you by your environment (return to our statement of Newton 2 and check this). This means that something outside of you, something in your environment, must exert the *drive force* that gets you moving. In other words, you can't exert a drive force on yourself. For example, you can't get yourself moving by pulling on your nose: if you pull on your nose, your nose pulls back on your hand (Newton 3), and the net force on your

FIGURE 6.8 You can't get anywhere by pulling on your nose.

body is zero, so you don't go anywhere (Fig. 6.8). So how do you get moving? What can provide the drive force?

As an example, consider the bicycle (Fig. 6.9). What provides the drive force? When a bicyclist "digs out" from rest on a dusty road, which way does the dust go? It goes backward. In which direction, then, is the back tire pushing against the road? Backward. Now enter Newton 3: Which way is the road pushing against the tire? Forward, right? We have discovered the drive force! You might have thought that *you* were pushing the bike forward when you pedaled, but actually the road was doing the job. You pump the pedals merely to get the tire to push backward on the road, so that the road will push forward on the tire. According to Newton 3, you can't have one force without the other, so it is necessary for the tire to push backward on the road in order for you to go anywhere.

What about the automobile? The idea is the same. The drive wheels push backward on the road, and the road pushes forward on the drive wheels. The motor doesn't push the car forward, the road pushes the car forward. The purpose of the motor is simply to turn the drive wheels so that they'll push backward against the road.

Exercise 11

A car travels along a straight, level road at a constant speed of 90 km/hr. How great is the net forward force on the car?

FIGURE 6.9 Find the drive force that keeps the bicycle (and Ziva) moving down the road.

Exercise 12

Since the net force on a car moving at constant speed is zero (Exercise 11), why do we need a drive force while the car is moving along a level road at constant speed, that is, why can't we just turn off the motor once we have attained the desired speed?

Exercise 13

In terms of Newton 3, explain why it is difficult to start a car moving on ice. Why is it difficult to stop a car on ice?

What about walking and running? Again, the idea is the same. Your foot pushes backward on the ground, so the ground pushes forward on your foot, providing the drive force.

What about swimming? Your hands and feet push backward against the water, and the water responds by pushing forward against you. A propeller-driven boat operates on the same principle.

More examples: An airplane pushes backward on the air, and the air pushes forward on the plane. A propeller-driven airplane operates in the same manner as a propeller-driven boat. The propellers are shaped so that when they spin they push large quantities of air backward. If you've ever stood behind a propeller-driven plane you may have experienced the gale produced by the propellers. A jet-driven airplane draws outside air into the engine, then heats it by combustion of fuel. Heating raises the pressure of the air inside the combustion chamber, and this high-pressure air then escapes through a nozzle at the rear of the engine (Fig. 6.10). The escaping high-velocity stream of air has been pushed backward by the engine, so the air pushes the engine (hence, the entire plane) forward.

Exercise 14

Judging from the above discussion, can a propeller-driven plane operate above Earth's atmosphere? Why? Can a jet-driven plane operate above Earth's atmosphere?

Suppose that some day in the winter you find yourself in the middle of a perfectly smooth frozen pond. It's cold, and you would like to go home. How are you going to get off the pond? Just walk off, you say? Sorry, that won't work. The ice is perfectly smooth, so there is no way your feet can push backward in order for the ice to push forward. Your feet just slide backward without pushing.

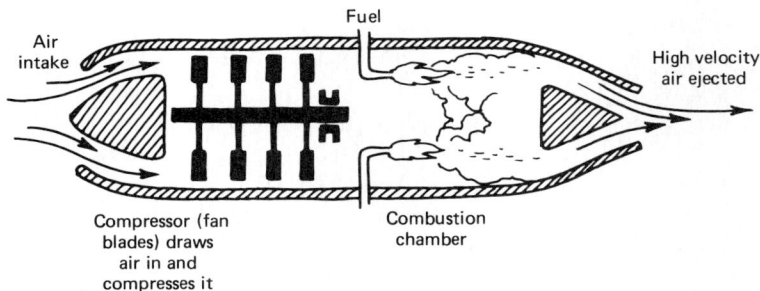

FIGURE 6.10 Schematic diagram of a jet engine.

Exercise 15
How are you going to get off the ice? (Answer below.)

87

Force: Why
Things Accelerate

As one solution to this problem, you could push the surrounding air backward by fanning, causing the air to push you forward. This solution is analogous to swimming or to the propeller-driven airplane.

To make things more difficult, imagine that this frozen pond is on the moon, where there is no air. Of course, on the moon there is not only no air, but also no water and hence no ice. For the sake of this exercise, though, imagine that there is ice on the moon and that you are standing on a perfectly smooth sheet of it. Because there is no air, fanning will get you nowhere. Now, how are you going to get off the ice?

Have you thought about it? In order to get off the ice, you need to push against something so that it can push back against you; you can't push against the ice or against the air. So, remove a boot and throw it away! As you accelerate the boot (while it is still in your hand), you exert a forward force on it. Thus the boot exerts a backward force on you during this time. This backward force will start you moving backward, so that by the time you release the boot you have acquired a certain backward speed. The same lack of friction that prevented walking now acts in your favor, because once you have started sliding backward, there is no friction to slow you down until you reach the edge of the pond. Neat.

Exercise 16
Is it possible to perform this stunt without actually releasing the boot? That is, could you hold on to the boot at the last minute and still get off the pond? (Answer below.)

The answer is no. You've got to release the boot. In order to hold on to the boot, you must de-accelerate it during the final part of your arm motion. This means that you must exert a backward force, toward you, on the boot (remember Newton 2), and according to Newton 3, the boot would then exert a forward force on you. A more detailed analysis of this situation would show that this forward force would exactly cancel any backward motion you are beginning to develop.

The "boot drive" in our discussion is an example of the principle of the **rocket drive.** When you throw the boot away, you are actually rocketing yourself across the ice. Whereas most drive forces operate on the basis of pushing against some object in the environment (against the ground, the water, the air), a **rocket** is defined as any device that carries along its own material to push against.* This rocket fuel might be a pile of shoes or hot gases ejected rapidly out the back end. The rocket engine pushes backward on the shoes or on the hot gases. The shoes or gases return the favor, as they must according to Newton 3, by pushing forward on the engine, providing the thrust to accelerate the rocket.

We may have come up with a new form of rocket propulsion: shoe power. Just put several strong-armed astronauts in the back end of your rocket ship, equip them with a pile of shoes, and tell them to start throwing! If enough shoes were thrown away fast enough, the rocket would slowly lift off the pad. I can see it now (Fig. 6.11).

*Although this sounds a little like pulling on your nose (Fig. 6.8), the difference is that you throw the rocket fuel away, whereas you retain your nose.

FIGURE 6.11 Shoe power.

NASA rockets and all ballistic missiles are powered by rocket engines that develop their huge thrusts by ejecting heated gases out the back end. These vehicles can accelerate even in the vacuum of space by pushing against their own ejected gases.

Checklist

force	inertia
pound	net force
newton	Newton's third law
mass	action-reaction pair
kilogram	drive forces
weight	propeller drive

acceleration proportional to force

acceleration inversely proportional to mass

Newton's second law

jet drive

rocket drive

Further Thought Questions
2. Some people feel that we are so surrounded by technology today, so removed from our natural environment, that we feel lonely, alienated, insecure. Is there any truth to this? Does an understanding of the principles behind such devices as the automobile, the airplane, and the rocket, help reduce such feelings of alienation?

Further Exercises
17. As a rocket rises, it loses fuel. Assuming that the rocket's thrust remains constant, how does this loss of weight affect the motion?
18. Suppose the rocket of the preceding exercise runs completely out of fuel while it is still pointed straight up. In what direction is the net force on the rocket right after the motors shut off?
19. Will the rocket ship of Exercise 18 stop moving upward and begin to fall right after the engines shut off or will it continue moving upward? Why?
20. An airplane weighing 20,000 N is in level flight at constant speed. The drag force caused by air resistance is 5000 N. (a) How large is the net force on the plane? (b) How large is the upward lift force provided by the wings? (c) How large is the thrust, that is, the forward drive force? (*Hint:* The net force is zero.)
21. In the preceding exercise, how large would the thrust have to be if the plane could be streamlined to reduce drag to 500 N? What if we could somehow reduce drag to 0?
22. As a shotgun fires a shell, which of the following pairs of forces form action-reaction pairs? (a) The expanding gases inside the gun pushing forward on the shell and the gun pushing backward against your shoulder; (b) the gun pushing backward against your shoulder and your shoulder pushing forward against the gun; (c) your shoulder pushing forward against the gun and the expanding gases pushing forward against the shell; (d) the expanding gases pushing forward against the shell and the shell pushing backward against these gases.
23. In light of the preceding exercise, explain in detail why your shoulder feels a force when you fire a shotgun.

Recommended Reading

1. George Papallo, *What Makes It Work?*, Arco Publishing Company, New York, 1972. This volume is similar to *The Way Things Work*, only smaller and simpler. Both references should be in every high school library.
2. *The Way Things Work,* in two volumes, Simon & Schuster, New York, 1967. If you are interested in rocket engines, jet planes, automobiles, and so on you will enjoy this encyclopedia designed as a reference for people who are curious about technological devices. No scientific or mathematical background is assumed, although some of the explanations are rather detailed.

For references on Newton and his physics see the Recommended Reading for Chapter Seven.

7

THE UNIVERSE AND NEWTON'S GRAVITY

Aristotle didn't need to ask why objects fall. He needed no special cause or force to explain the fall of, for example, an apple, because to him, downward motion was simply the natural motion of the apple.

You'll recall, however, that Newton had a different point of view. For him, straight-line motion at constant speed was natural motion; any other motion was unnatural. Thus a falling apple, which is accelerated, must be caused or forced to fall. This idea made it necessary to invent *gravity,* a force that reaches up from Earth and pulls objects downward.

Isaac Newton developed a quantitative theory of gravity and extended the idea from Earth to that ethereal region beyond the moon which had been supposed to be essentially different from our own planet. Thus Newton unified the heavens with Earth and altered forever our perception of humanity's place in the universe.

The Idea of Gravity

In 1665, when England was suffering from the great plague, authorities at Cambridge University closed the school for the duration of the disease. Isaac Newton (see Fig. 7.1), then in his fourth year of studies, returned to his boyhood home on a farm in the Lincolnshire village of Woolsthorpe. With a lot of time and few responsibilities, the 23-year-old student whiled away the hours by inventing the differential calculus, formulating the principle of gravity, and discovering a few other details such as the widely used binomial theorem of algebra.

Newton said that he gained important insight into gravity when he noticed an apple falling from a tree. In Newton's thinking (contrary to that of Aristotle), the downward, accelerating fall of an apple was not natural motion. After all, Galileo had determined that natural motion was nonaccelerated. Thus Newton reasoned that some force must be pulling the apple downward toward Earth and that this downward *gravitational force* acts on every object near Earth.

Exercise 1
(a) Does Earth exert a gravitational pull on objects that are at rest, such as an apple attached to a branch or a book at rest on a table? (b) Do any forces other than the gravitational force act on the two objects in (a)? (c) Identify these other forces. (d) What can you say about the net force on each of these objects?

In the course of his reflections on the fall of an apple, Newton was apparently struck by a fundamental new idea: Perhaps the force that causes the fall of an apple extends all the way to the moon. And if Earth exerts a gravitational pull on the moon, then perhaps all the heavenly bodies exert gravitational forces on each other: the sun

FIGURE 7.1 Isaac Newton. (Yerkes Observatory Photo,
University of Chicago.)

on each planet; the planets on one another, the planets on the sun, and so on.
Newton thus conceived the idea that the gravitational force is universal, that it reaches
throughout the universe to guide the motions of the planets and the stars just as it
guides the fall of an apple.

And thus our perception of the heavens was permanently altered. Before Newton,
it was generally believed that the astronomical bodies were essentially different from
this base Earth; that, being made of Aristotle's ethereal quintessence, they were
perfect and were not subject to the earthly laws of change and decay.* Now Newton
announces that the stars are subject to the same law of gravity that governs the fall
of an apple. The stars are no longer special.

Let's work through a portion of Newton's thinking about gravity and the moon.

Exercises 2

Consider the moon as it moves in a circle around Earth. Does the moon have an
acceleration? Is a force exerted on the moon? How do you know? What is the name
of this force? (Answers below.)

The moon moves in a circle and is therefore accelerated because its direction of
motion keeps changing. According to Newton's physics, acceleration can only be
produced by forces, so there must be a force acting on the moon. Newton then made
the mental leap, the analogy that perhaps the force that bends the moon's path into
a circle is the same force that pulls the apple to the ground, namely the force of
gravity produced by Earth. Creative scientific work is largely a matter of finding

*A few pre-Newtonian scientists, such as Galileo, disagreed with this belief.

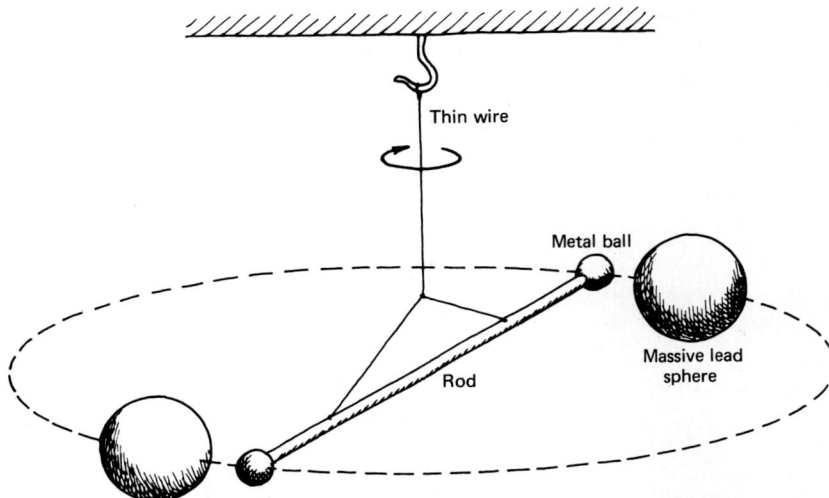

FIGURE 7.2 An experiment to directly measure the gravitational force between laboratory-scale objects. The gravitational attraction between the spheres and the balls causes the thin wire to twist by a measurable amount.

(inventing?) just such analogies, of working out connections between dissimilar phenomena, of finding a unifying point of view.

As if this grand synthesis wasn't enough, Newton took the further step of proposing that *every* object, and not only astronomical objects like the earth, moon, and sun, exerts a gravitational pull on *every other* object. Even human-scale objects such as apples and books exert gravitational pulls on each other. It might be difficult to observe the tiny gravitational pull that an apple exerts on a book, especially when this force is so overshadowed by the much larger pull of Earth on either one of these objects, but, according to Newton's principle of gravity, this force is present.

The small gravitational force among human-scale objects such as books or apples has been directly observed in laboratory experiments like the one shown in Figure 7.2. In this experiment, first performed about a century after Newton's work on gravity, a thin horizontal rod with a metal ball attached to each end is suspended from the ceiling by a thin wire. Large lead spheres, placed near the two balls on opposite sides of the rod, attract the balls, twisting the wire by a measurable amount.

Exercise 3
The gravitational force between two 1-kg lead balls is the same as the gravitational force between two 1-kg wooden balls. The gravitational force doesn't depend on the substances of which the objects are made. Can you think of any forces that do depend on the substances of which the objects are made?

The Law of Gravity

From his study of falling objects and the motion of the moon, Newton formulated the following quantitative description of the gravitational force.

Newton's Law of Gravity. Every object attracts every other object with a gravitational force that is proportional to the product of the masses of each body and that is inversely proportional to the square of the distance* between them. More briefly:

$$\text{Gravitational force} \propto \frac{(\text{mass of 1st object}) \, (\text{mass of 2d object})}{(\text{square of distance between them})}$$

This statement goes beyond the statement that all objects exert gravitational pulls on all other objects. It tells you about the size of this force. For instance, the force is larger if one or the other or both of the objects have a large mass than if both have a small mass. This is why the force by Earth on an apple is much larger than the force between two apples. The law of gravity also tells you that the force is smaller if the objects are farther apart, since the force is inversely proportional to the square of the distance. For example, the force exerted by Earth on our moon is much larger than the force exerted by Earth on one of Jupiter's moons.

Let's consider the gravitational force between a couple of human-scale objects, a grapefruit and an orange. Suppose the grapefruit and orange are 1 m apart. At this distance, the gravitational force by either one of these objects on the other is about 10^{-12} N—an extremely small force. Suppose that we somehow managed to double the mass of the grapefruit. What would this do to the gravitational force? The law of gravity tells us that the force is proportional to the mass of the grapefruit, so doubling this mass would double the force. Thus the two would attract each other twice as strongly. What if, instead of doubling the grapefruit's mass, we moved the grapefruit and orange an extra meter apart, from 1 m to 2 m? This would decrease the force, but by how much? The law of gravity tells us that the force is inversely proportional to the square of the distance. Since the distance doubles, the square of the distance quadruples, which means that the force is reduced by a factor of four. Thus the grapefruit and orange would attract each other one-fourth as strongly.

Perhaps you have already thought of the following question about the gravitational force between our grapefruit and orange. What is the correct distance between the two to use in Newton's law of gravity? Is it the distance between the near sides of the grapefruit and orange, or between centers, or between the far sides? If you guessed that it is the distance between centers, you were right. In the law of gravity, distances are supposed to be measured from the centers of any spherical objects such as grapefruits, oranges, moons, planets, and stars. For two nonspherical objects, it is more difficult to precisely define the distance to be used in Newton's law of gravity; we won't pursue this detail here.

Exercise 4

(a) Suppose we had doubled the mass of the orange instead of the mass of the grapefruit. What would this do to the gravitational force? (b) What if we doubled the mass of both the apple and the orange? (c) What if we tripled the distance between the two? (d) What if we halved the distance between the two?

Exercise 5

How big, numerically, is the gravitational force exerted by Earth on your body? (Answer below.)

---

*The square of a number means the number multiplied by itself. For example, the square of 2 is 4, and the square of 3 is 9.

The size of the gravitational force acting on your body right now is simply your weight, according to our definition of weight in Chapter six. Using your body and Earth as the two objects referred to in Newton's principle of gravity,

$$\text{Your weight} \propto \frac{\text{(your mass) (mass of Earth)}}{\text{(square of distance from you to center of Earth)}}$$

If you increase your mass (in kilograms), your weight (in newtons or pounds) increases by the same percentage. Weight is proportional to mass. If the mass of Earth were somehow increased, your weight would also increase.

Exercise 6
Earth's radius is about 6000 km. How much would you weigh at 6000 km above Earth's surface? At 12,000 km above Earth's surface? At 18,000 km above Earth's surface?

Satellites and Circles

Artificial Earth-orbiting satellites provide fine examples of many physical principles. In Chapter six we discussed the rocket engines that boost artificial satellites into orbit. As you may be aware, large rocket engines use up all their fuel and are separated from the rest of the craft after a few minutes, so satellites use no rocket power as they circle Earth.* What keeps them up? As another example, the moon obviously has no rocket engines on board. What keeps it up?

Exercise 7
Consider a 50,000 N (about 6-ton) satellite in low circular orbit, say 100 km above Earth's surface. (a) Approximately how big, and in what direction, is the gravitational force on this satellite? (b) Why do you suppose part (a) used the word *approximately?* (c) Other than gravity, what forces are acting on this satellite? (Answer below.)

Have you thought about it? The gravitational force on the satellite is roughly its 50,000 N weight directed downward toward the center of Earth. At an altitude of 100 km, the precise force is about 3 percent less than this, because the gravitational force on an object decreases as the object gets farther from Earth's center. In other words, a satellite that weighs 50,000 N on Earth will weigh a little less at 100 km above Earth. Besides Earth's gravity, *no* forces are acting on this satellite (except for the much smaller gravitational forces exerted by the sun and the moon). There is no upward force on the satellite to hold it up by balancing gravity; there is an unbalanced, net force of nearly 50,000 N acting downward on the satellite.

With an unbalanced force acting downward on the satellite, why doesn't the satellite fall to the ground? Briefly, the answer is that the satellite is falling, but not to the ground. Instead, its large forward speed keeps it falling around Earth. Let me explain.*

Imagine you are standing on the cliffs of Dover overlooking the English Channel. You pick up a stone and throw it horizontally, straight east, toward the sea. The stone

---

*Except for the occasional use of small rocket bursts to keep the satellite properly oriented.

*Three centuries ago Newton gave a similar explanation in connection with the moon's motion around Earth.

will travel a short distance and fall into the water (Fig. 7.3). Let's take a close look at the path of thrown objects such as this stone.

Exercise 8

(a) What is the direction of the force on the stone, neglecting air resistance? (b) Recalling Newton 2, what must be the direction of the stone's acceleration? (c) Recalling Galileo's work on falling objects, what is the magnitude of this acceleration? (Answer below.)

Newton's principle of gravity tells us that the stone has a downward force on it caused by gravity. According to Newton 2 the acceleration is in the direction of the force, so the acceleration must be downward, despite the fact that the stone isn't moving directly downward. That is, the acceleration is downward, but the velocity is partly eastward. Galileo said that acceleration caused by gravity is 10 m/sec², so this must be the magnitude of the acceleration of the stone.

Let's be more precise. The stone is thrown horizontally. At any later time it has moved a certain distance in the horizontal direction and it has also dropped a certain distance below the horizontal, as shown in Figure 7.3. According to the above exercise, the stone has a downward acceleration of 10 m/sec² but no horizontal acceleration (provided we neglect air resistance). In other words, our stone's downward speed keeps increasing by 10 m/sec during every second, but its horizontal speed remains constant.

For example, if you throw the stone horizontally at 15 m/sec, then during every second the stone moves 15 m horizontally, but it also moves vertically. In fact, while moving horizontally it also moves 5 m downward during the first second—this is the distance downward traveled by any falling object during the first second after release (Chapter Five).

Suppose the stone hits the water 2 sec after release (Fig. 7.3). How high are the

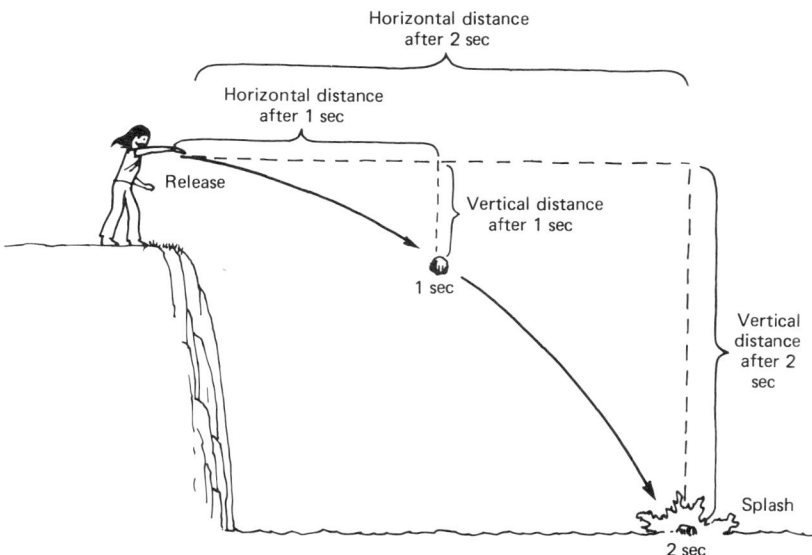

FIGURE 7.3 Throwing a stone from the cliffs of Dover.

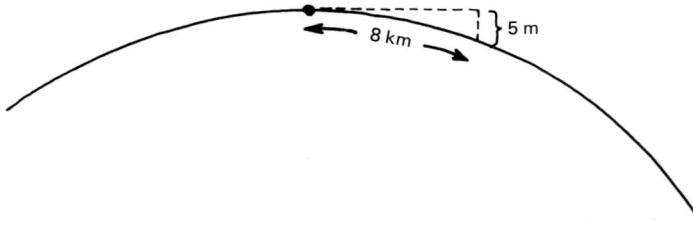

FIGURE 7.4 Our spherical Earth curves away from the horizontal by 5 m for every 8 km.

cliffs of Dover, and how far is it from the foot of the cliffs out to the splash? In 2 sec, any falling object drops a total of 20 m (see Chapter Five), so the cliffs must be 20 m high. Since the stone was thrown with a horizontal speed of 15 m/sec, it must travel horizontally 30 m during the 2 sec it is in the air. Thus it is 30 m from the foot of the cliffs to "splashdown."

Exercise 9

Suppose you throw a stone horizontally from the cliff and drop a second stone over the edge of the cliff from the same initial height, both at the same time. Which stone lands first? (Answer below.)

The two stones land at the same time. As we've seen above, any dropped or horizontally thrown object falls 5 m downward during the first second, a total of 20 m downward during the first 2 sec, and so on, regardless of the object's horizontal motion. Therefore, both objects fall from the release point to the ground in the same time.

If you throw the stone faster, say at 30 m/sec, it will travel farther during each second but it will still fall 5 m below the horizontal during the first second, a total of 20 m below the horizontal in 2 sec, and so on. The distance to the stone's splashdown is now greater, however, because it was thrown faster. In fact, if the cliffs are 20 m high, splashdown will still occur in 2 sec and the stone will hit the water 60 m from the foot of the cliffs.

Suppose you give the stone a brief burst of rocket power, launching it horizontally at several thousand meters per second. During the first second, the stone still falls just 5 m below the horizontal, but it travels several thousand meters horizontally.

Wait a minute, though. Over a distance of several thousand meters, the earth is significantly curved. For every 8 km measured along the ground, Earth curves inward (below the horizontal) by 5 m (Fig. 7.4). If you were to watch a boat 5 m in height sail out to sea, its top would disappear below the horizon when it had reached 8 km out.*

If you throw the stone at several thousand meters per second, it will still drop by 20 m during the first 2 seconds. However, it will not yet have hit the water, despite the fact that the cliffs of Dover are only 20 m high, as the ground has "fallen away" a little because of Earth's curvature.

---

*Assuming that your eye is at ground level. If you are standing upright, your eye is about 2 m above the ground, and the ship won't disappear until it is about 11 km out to sea.

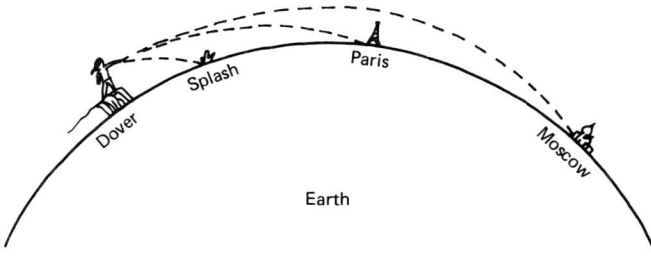

FIGURE 7.5 Throwing stones faster and faster from the cliffs of Dover.

Exercise 10

Using the facts noted above, what horizontal launching speed must you give the stone in order for it to orbit Earth at a constant altitude? (Answer below.)

Earth curves inward by 5 m for every 8 km measured along the ground, and the stone falls 5 m below the horizontal during the first second. Therefore, if the stone is thrown at 8 km/sec (8000 m/sec), it will travel 8 km during the first second and will still be at its original height, because the stone fell by 5 m while Earth curved inward by 5 m. As shown in Figures 7.5 and 7.6, the stone will continue "falling" around Earth at 8 km/sec, never getting any closer to the ground!

And *that* is why satellites stay up.

Satellites are falling all the time in the sense that they keep dropping below their initial direction of motion. But while the satellite drops, Earth's surface keeps curving inward, so the satellite stays at a constant height above the ground.

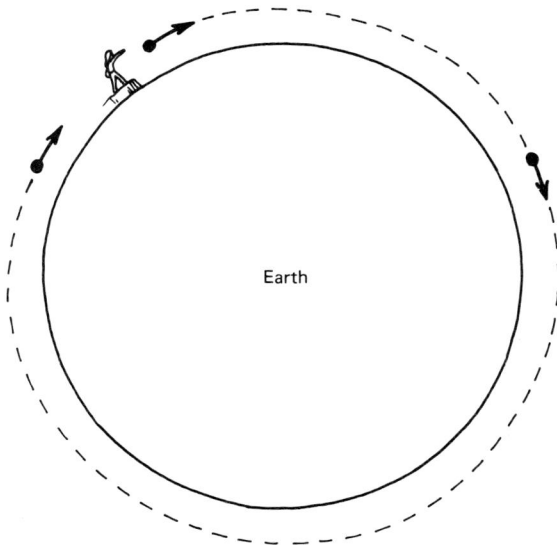

FIGURE 7.6 Throwing a stone *really* fast from the cliffs of Dover.

Exercise 11

Earth's radius is about 6000 km. How long does it take a low-orbit satellite to circle Earth? *Note:* The circumference of (distance around) a circle is 2π times its radius, where π is approximately 22/7, or 3.14.

Exercise 12

(a) Would the orbital speed be 8 km/sec for a high-orbit satellite at, for example, 1000 km above Earth? (b) Would the 80-minute orbital time found in the preceding exercise be correct for a high-orbit satellite? (c) What is the orbital time (time to complete one orbit) for satellites that orbit at 400,000 km (240,000 mi) above Earth? *Hint:* This is the distance to the moon.

Communication satellites transmit signals across large distances by receiving these signals from one point on Earth's surface and beaming them down to another point. In order to provide a stationary target for the transmitted signals, these satellites must remain at a fixed point above Earth.

Exercise 13

How long must it take for a communication satellite to carry out a complete orbit? (Answer below.)

The orbital time must be 24 hours, if a communication satellite is to stay above a fixed point on Earth. Such an orbit is said the be **synchronous** with Earth's spin.

We can get a rough idea of the altitude of these satellites as follows: Very near Earth the orbital time is 80 minutes (exercise 11), whereas the moon's orbital time is 27 days. Thus the radius for a 24-hour orbit must be considerably larger than the

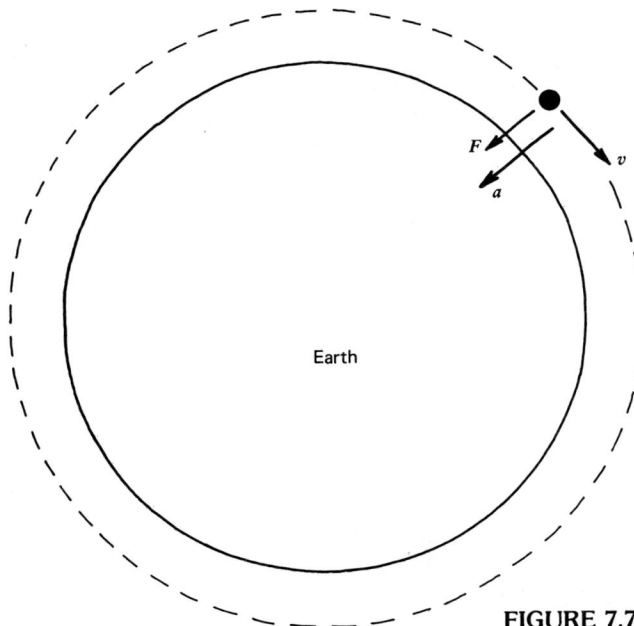

FIGURE 7.7 Analysis of an orbiting satellite.

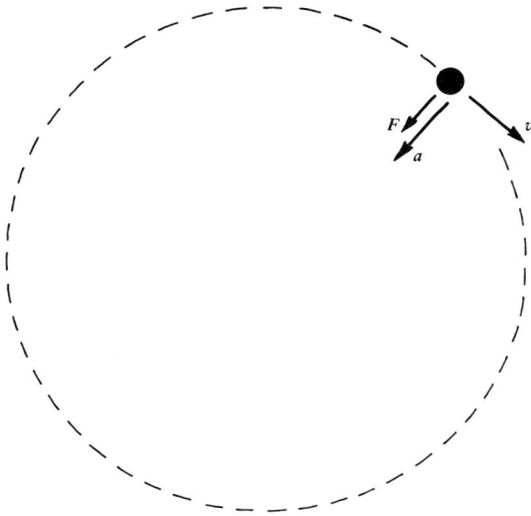

FIGURE 7.8 Analysis of circular motion.

6000-km radius of Earth but considerably smaller than the 400,000-km radius of the moon's orbit. The actual altitude turns out to be 47,000 km. This is 7 Earth radii, or 12 percent of the distance to the moon.

Figure 7.7 summarizes circular motion under the influence of gravity. The arrows marked v, a, and F show the directions of the velocity, acceleration, and force. Note that the acceleration and force are in the same direction, as demanded by Newton 2.

There are many examples of circular motion, or at least near-circular motion: the Earth satellite discussed above, an electron orbiting a nucleus, a bicycle racer on a circular track, a planet orbiting the sun, a ball on a string swung in a circle. Our analysis in this section applies to all of these examples. In every case, the velocity, acceleration, and force are arranged as in Figure 7.8. In every case, there is an unbalanced force acting toward the center of the circle to hold the object into the circle, that is, to bend the object's path into a circle. Without this force, the object would move in a straight line (Newton 1).

Exercise 14
Identify or describe the force that holds the object into its circular path in the case of (a) a car moving around a level, circular speedway; (b) a planet orbiting the sun; (c) a ball attached to a string and swung in a circle.

Weightlessness

As you can see from Figure 7.9, astronauts in orbit feel weightless. Recall, however, that *weight* means "the force of gravity." An astronaut orbiting Earth isn't really weightless, because the force of Earth's gravity is still acting. In order to be truly weightless, the astronaut would have to be in deep space, far from any planet or star, and even then the astronaut's weight would not be exactly 0 because of the small gravitational forces from distant stars.

FIGURE 7.9 Astronauts in orbit.

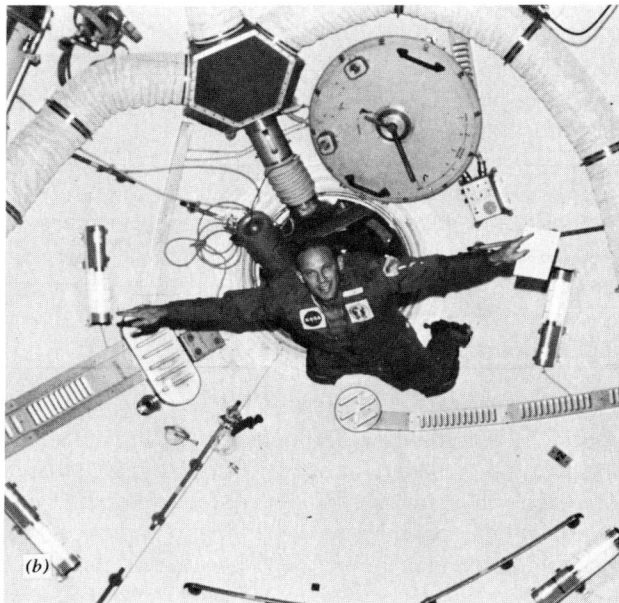

So astronauts in orbit are not weightless, yet they feel weightless. How can we explain this apparent weightlessness?

Let's begin by considering Dick, who is riding in an elevator. Suppose that Dick's actual weight, the force of Earth's gravity on him, is 600 N (about 150 lb). You have probably noticed that when you ride an elevator accelerating upward from the bottom floor, you get pressed down a little, you feel heavier. So does Dick. In fact, if he were standing on a spring scale during this operation, the scale would register more than his actual 600-N weight.

On the other hand, if Dick's elevator accelerates downward from the top floor, he feels lighter than usual. A spring scale would register less than his true 600-N weight.

Exercise 15

What would a spring scale register during the middle of the elevator's run, when the elevator's speed is constant? (Answer below.)

If the elevator is not accelerating (Exercise 15), Dick's spring scale reads his true 600 N. With no acceleration, there must be no net force (Newton 2), and so the 600 N downward gravitational force on Dick must be balanced by a 600 N upward force of the scale pushing on Dick's feet. Dick's feet, then, must push back on the scale with a force of 600 N (Newton 3), so that's what the scale reads.

Downward accelerations give Dick a feeling of reduced weight. The floor is accelerating downward, thus Dick doesn't press as hard as usual on the floor and the floor doesn't press as hard as usual on him (Newton 3 again). Anyone's feeling of weight is determined by the force with which the floor pushes against his or her feet.

Exercise 16

Is there any motion of the elevator that would make Dick feel weightless? Describe this motion. What is an easy way to achieve this motion? (Answer below.)

Dick feels weightless whenever the elevator's downward acceleration is 10 m/sec^2, the acceleration of a freely falling object. The reason is that, at this acceleration, Dick falls freely also, right along with the elevator. Dick and the elevator just keep pace with each other. From Dick's perspective, he is floating inside the elevator. He could be completely unaware that, from the perspective of a person standing on solid ground, he is zooming downward at a great rate, speeding up another 10 m/sec during every second. Free fall (a downward acceleration of 10 m/sec^2) reduces the apparent weight to 0.

There is an easy way to achieve this effect:—cut the elevator cable (Fig. 7.10.) Unfortunately, this action produces apparent weightlessness only until Dick reaches the bottom floor.

Now let's consider Dick's friend Jane, who is in orbit (Fig. 7.11.) Jane's space capsule is falling in a circle around Earth with a forward speed of 8 km/sec and a downward acceleration of 10 m/sec^2 (see the preceding section). Just as in the elevator example, Jane's downward acceleration of 10 m/sec^2 reduces her apparent weight to 0.

Look at it another way: Jane must be doing the same thing as the capsule, namely falling in a circle around Earth with a forward speed of 8 km/sec and a

FIGURE 7.10 Dick in a freely falling elevator.

downward acceleration of 10 m/sec². Jane and the capsule are both in orbit. The two orbits have the same radius and the same speed, so Jane just keeps pace with the capsule. Just as Dick and the elevator were freely falling after the cable was cut, Jane and the capsule are freely falling around Earth.

This analysis also applies to an orbit around the moon, for example. The orbital speed and acceleration are different for a moon orbit than for an Earth orbit, but Jane is still freely falling and hence feels weightless.

In fact, Jane feels weightless no matter where she is in the solar system or in the universe, so long as no rockets or other drive forces act on the capsule. The reason for this phenomenon is that gravity is then the only force on Jane or on the capsule; because Jane and the capsule both fall freely, they just keep pace with each other.

Jane makes a trip to the moon. During which portions of the trip does she feel
weightless? At any point along the way is she truly weightless?

Other Orbits

We've seen that an object launched horizontally at 8 km/sec will fall around
Earth in a circular orbit. An object launched more slowly will move some distance
around Earth, getting closer and closer to Earth until it hits the ground (Fig. 7.12).
What about objects launched faster than 8 km/sec?

An object launched horizontally at more than 8 km/sec is moving too fast for its
orbit to be bent into a circle by Earth's gravity, so the object moves out into a wider
curve as shown in the two faster orbits in Figure 7.12. In the wider curve, however,
the force of Earth's gravity pulls partially backward on the object, rather than entirely
sideways as in the circular orbits, slowing it down. The object thus slows down as it
gets farther above Earth's surface, just as a stone thrown upward slows as it rises. If
the object was not started too fast (less than about 11 km/sec), it will eventually slow
down enough to turn around and return toward Earth as shown in Figures 7.12 and
7.13. The object speeds up as it approaches Earth, just as a stone speeds up as it falls.
Eventually, the object gains enough speed to begin rising again, and the whole process
starts over (Fig. 7.13).

A detailed mathematical analysis shows that the resulting orbit is an ellipse.
Similar reasoning shows that any object moving in free fall around some central
object must move in an elliptical orbit. Newton's laws of mechanics and of gravity
thus explain Kepler's ellipses! In fact, Newton's laws explain all three of Kepler's
principles of planetary motion, an important achievement for Newton's physics.

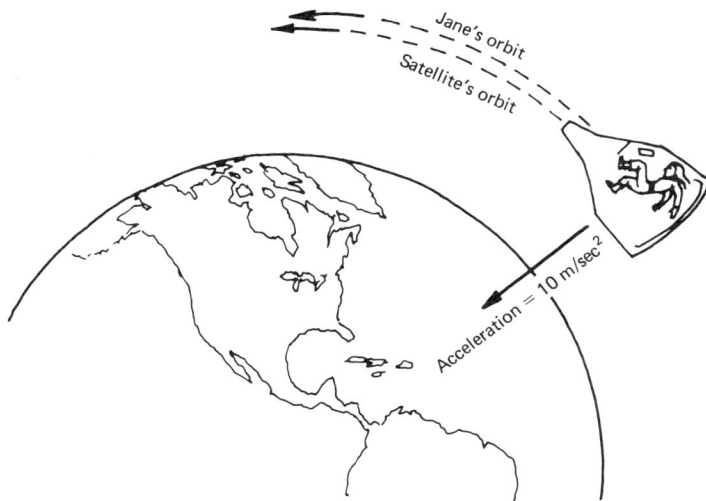

FIGURE 7.11 Jane and her satellite, both in orbit.

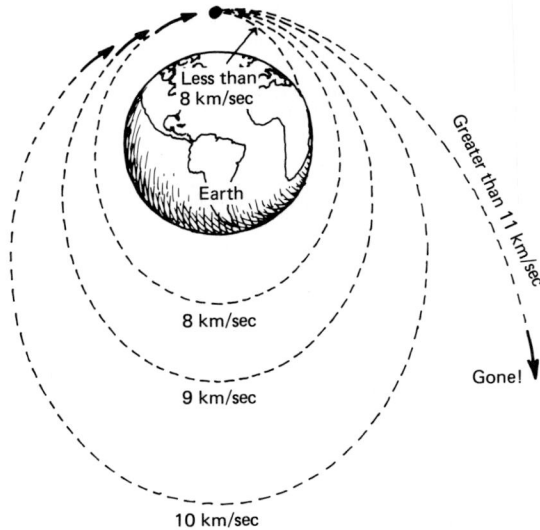

FIGURE 7.12 Several orbits.

If an object is launched fast enough, Earth's gravity will not be able to turn it around; the object escapes Earth (Fig. 7.12). Unless nongravitational forces, such as a rocket drive, come into play, the object will never return. The initial speed required for escape is called the **escape velocity.** The escape velocity from Earth happens to be 11 km/sec (25,000 mph). If you managed to throw a rock faster than 11 km/sec, it would never return, (except for the effects of air resistance, which would slow it considerably.) What goes up *doesn't* always come down!

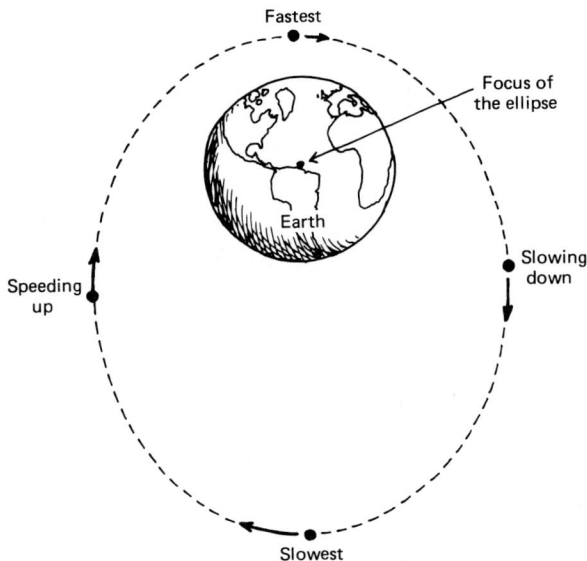

FIGURE 7.13 The elliptical orbit.

gravitational force

Newton's law of gravity

satellite motion

falling around Earth

velocity, acceleration, force during circular motion

communication satellite

synchronous orbit

apparent weight in elevators

apparent weightlessness

elliptical orbit
escape velocity

Thought Questions

1. In Newton's day, the idea of an invisible force that could act even across empty space seemed farfetched. Does it seem farfetched to you that Earth can exert a gravitational force on the moon, across a distance of some 400,000 km?

2. Is the force of gravity real or is it merely a useful fiction invented by scientists?

Further Exercises

18. Some objects, for instance, helium-filled balloons, rise upward when you release them. Does Earth exert a gravitational force on a helium-filled balloon? In what direction is this gravitational force?

19. Continuing the preceding exercise, in what direction is the net force on the balloon? Is there any other force acting on the balloon besides the gravitational force? What is its direction? This force is called the *buoyant force*, caused by the surrounding air. It is similar to the buoyant force caused by surrounding water that keeps a boat afloat. In the case of a helium-filled balloon, the buoyant force is greater than the force of gravity.

20. A falling leaf floats slowly to the ground, moving at constant speed. How big is the net force on the leaf? Is there a force other than gravity acting on the leaf? What do we call this force? Is this force larger than, or less than, or equal to, the weight of the leaf?

21. (a) Suppose you were standing on a planet having the same radius as Earth but only half of Earth's mass. Find your new weight. (b) Suppose another planet had the same mass as Earth but only half of Earth's radius. Find your weight on this planet. (c) Finally, suppose a third planet had half of Earth's mass and half of Earth's radius. Find your weight on this planet.

22. If you were standing on the moon, your weight would be about one-sixth of your normal weight, despite the fact that the moon's radius is much smaller than Earth's radius. This result appears to conflict with Newton's law of gravity, according to which gravitational forces are larger for smaller distances. Resolve this difficulty.

23. Suppose you throw a stone horizontally eastward at 30 m/sec. How far does the stone travel horizontally during the first second after release? During the first 3 sec?

24. Suppose that, in the preceding exercise, you throw the stone horizontally from the top of a building that is 80 m high. How many seconds will the stone remain in the air? (Consult Chapter Five.) How far from the foot of the building will the stone land?

25. An Earth-orbiting weather satellite photographs hurricane patterns over the Gulf of Mexico. In order to return the film to Earth, the film is released through a small door on the bottom of the satellite and allowed to drop to Earth. What is wrong with this scheme?

Recommended Reading

1. Barry M. Casper, and Richard J. Noer, *Revolutions in Physics,* W. W. Norton & Company, New York, 1972. A large portion of this textbook about the conceptual revolutions in physics is devoted to the Newtonian synthesis.
2. I. B. Cohen, *The Birth of a New Physics,* Doubleday, Anchor Books, Garden City, N.Y., 1960. A brief, readable account that covers many of the topics of Chapters Two, Five, Six, and Seven of this book.
3. F. Manuel, *A Portrait of Isaac Newton,* Harvard University Press, Cambridge, Mass., 1968. This biography emphasizes Newton's psychological makeup.

8

OF SPACE COLONIES AND BLACK HOLES

This chapter is a random sampler in Newtonian physics. It presents two applications that I find intriguing—space colonization and gravitational collapse—and it puts gravitational force into its modern perspective as one of the four known fundamental forces. This chapter also contains some philosophizing about the validity of Newton's physics as we see it today and about the mechanistic world view frequently associated with Newtonian ideas.

Project L-5

Picture this: Colonists from Earth establish an excavation and launch site on the moon. Soil mined from the moon is catapulted from the launch site across 500,000 km to a location, called L-5, in the Earth-moon system. Other colonists on an orbiting station at L-5 catch the moon soil and process it into such useful substances as oxygen, iron, glass, and aluminum. Using the processed materials, the colonists construct a giant space habitat that is several kilometers long and that houses thousands of people. Once the habitat is finished, the community turns its attention to manufacturing large solar collectors. As each collector is completed, it is towed into synchronous orbit above Earth. The collectors convert solar energy into electrical energy and then into radiant energy in the form of microwaves, which are beamed down to receiving stations on Earth. These microwaves are reconverted to electrical energy, thus solving Earth's energy problems (Fig. 8.1).

Carry the scenario a step further: When they are not manufacturing solar collectors, the habitat community builds another habitat. Then there are two habitats. Still using materials from the moon, each habitat builds another habitat. Then there are four habitats. And so on. As the habitats continue reproducing themselves, the new habitats are moved to other locations in the solar system. Eventually, the number of habitats being reproduced each year is sufficient to house the yearly increase in Earth's population, thus solving Earth's overpopulation problem. Earth's food problem is solved as well, because the habitats are self-sufficient in food as in everything else. Now that the energy problem, the population problem, and the food problem no longer exist, international tensions subside. Humans bend their swords into plowshares and practice war no more.

Science fiction? Utopian? Perhaps. But various versions of this scenario have been the subject of serious scientific study. NASA has sponsored several interdisciplinary studies of the project, and it is beginning to appear that the project may be "technologically feasible." In other words, our present scientific and technological capabilities are such that the human race could actually play out this scenario. Whether we will play it out, or should play it out, is, of course, another question.

The primary justification for the project is Earth's need for energy. The only long-

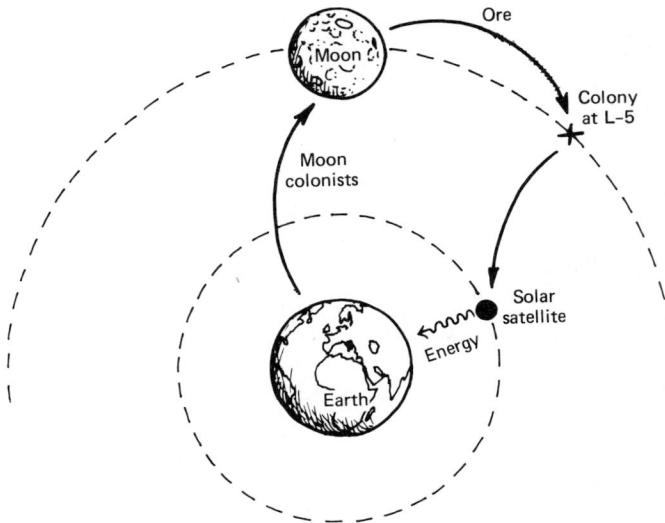

FIGURE 8.1 Project L-5

term, large-scale sources of energy appear to be nuclear fission, nuclear fusion, geothermal, and solar power. In the opinion of many, solar energy is the most desirable option. Solar energy can be received on land, on the sea, or above the surface of Earth. Collectors above Earth's surface would be placed in synchronous orbits, discussed in Chapter Seven. According to one popular proposal, a single solar satellite would combine a large number of solar cells into a raftlike array the size of Manhattan Island.

The solar cells, also known as **photovoltaic cells,** would convert the sun's radiant energy into electric current. This electric current would be used to produce **microwaves,** which are similar to radio waves. The production of microwaves from an electric current is similar to the production of radio waves from an electric current in a radio antenna. The microwaves would be beamed down to a microwave receiver (like a radio receiver antenna) on Earth. A typical receiving region would be approximately circular and would measure about 10 km in diameter. The microwaves would then be converted back into electric current for use on Earth.

One such satellite would provide 5,000 megawatts (MW) of electric power at the ground—equivalent to the power output of five large present-day nuclear generating plants.

A growing number of corporations and individuals are interested in such solar-satellite schemes and are pushing for government-sponsored development studies. One bill in Congress sought $200 million in funding over 5 years.

But why go all the way to the moon for the materials for such a satellite, and why manufacture it in outer space, when the materials and manufacturing capability exist right here on Earth? There are two reasons.

(1) It is easier to boost material into orbit from the moon than from Earth. Recall that the moon's gravitational pull is one-sixth that of Earth. The escape velocity from the moon is therefore much less than the escape velocity from Earth. Furthermore,

air resistance is not a factor because of the near-perfect vacuum on the moon. One kilogram can be lifted off the moon with only 5 percent of the energy required to lift it off Earth.

(2) Recall (Chapter Seven) that unpowered orbital motion produces apparent weightlessness. Because the habitat at L-5 would be a freely orbiting satellite, the environment on board would be apparently weightless. Many industrial processes are enormously simplifed in such a weightless environment. An important example is the fabrication of the delicate crystals that are the main component of photovoltaic cells.

On the moon, large quantities of soil would be mined. The soil would be put into sacks and accelerated to the moon's escape velocity. In one proposed scheme, each bundle would be placed in a bucket, which would be accelerated by an electric motor along a straight, 4-km-long ramp. A solar-powered steam-generating plant on the moon would provide the electric current for the motor. At the end of the ramp, the bucket would stop and the bundle would catapult into space. A steady parade of such soil bundles would streak toward their target at L-5. The habitat would catch the bundles and process them into their useful constituents.

The point L-5 is one of the five balance points known as **Lagrange points** associated with the Earth-moon system (Fig. 8.2). It is distinguished by the fact that

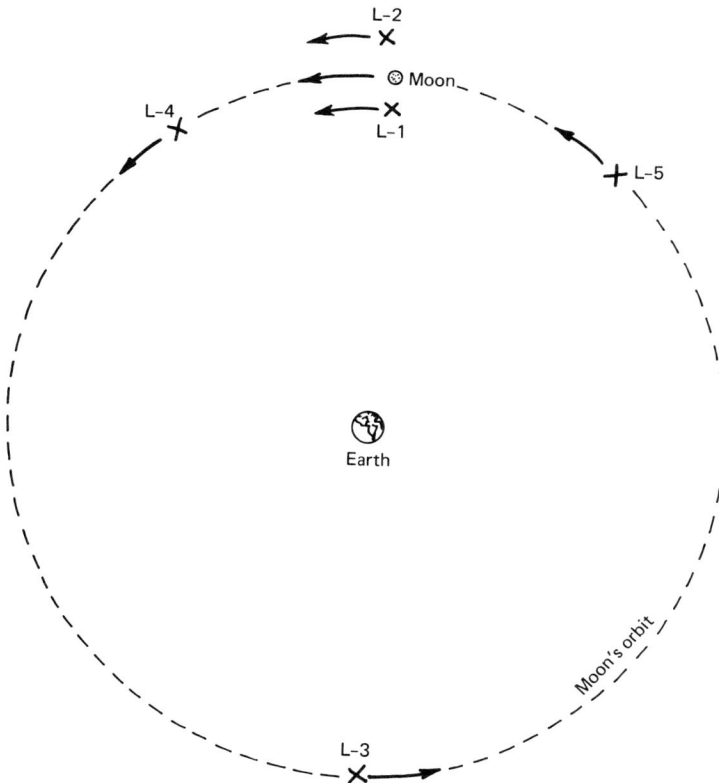

FIGURE 8.2 The Lagrange points.

an object displaced slightly from L-5 will experience forces that bring it back to L-5, an obvious advantage for the location of a space habitat.

Most of the industrial operations for this project would be carried out at L-5. According to one scheme, the habitat would be a huge cylinder several kilometers long and perhaps 1 km across, housing 10,000 to 100,000 people (Fig. 8.3). Of course it would be inconvenient for all these people to be floating around in a weightless (*apparently* weightless, really) environment inside the cylinder. The solution is to have the cylinder spin around its axis (Fig. 8.4).

Why does the spin of the cylinder solve the weightlessness problem? You can answer this question by working through the following exercises, checking your answers as you go.

Exercise 1

According to Chapter Seven, an object in circular motion must have an unbalanced force on it directed (*choose one*) (a) toward the outside of the circle; (b) toward the center of the circle; (c) in the direction of motion (forward); (d) Nonsense—an object in circular motion has no unbalanced force on it.

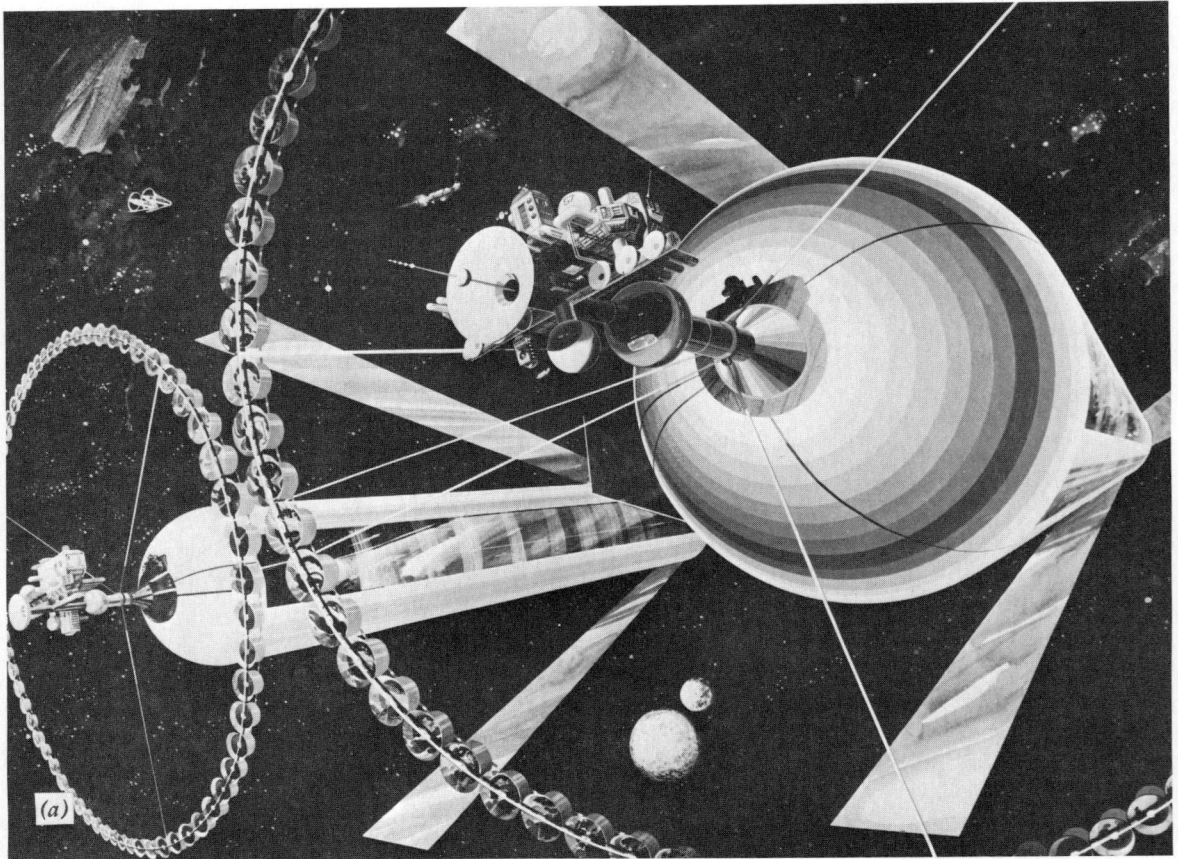

FIGURE 8.3 (*a*) Artist's concept of a pair of cylindrical habitats.

Exercise 2

Rex Stargazer stands on the inside surface of the spinning cylinder. Is there an unbalanced force on him? If so, identify or describe this force and give its direction.

Exercise 3

Is Rex exerting a force on the floor (the inside surface) of the cylinder? If so, give the direction of the force that he exerts on the floor. What principle of physics tells you that this force must be present?

Exercise 4

How do things feel from Rex's point of view, inside the spinning cylinder, and why does Rex feel this way?

So Rex, or any inhabitant of the spinning cylinder, feels that an outward force is exerted on him in response to the actual inward force exerted by the inside surface of the cylinder. From the point of view of Newtonian physics an outward force doesn't really exist, there is only an apparent outward force resulting from the spin of the cylinder. Rex would naturally interpret this apparent force as gravity.

(b)

FIGURE 8.3 continued. (b) Interior view of a cylindrical habitat, artist's concept.

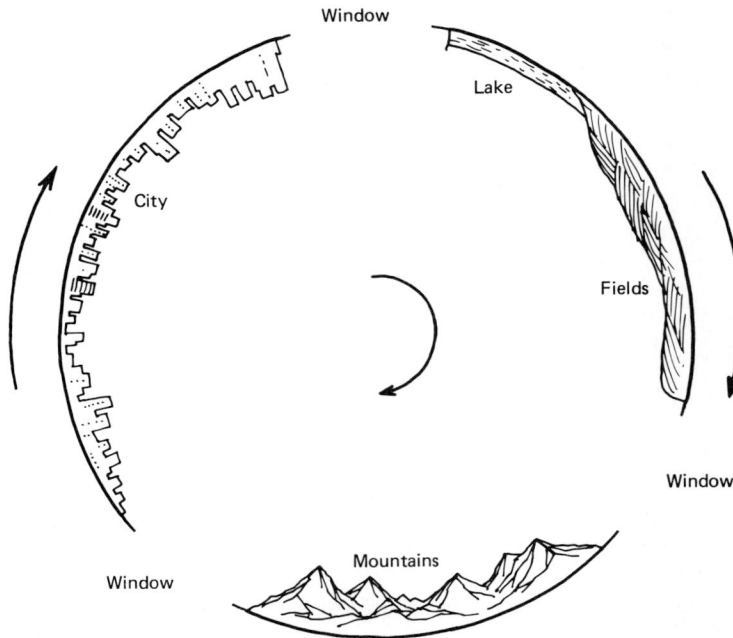

FIGURE 8.4 Cross-section of the space habitat, showing spin.

Rex's apparent weight depends on the rate at which the cylinder spins, because at higher spin rates the floor must press harder against Rex to hold him into his circular motion. Apparent weight can be adjusted by adjusting the rate of spin. For a cylinder with a radius of 1 km, all apparent weights equal ordinary weights on Earth if the cylinder spins through a complete revolution every minute. At this spin rate, Rex feels his weight is normal as he stands on the floor of the cylinder.

The habitat at L-5 could be made quite livable. Sunlight reflected from large mirrors outside the cylinder and directed into the cylinder through windows would supply light, heat, and energy, and the cylinder would be filled with air. Such earthly features as lakes, fields, hills, clouds, and rain could be reproduced. Because the apparent gravitational force becomes weaker as one moves upward from the floor toward the cylindrical axis, and reduces to 0 at the axis, skydiving from a point near the axis would be very interesting. Human-powered flight, a near impossibility on Earth, would be quite feasible near the axis.

The entire project would cost several *trillion* dollars and require 50 years for completion. The NASA space shuttle, a reusable rocket ship, would play an important role in the project. There are, however, no definite plans to go ahead with project L-5.

Implications of Space Colonization

The space colonization project is a good example of the complicated interactions between science and society. The primary justification for the project is social—conventional energy resources are becoming inadequate for our tremendous energy created the wide array of energy-consuming devices in use today.

The space colonization project is not only a consequence of social and scientific developments; if carried out, it will also cause such developments in the future. Any technological project that costs trillions of dollars will have a noticeable effect on both science and society.

Thought Question 1
List some effects that the colonization project will have on science and technology.

Thought Question 2
List some beneficial social and environmental effects.

Thought Question 3
List some harmful social and environmental effects.

Here are a few of the scientific and technological effects that have been suggested: The project would stimulate our exploration of the solar system; large telescopes, located outside Earth's atmosphere, could be erected on the habitat. Many industrial processes could be carried out more efficiently in the 0-gravity and vacuum environment of space. The crystals that are the central component of photovoltaic cells could be fabricated more easily in space; these crystals would be useful in land-based solar projects as well as in the orbiting solar collectors. On the other hand, the project would take money, attention, and expertise away from other scientific and technological projects.

Suggested social and environmental benefits include the following: The project might go a long way toward solving Earth's energy problems. As more habitats are built, a significant fraction of Earth's population could be housed in them, reducing the population pressures on Earth. Industrial activity could eventually be moved from Earth to the habitats, reducing pollution on Earth. Endangered species could be preserved on specially constructed habitats. Many nations could work together on the project, stimulating international cooperation and understanding. The habitats could become experimental communities in which cultural diversity would flourish. Land would be plentiful for agriculture, for industry, and for living.

Many social and environmental drawbacks have also been suggested. Cost is an obvious drawback. Those living in the vicinity of the receiving area for the high-energy microwave beam from each solar satellite might suffer health hazards. The microwave beam might disturb Earth's atmosphere or Earth's radiation belts. Moon colonies, solar satellites, and space habitats would present unprecedented international legal problems—United Nations representatives have yet to agree on a suitable definition of *space*. What would the legal status be of the residents of a habitat? Could they declare independence? Colonies could become involved in Earth-based tensions: weapons could be mounted on them, and conflicts among them might occur. Perhaps world government must be established before the project can become politically feasible.

This brief listing of pros and cons of the colonization project illustrates an important question facing all of us in the closing decades of the twentieth century. The question is: To what extent can modern science and technology help us solve our problems, and to what extent are science and technology *part of* the problem? What are the human pros and cons of the scientific enterprise? This question will recur at many points in this book.

Nature's Four Sub-atomic Forces

Exercise 5
List a dozen or so different types of forces (pushes or pulls). (Answer below.)

Newton's principle of gravity describes one of the fundamental forces in nature: the force of gravity. There are many other types of forces: direct pushes, such as a person pushing a crate or a bat hitting a ball; direct pulls, such as a car pulling a trailer; and less obvious forces, such as friction, air resistance, the magnetic force that turns a compass needle, and the buoyant force that supports floating objects. Recall that Newton unified a wide range of forces with the concept of gravity. A person's weight; the force holding Earth into its orbit around the sun; the force holding the moon in its orbit; the force pulling a falling apple to Earth—Newton saw all these and many more as aspects of a single force, the force of gravity.

It is possible to unify many of the other forces, to see them as different aspects of just a few fundamental forces, provided you look at nature from the point of view of individual subatomic particles, such as protons, electrons, and neutrons. From this *subatomic* point of view, there are only four different types of forces.

For now, I'll list these four forces and give a few examples. Gravity is one of the four. We have already discussed the gravitational force in detail. The other three will receive more detailed treatment later.

1. The Gravitational Force

From the subatomic point of view, this is the weakest of the four forces. That is, the gravitational attraction between two subatomic particles (e.g., two protons) is smaller than any of the remaining three forces acting between the same two particles. The gravitational force exerted by one subatomic particle on another is so weak that it is nearly negligible compared to the other three forces. This weak gravitational force is simply a result of the fact that the two masses are so small. Gravitational forces are large only when at least one of the objects has a large mass.

2. The Weak Nuclear Force

This force is stronger than gravity but weaker than the remaining two forces. The weak nuclear force is not well understood, partly because it has a short range and is thus hard to detect. Although the gravitational force is weak, it has a long range; that is, if we strengthen it by putting a lot of matter together in one place, the gravitational force will become large even at great distances. For example, the sun's gravitational force reaches across 93 million miles to hold Earth into its orbit. The weak nuclear force, on the other hand, has a range roughly equal to the size of an atomic nucleus. That is, two objects separated by more than this distance cannot exert the weak nuclear force on each other.

The weak nuclear force has practically no effect outside a nucleus because particles outside a nucleus are usually so far from all other particles that they cannot feel the weak nuclear force from any other particle. For instance, the electrons orbiting an atomic nucleus do not feel the weak nuclear force. The weak nuclear force does have important effects within the nucleus, however. We will study one of these effects, known as beta decay, in connection with nuclear physics.

3. The Electromagnetic Force

This force is stronger than the preceding two forces but weaker than the strong nuclear force (discussed below). In addition to being fairly strong, it is a long-range force, like gravity, and thus has important and easily observed effects. Its strength and its long range help make it the best understood of the four forces.

A few familiar examples of the electromagnetic force are the forces between simple bar magnets; the forces that produce the electric shock you feel when you shuffle across a rug on a cold day and then touch a piece of metal; the force on a compass needle; the forces that make your hair stand up when you brush it on a cold day; and the forces that produce the effects in such electrical devices as light bulbs and electric motors.

4. The Strong Nuclear Force

As its name implies, this is the strongest of the four forces. However, it has an extremely short range: it acts only within the nucleus. More precisely, the range of this force is roughly equal to the distance across a nucleus. Partly because of its short range, this force is also rather poorly understood.

The strong nuclear force holds the nucleus together. The great strength of this force makes it difficult to dislodge a particle from a nucleus or to split a nucleus into fragments.

Many familiar types of pushes and pulls do not seem to fit into the above four categories. For example, what about the force by a bat hitting a ball? By a chain pulling on a trailer? By a table holding up a book? What about air resistance? What about friction?

To answer these questions, let's focus our attention once more at the atomic and subatomic level. Consider the force by a bat on a ball. As the bat smashes into the ball, the bat's atoms are very close to the ball's atoms. Recall (Chapter Three) that the atom consists of a small nucleus buried in the middle of the much larger orbits of the electrons. Thus as the bat's atoms and the ball's atoms get closer, the electrons approach each other first. When they are close together, the bat's electrons and the ball's electrons exert large electromagnetic forces on each other. These forces distort the electron orbits in many of the atoms, and they repel the bat and ball from each other, pushing the ball out into center field.

A similar subatomic explanation applies to a chain pulling a trailer, a table supporting a book, air resistance, friction, and, in fact, to every force in your everyday environment other than the gravitational force. All of these forces result from one thing touching another thing: bat touches ball, table touches book, air touches moving object, two objects exert frictional forces as they slide against each other. When viewed at the subatomic level, all of these **contact forces** are electromagnetic. From the atomic point of view, what we mean when we say that one thing is touching another thing is that the electron orbits around the atoms of the first thing are so close to the electron orbits around the atoms of the second thing that the electrons are exerting large repulsive electromagnetic forces on each other. Think about that the next time you touch someone.

The tremendous variety of forces around us can be reduced to just four fundamental forces. Such a reduction, or unification, of apparently different phenomena is one of the goals of the scientific enterprise.

Newton unified the falling apple and the "falling" moon. He found a point of view from which the two phenomena could be viewed as manifestations of the same principle: the law of gravity. After Newton, the apple and the moon were no longer as different as they had been.

Kepler unified the voluminous planetary data collected by Tycho Brahe. Kepler found that the data could be reduced to, or explained by, three principles of planetary motion. And Newton built on the unification achieved by Kepler and Galileo and others to achieve a grand synthesis of apples, moons, planets, and stars.

As another example of unification, the billions of different objects in your environment reduce to millions of types of molecules, which reduce to 105 types of atoms, which reduce to just three types of subatomic particles: protons, neutrons, and electrons.

Albert Einstein spent most of his life searching for that **unified field theory** which would tie together the gravitational force and the electromagnetic force. He never reached his goal, but physicists today continue the search.

One goal of contemporary high-energy physics research is deeper understanding and further unification of the fundamental forces. Perhaps the day will come when we will view all forces as manifestations of a single, underlying universal force.

Limitations of the Newtonian Scheme

Newtonian physics represents one of the greatest unifications of natural phenomena that has ever proceeded from the human mind. From moons and meteors to apples and automobiles, Newton's physics explains all.

Thought Question 4
Does the validity of Newton's predictions prove for certain that Newton's theories are true?

Recalling the lessons about the methods of science in Chapter Two, the answer to this question is no—we can never prove a theory.

Recall, however, that we can disprove a scientific theory. Many principles of Newtonian physics actually have been disproved in the twentieth century. To put it more gently, Newton's ideas have been found to have only a limited range of validity. Newton's physics has been found to be invalid in the following situations:

(1) Newtonian physics does not apply to objects moving at a significant fraction of the speed of light. Now, light moves really fast—a light beam travels a distance equal to eight times around Earth in 1 sec! For all "ordinary" speeds, then, Newton's ideas work just fine. Around 1905, Albert Einstein invented a theory known as special

relativity that works at high speeds as well as at low speeds. For slower objects (slower than about 1 percent of the speed of light) the predictions of special relativity and of Newtonian physics are essentially identical, so either theory may be used in this range.

(2) Newtonian mechanics does not apply to objects consisting of only a small number of atoms. For larger objects (larger than, for example, the smallest object visible to the unaided eye), Newtonian mechanics is valid. During the period from 1900 to 1930, a group of physicists worked out a theory known as quantum mechanics, which seems to work for objects of all sizes, from large objects down to and including subatomic objects. For larger objects, the predictions of quantum mechanics and of Newtonian mechanics are essentially identical,* so either theory may be used.

(3) Newtonian gravitational theory does not apply to very strong or very extended gravitational forces. For ''ordinary'' gravitational forces, such as the weight of your body or the force holding the moon into its orbit, Newton's theory of gravity works. But Newton's theory is inapplicable to the strong gravitational forces found in the vicinity of collapsed stars (see below) or to the extensive gravitational forces reaching across large expanses of the universe. In 1916, Einstein published the theory of general relativity, which appears to be valid for all gravitational forces. For ''ordinary'' gravitational forces, the predictions of general relativity and the predictions of Newton's theory are nearly identical.

To sum up: Newtonian physics is valid provided the objects under study aren't moving too fast, aren't too small, and aren't subject to very strong or very extensive gravitational forces.

Despite these limitations, Newtonian physics is still the basis for most scientific and technological work, because most phenomena of interest to humans involve neither high speeds nor small objects nor unusual gravitational forces. The theories of special relativity, quantum mechanics, and general relativity are mathematically more difficult than Newtonian physics, so we resort to these more modern theories only when forced to do so by the limitations on Newton's physics.

Although most scientific theories have their limitations, we don't necessarily discard a theory just because it is limited. We are more likely to note the limitation and to apply the theory only within its range of validity, for the theory might still be useful within its limited range. The hallmark of a good scientific theory is not that it is universally true, but rather that it is fruitful: fruitful in making specific predictions, in suggesting new ideas, in unifying a wide range of phenomena.

The World as Machine

During the eighteenth and nineteenth centuries, Newton's ideas were so successful that they took on an aura of certainty. Many scientists and philosophers felt that a thing was not really understood until it had been explained in Newtonian terms. The very word *understand* came to mean ''to explain in terms of Newtonian physics.''

Newtonian physics thus spawned the ***mechanistic world view:*** the idea that the universe is a grand machine, running like clockwork according to predetermined laws.

*Exception: The large-scale effects known as superfluidity and superconductivity are explainable by quantum mechanics but not by Newtonian mechanics.

In what sense is Newtonian physics mechanistic? Consider a collection, or *system,* of particles, all exerting specific types of forces on each other. (Scientists often use the word *system* to mean "a portion of nature toward which we wish to direct our attention.") Suppose that this system is isolated from all external influences, so that the forces on any particle of the system arise only from the other particles of the system. Further suppose that Newtonian physics is valid for these particles. Then, according to Newton's laws of mechanics, the behavior of the entire collection is entirely and precisely predictable, for all time. To state this important idea more completely: Given the positions and velocities of every particle of the system at any one instant, it turns out to be possible to use Newton's laws to calculate the positions and velocities of every particle at every later instant. The collection of particles operates in a precise and predictable manner, *just like a machine.*

Thus if everything is made of particles and if Newton's laws are precisely and universally correct, then everything operates in a completely predictable manner. If we take this view seriously, then the universe and everything in it, including you and me, are just so many machines. For example, you might suppose that you decided, of your own free will, to scratch your ear just now. According to the mechanistic view, the scratching of your ear was entirely determined aeons ago, before your birth and before the birth of the solar system, by the Newtonian laws of the universe. If you should decide, impulsively, *not* to scratch your ear, hoping perhaps to thwart the laws of the universe, well, that decision was predictable as well. In this view, free will is a figment of our imaginations.

Over a span of nearly three centuries, predating Newton and extending well into the twentieth century, many of the central figures of western civilization have subscribed to this view of the world as machine.

Consider Kepler:

My aim is to show that the heavenly machine is not a kind of divine, live being, but a kind of clockwork—insofar as nearly all the manifold motions are caused by a most simple, magnetic, and material force, just as all motions of a clock are caused by simple weight.

The Scottish philosopher and historian David Hume:

Look around the world, contemplate the whole and every part of it: You will find it to be nothing but one great machine, subdivided into an infinite number of lesser machines, which again admit of subdivisions to a degree beyond what human senses and faculties can trace and explain.

To the French philosopher Voltaire it seemed unbelievable

that all nature, all the planets, should obey eternal laws, and that there should be a little animal, five feet high, who in contempt of these laws, could act as he pleased, solely according to his caprice.

The French scientist and social leader Laplace:

An intelligence knowing, at a given instant of time, all forces acting in nature, as well as the momentary positions of all things of which the universe consists, would be able to comprehend the motions of the largest bodies of the world and those of the smallest atoms in one single formula."

The nineteenth-century physicist Lord Kelvin:

> I never satisfy myself until I make a mechanical model of a thing. If I can make a mechanical model I can understand it. As long as I cannot make a mechanical model all the way through I cannot understand.

And the twentieth-century philosopher and mathematician Bertrand Russell wrote in 1902:

> That Man is the product of causes which had no prevision of the end they were achieving; that his origin, his growth, his hopes and fears, his loves and his beliefs, are but the outcome of accidental collocations of atoms;—all these things, if not quite beyond dispute, are yet so nearly certain, that no philosophy which rejects them can hope to stand."

We will find in later chapters that twentieth-century physics leads to world views that are quite at odds with the preceding philosophy. Nevertheless, it seems likely that after several centuries of viewing the universe as a machine, the western mind is still dominated by mechanistic ways of thinking.

Thought Question 5
Are we still influenced by this mechanistic way of thinking? In what ways?

Thought Question 6
If you had lived in the nineteenth century and had been presented with the arguments of Hume, Voltaire, Laplace, and others, what would your response have been?

Black Holes

The large-scale universe seems to come in clumps. We find matter clumped into galaxies, into stars, into planets, into moons. These aggregations of material are caused by gravity always pulling matter inward toward other matter. Great extended bodies of gas and dust are thus pulled together by their own gravity into such compact objects as stars, planets, and moons. This process of gathering matter together by gravitational attraction is called ***gravitational collapse.***

A typical star, such as our sun, is a large, gaseous object, gathered together by gravity and powered by a nuclear energy source (see Chapter Twenty-One) at its center. This nuclear energy source produces thermal energy that produces outward-directed forces throughout the star. The outward-directed forces resist the inward pull of gravity and enable the star to remain in its extended, gaseous state. Every star eventually runs out of nuclear fuel, at which point there are no outward-directed forces to resist the inward pull of gravity. So the star shrinks. In fact, it shrinks drastically, because the star's large mass (compared to, for example, Earth's mass) produces very large inward-pulling gravitational forces throughout the star.

The process of shrinking usually causes the star to collapse to a much more compact state in which its entire mass (perhaps a million times the mass of Earth) is squeezed together until the star is about the size of Earth. Stars that have reached this state are called ***white dwarfs.*** If you could somehow transport a single cupful of material from a white dwarf to Earth, it would weigh several tons.

Exercise 6
Suppose a star collapsed to 1/100 of its original radius. By what factor would the force of gravity on the surface be multiplied?

Our sun will someday run out of its nuclear fuel and collapse to become a white dwarf. But don't despair; it still has some 5 billion years to go.

Suppose that an extremely massive star, one whose mass is perhaps ten times the sun's mass, runs out of nuclear fuel and collapses. The force of gravity then collapses the star until it reaches the white dwarf state. However, not even the dense white dwarf star can stand up under the intense gravitational pull of so much mass. The star continues to collapse, until it is much more compact than a white dwarf. It is thought that this process occurs in any collapsing star that has a mass greater than about 1.5 times the sun's mass.

Many of these more massive stars eventually halt their collapse when they reach a configuration known as the **neutron star state.** This object, called a **neutron star,** is, by earthly standards, an unusual object indeed. A neutron star may have the mass of several suns squeezed into a sphere a few kilometers across. A thimbleful of this material would weigh millions of tons on Earth. The pressure inside such a star is so great that ordinary atoms are squeezed out of existence, and electrons and protons are squeezed together to form neutrons, so that only neutrons remain. Such an object can be thought of as a very large nucleus a few kilometers across, composed only of neutrons.

It is thought that neutron stars have masses less than about three times the sun's mass. For more massive objects, the ever-present pull of gravity is so strong that even neutron stars cannot withstand it. Such objects continue collapsing past the neutron star state. But there is no known stable state for a star that is more compact than a neutron star. Once the force of gravity has compressed a star beyond the neutron star state, there is literally no stopping the collapse. No known force is strong enough to resist the enormous pull of gravity within the star. The object just keeps collapsing, forever. It becomes a **black hole.**

You can't throw anything out of a black hole. In order to do so you would have to get it moving faster than the speed of light, because the escape velocity (Chapter Seven) from a black hole exceeds the speed of light. Since no known object can exceed the speed of light, nothing, not even light, can escape a black hole. Thus the object is called *black.**

It is difficult to assign a size to a black hole, because it collapses forever. Any object within 30 km of a typical black hole (having a mass of, let's say, 10 solar masses) is caught in it, because inside this radius the escape velocity exceeds the speed of light. In this sense, then, the radius of a typical black hole is about 30 km.

Nobody was much concerned with black holes until 1967, when the first neutron star was discovered. J. Robert Oppenheimer and others described the neutron star as a theoretical possibility as early as 1934, but none of these objects were observed for 3 decades after this prediction. Because no neutron stars had been discovered, nobody worried about the even more esoteric prediction, also made by Oppenheimer

*More precisely, according to the general theory of relativity (Chapter Twenty-Four), the object is black because space is warped back on itself near a black hole. The above explanation, in terms of Newtonian physics, will do for now, however.

in 1934, of the black hole.* Since 1967, however, more than 100 neutron stars have been discovered. Today's scientists take the possibility of black holes very seriously indeed.

Several black holes may already have been discovered. At least, black holes seem to provide the most reasonable interpretation of several puzzling sources of high-energy radiation. For example, the X-ray source known as Cygnus X-1 is possibly a double star consisting of a large and luminous giant star and a tiny but massive (10 solar masses) black hole. The visible star and the black hole orbit each other. The X-rays received from this source are apparently emitted from rapidly moving gas that is torn off the giant star by the black hole. Some of this gas must be disappearing "down" the black hole, never again to be seen in our universe.

Recently, scientists have indulged in some weird and wonderful speculations about black holes. These ideas, discussed below, are only theoretical possibilities, and have not, as of this writing, been confirmed by observation.

The mysterious objects known as quasi-stellar objects, or **quasars,** might be powered by gigantic black holes. Astronomers have observed a large number of quasars recently. According to prevailing opinion, quasars are highly energetic objects residing at great distances from our galaxy, comparable in mass to an individual galaxy but with a vastly greater energy output than typical galaxies. It has been suggested that the energy source for these creatures might be a giant black hole, comparable in size to our solar system (recall that an "ordinary" black hole is just 30 km across) and with the mass of millions of suns. The idea is that the black hole inhabits the center of the quasar and provides energy to the quasar by devouring the surrounding stars. If roughly one star per year were sucked into the black hole, the energy released by the accelerated gases just before they disappeared down the black hole would suffice to power the quasar.

Giant black holes might inhabit the centers of many, or most, galaxies in the universe. In fact, observations of radio waves coming from the centers of some galaxies suggest the presence of giant black holes. Perhaps the bright centers found in all galaxies are powered by massive black holes consuming the stars around them (Fig. 8.5).

Some evidence suggests that the center of our own Milky Way galaxy is the home of a giant black hole. At any rate, the radio waves emitted by our galactic center are consistent with the presence there of a black hole having a mass of 100 million suns and a size comparable to the distance from Earth to sun. It is difficult for us to know exactly what is going on at the center of our own galaxy because our view is obscured by great clouds of gas and dust. We do know that the position of the center lies in the Milky Way* in the direction of the constellation Sagittarius.

The next time you are out in the country on a starry summer or autumn night, find the constellation Sagittarius (Fig. 8.6). It is visible in the southern sky during the summer and early autumn months, close to the band of light known as the "Milky Way." The Milky Way itself is actually composed of those stars in our galaxy which

---

*To be more precise, the French mathematician Pierre-Simon de Laplace was the first to demonstrate theoretically that a mass could be so large that not even light could escape from its surface. His paper, based on Newton's physics, appeared in 1798; his idea, based this time on Einstein's physics, was rediscovered in 1934.

*The galactic center must lie in the band of light known as the Milky Way, since the Milky Way actually outlines the plane of our dish-shaped galaxy.

FIGURE 8.5 (a) and (b) Two galaxies. Note the bright
centers. (b) This galaxy appears edge on as seen from
Earth. (Palomar Observatory, California Institute of
Technology.)

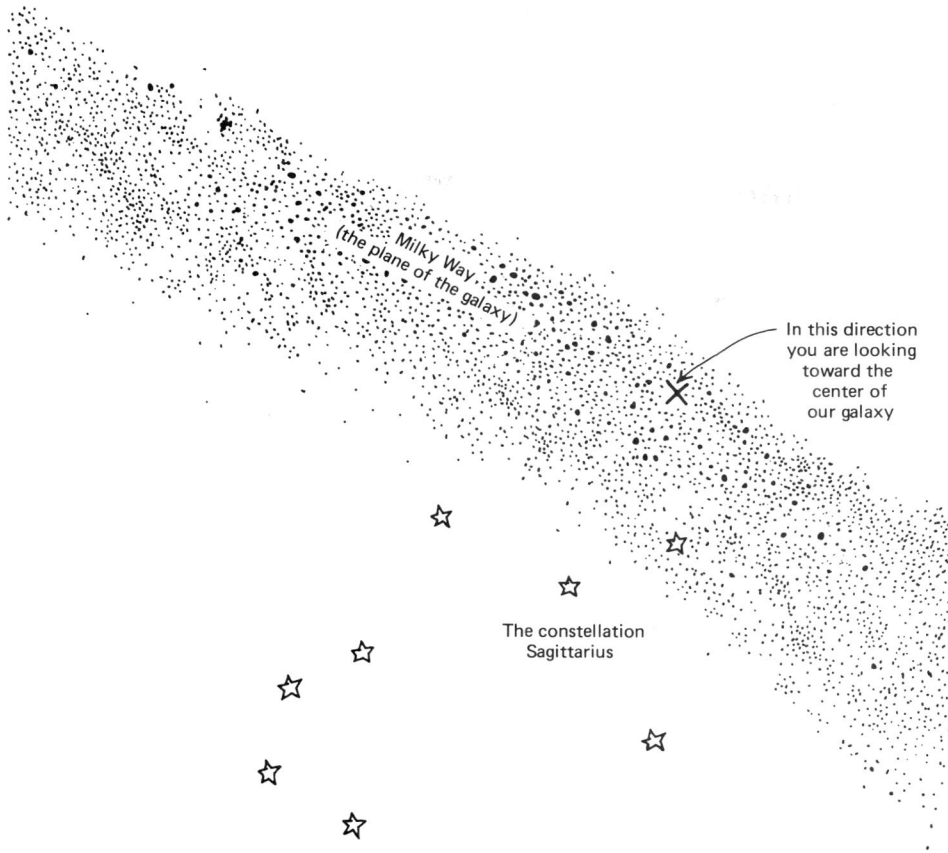

FIGURE 8.6 The constellation Sagittarius. Far beyond these stars, far beyond the visible Milky Way, lies the center of our Milky Way galaxy. The direction toward the center is shown.

lie relatively close to us. Far beyond Sagittarius, far behind the Milky Way, is the center of our galaxy.

As if giant black holes weren't enough, current theories of general relativity, quantum mechanics, and cosmology (the study of the structure and evolution of the universe) suggest the existence of **small black holes.** Such objects might have the mass of a good-sized meteoroid (a rock about a kilometer across) but the size of a proton! Mini black holes could have been created when our universe was created. Furthermore, unlike ordinary black holes, small black holes can lose significant amounts of mass by an esoteric microscopic process. The smaller the black hole, the faster it would lose mass. Eventually, the black hole would vanish entirely. Small black holes created when our universe was born might be completing this "evaporation" process right now.

In a sense, the material that enters a black hole leaves our universe. It can never again be observed in our universe. What happens to it? Does it reappear in some other universe? Does it reappear in our own universe at some other point in space and time, emerging as a "white hole," a sort of reversal of a black hole? Perhaps black holes are connected to white holes in this or another universe by "worm holes."

Perhaps our own universe is a black hole in some other universe.
Such are the possibilities of modern astrophysics!

Checklist

photovoltaic cells	strong nuclear force
microwaves	contact forces
Lagrange points	limitations of Newtonian physics
habitat at L-5	the mechanistic world view
four subatomic forces	gravitational collapse
gravitational force	white dwarf
weak nuclear force	neutron star
electromagnetic force	black hole
	quasar

Further Thought Questions

7. Would the space colonization project be beneficial for the human race or would it create more problems than it would solve?

8. Many people feel that the human race has no business venturing into space until we can solve our problems here on Earth. How do you feel about this?

9. Many phenomena are unified at the subatomic level. For instance, we saw in this chapter that all of nature's many forces can be explained by just four different types of forces acting among subatomic particles. Give other examples of the unification of natural phenomena at the atomic or subatomic level.

Further Exercises

7. You may have played with a toy top or gyroscope and observed that it spins for a long time once it is set into motion. The habitat at L-5 will need to be set into spinning motion in order to create the effect of gravity on the inside. Unlike your top or gyroscope, the habitat will be isolated in the vacuum of space. Judging from your experience with tops, would you expect that external forces will need to be exerted on the habitat in order to start it spinning? Once it is spinning, would you expect that external forces will need to be exerted to keep it spinning?

8. Suppose that the habitat at L-5 suddenly stopped spinning. What would happen to the inhabitants? Would they fall to the center? Would they fall to the floor (the inner surface of the cylinder)? Would they be aware that anything had happened?

9. Which of the four subatomic forces is responsible for lightning? Which is responsible for weight? Which is responsible for the bouyant force that supports floating objects?

10. Of the four subatomic forces, which would you guess is responsible for an atomic bomb explosion? Which is responsible for the operation of a nuclear reactor?

Recommended Reading

1. Arthur C. Clarke, *Rendezvous with Rama,* Ballentine Books, New York, 1973. Science fiction about the human exploration of someone else's cylindrical space habitat.

2. T. A. Heppenheimer, *Colonies in Space,* Warner Books, New York, 1977. A complete treatment of space colonization by a planetary scientist.

3. William J. Kaufmann, III, *Relativity and Cosmology,* Harper & Row, New York, 1973. A book for nonscientists about black holes, quasars, and the shape of the universe.

4. *L-5 News,* published monthly by the L-5 Society, 1620 North Park Ave., Tucson, Ariz. 85719.

5. G. K. O'Neill, *The High Frontier,* William Morrow & Co., New York, 1976. Project L-5 originated with physics professor Gerald O'Neill of Princeton University.

6. *2001,* a film (released in 1968 by MGM Studios and directed by Stanley Kubrick) based on the novel by Arthur C. Clarke, features a torus-(doughnut-) shaped space colony, artificial gravity produced by spin, and many other examples of the principles of physics.

7. Allen Wheelis, *The End of the Modern Age,* Harper Torchbooks, Harper & Row, New York, 1973. An elegant discussion of the rise (with Copernican cosmology), triumph (with Newtonian physics), and fall (caused by relativity and quantum mechanics) of the mechanistic world view.

9

THE LAWS OF ENERGY

Energy is important because it can be used to make things go and because the world is suffering from an energy crisis. The world energy problem is one of the major themes of this book. To deal intelligently with the social aspects of energy, we must learn what energy *is* and we must learn nature's rules regarding energy. This chapter is devoted to the physics of energy.

What Is Energy?

Because physicists define energy in terms of the concept of work, let's discuss work before defining energy.

People use the word **work** in several ways in everyday language. In order for this word to be useful to the scientist, though, it must have a single, precise meaning. Physicists say that work is done whenever a force is exerted on an object while the object moves through some distance. This definition is fairly close to one of the everyday meanings of the word. Most people would say, for instance, that they are doing work on a cement block as they exert a force with their arm to slide the block across a table. As some of the following exercises show, however, other common meanings of *work* do not fit the physicist's definition.

Exercise 1
I push against a wall for 5 minutes. I push hard. My muscles get very tired. I work up a terrific sweat. Have I done any work on the wall? (Answer below.)

Exercise 2
For 1000 years a meteoroid traverses a part of the Galaxy within which gravitational forces are essentially zero. Was any work done on the meteoroid? (Answer below.)

Work requires both force and motion: no motion (Exercise 1) means no work; no force (Exercise 2) means no work.

For the simplest case, when the force is constant and the motion is in a straight line along the direction of the force, physicists use the following quantitative definition of work:

The **work** done by a force acting on an object is the product of the force and the distance through which the object moves.

Briefly,

$$\text{Work} = \text{force} \times \text{distance}$$

Exercise 3

In the metric system what units can be used to measure work? In the English system what units can be used? *Hint:* You can derive the answer from the definition of work.

Exercise 4

A 2000-lb car is raised 4 ft by a car lift. How much work does the lift do on the car? The car is now lowered 3 ft. How much work does the car do on the lift? How much work does the lift do on the car while the lift is stationary?

Exercise 5

The expanding gases inside a rifle barrel exert a force of 2000 N on a bullet as it moves down the barrel. If the barrel is 1.2 m long, how much work do these gases do on the bullet?

Work can be measured in terms of the **newton-meter** (N·m), the work done by 1 N in moving through 1 m, or the **foot-pound** (ft·lb), the work done by 1 lb in moving through 1 ft. Several other units popular for measuring work are the **joule** (J), the **calorie** (cal), the **kilocalorie** (kcal), and the **British thermal unit** (Btu). These units are related as follows:

$$1 \text{ J} = 1 \text{ N·m}$$
$$1 \text{ cal} = 3 \text{ ft} \cdot \text{lb} = 4 \text{ J*}$$
$$1 \text{ kcal} = 1000 \text{ cal}$$
$$1 \text{ Btu} = 780 \text{ ft} \cdot \text{lb.}$$

Joule is simply another name for newton-meter. The joule is the standard metric unit for work. Because of an unfortunate mix-up, the **dieticians' calorie** is the kilocalorie defined above.

Exercise 6

How many foot-pounds are there in 1 dietician's calorie? How many joules? How many Btu's?

Power is a useful concept closely related to work.

The **power output** by a force is the rate at which the force does work, that is, the work done divided by the time to do it.

Briefly,

$$\text{Power} = \frac{\text{work}}{\text{time}}$$

Several units are used to measure power. The standard English unit is the *ft · lb/sec,* the power output when 1 lb acts on an object as it moves through 1 ft in a time of 1 sec. The standard metric unit for power is the **watt** (W), which is defined as a J/sec. That is, the watt is the power output when 1 J of work is done in 1 sec. The **horsepower** (hp) is another widely used unit. These units are related as follows:

$$1 \text{ hp} = 550 \text{ ft} \cdot \text{lb/sec} = 750 \text{ W}$$
$$1 \text{ W} = 1 \text{ J/sec} = 0.75 \text{ ft} \cdot \text{lb/sec}$$

---

*More precisely, 1 cal = 3.1 ft · lb = 4.2 J.

The horsepower, 550 ft · lb/sec, is a very optimistic figure for the rate at which a horse can do work. Real horses can do work at about half this rate.

An example: A person weighing 600 N walks up a flight of stairs in 5 sec. Suppose the distance between floors is 3 m. Let's determine the work done and the power expended by the person's muscles as they lifted the body. The muscles must exert a force of 600 N through 3 m just to lift the person, in addition to any forces needed in moving horizontally up the stairs. So, the work that goes into lifting the person is 600 N × 3 m = 1800 N · m = 1800 J. Since 1 cal = 4 J, this work can also be expressed as 450 cal or 0.45 kcal (0.45 dietician's calories). This work is done in 5 sec, so the power output is 1800 J/5 sec = 360 J/sec = 360 W, a little less than ½ hp.

Do not confuse work with power. Power is the rate of doing work, the work done divided by the time to do it. Equivalently, the work done is the power output multiplied by the time. The amount of work done is one thing, and the rate at which that work is done is something else. Try these exercises:

Exercise 7

A bat delivers a 200,000-N blow to a baseball. Bat and ball are in contact for only 0.001 sec, during which the bat (and ball) moves a distance of 0.02 m. How much work was done, and what was the power output by the bat?

Exercise 8

The ion-drive, a rocket-propulsion system that might power interplanetary spaceships in the future, operates by using electric forces to eject ions (electrically charged atoms) rearward. If a low-power ion drive having a 2 W output operates continuously for a week, how much work is done?

Exercise 9

The **kilowatt hour** (kW · h) is the amount of work done when a power of 1000W (1 kW) operates for 1 hour. A kilowatt hour is equivalent to how many joules? To how many calories? To how many foot-pounds? When you pay for kilowatt hours on your electric bill, you are paying for work that the power company did for you.

In Exercise 7, the power output is large even though the total work done is not large, because the time is short. A large power output is typical of brief, "impulsive" forces, like a bat hitting a ball. In Exercise 8, the total work done is large even though the ion drive has a low power rating. This situation is typical of low-power outputs operating continuously over long periods of time. Remember this the next time you go out of your house and leave the lights burning.

Now we can define energy.

Consider a parked car with its tank full of gasoline. Obviously, the gasoline can be used to power the car down the road. The gasoline gives the car the ability to do a certain amount of work, and this ability is what we mean when we say that the gasoline has energy. Any system (such as a tank of gasoline) that has the ability to do work is said to have energy. Quantitatively, the amount of energy possessed by the system is the amount of work the system can do. To summarize:

> **Energy** is the ability to do work. The amount of energy that any system possesses is the amount of work that the system is capable of doing.

Because energy is equal to the amount of work a system can do, it is measured with the same units used to measure work: newton-meter, joule, foot-pound, calorie, kilocalorie, and British thermal unit.

Exercise 10
A slice of bread contains about 60 dietician's calories. How much energy is in the slice of bread, in joules and in foot-pounds?

Exercise 11
A 100-hp automobile engine runs for 10,000 sec (about 3 hours) on a tank of gasoline. How much energy is in the gasoline, assuming that all the energy is used in running the engine? (This assumption is not really justified. The energy in the gasoline is actually several times the amount you calculated; most of it is wasted.)

You can think of energy as *stored work*, the amount of work a system is capable of doing sometime in the future. This work might never be done. For instance, in Exercise 11, the car might never move out of the driveway. Nevertheless, the full tank of gasoline contains 5.5×10^8 ft · lb of energy.

Types of Energy

Physicists classify energy into several different categories.

For example, a baseball moving toward home plate can do work by compressing the catcher's mitt. Since the baseball in motion has the capacity to do work it therefore has energy. This type of energy is called **kinetic** ("motion") **energy**.

We will distinguish among eight types of energy:

1. **Kinetic energy** results from motion.
2. **Gravitational energy** results from gravitational forces.
3. **Elastic energy** is the energy resulting from the ability of a stretched or distorted object to snap back to a less distorted shape.
4. **Thermal energy** (sometimes called *heat energy*) results from temperature.
5. **Electromagnetic energy,** or **electric energy,** results from electromagnetic forces.
6. **Radiant energy** results from electromagnetic radiation. (Examples of electromagnetic radiation are a light beam, an X-ray beam, and a radio wave.)
7. **Chemical energy** results from chemical structure, that is, the way in which atoms are arranged into molecules.
8. **Nuclear energy** is energy that results from nuclear structure, that is, the way in which protons and neutrons are arranged into atomic nuclei.

Thought Question 1
The moving baseball is one example of kinetic energy. Name one or two examples of each of the other types of energy.

A rock held above the ground has more energy than a rock at ground level because it can do work in getting back down to the ground. This energy is a result of the force of gravity pulling downward on the rock, so it is gravitational energy. Suppose that the rock weighs 40 N (about 10 lb) and that you hold it 2 m above the ground. If you lower the rock slowly to the ground, it will exert a 40-N downward force on your hand through the entire 2 m. Thus, at 2 m above the ground, the rock was capable of doing 40 × 2 = 80 N·m or 80 J of work in getting back to the ground. In other words, at 2 m high the rock had 80 J of gravitational energy (relative to its energy at ground level).

I think you can see from this example that any object above ground level has gravitational energy and that we can calculate the amount of this gravitational energy from the following formula:

Gravitational energy = weight of object × height above ground

Whereas gravitational energy depends on the object's position relative to the ground, kinetic energy depends on the object's speed. For instance, the faster a baseball moves, the more work it can do on the catcher's mitt, and the more energy it has. Kinetic energy turns out to be proportional to the square of the speed of the object. That is, doubling an object's speed quadruples its kinetic energy. For example, a car that doubles its speed quadruples its kinetic energy; it can do not just two, but four, times the work it could do at the slower speed and four times the damage if it runs into something.

An object's kinetic energy also depends on its mass. A more massive ball does more work on a catcher's mitt than a less massive ball moving at the same speed.

A detailed analysis shows that

Kinetic energy = 1/2 × mass of object × square of speed

We'll study all eight forms of energy at appropriate points in the book.

The Principle of Conservation of Energy

Suppose you have a system (for example, a moving baseball or a tank of gasoline) possessing 20 J of energy. Having 20 J of energy, the system can do 20 J of work. Now suppose that the system actually does 5 J of work on some second system (the ball could do work on a catcher's mitt; the gasoline could do work on a car). How does this affect the energy of the two systems?

Experiments show that the first system loses precisely 5 J and the second system gains precisely 5 Joules of energy.

Experiments involving every known form of energy have tested this idea again and again. To date, no exceptions have been found. We'll call this idea the

Work-Energy Principle. If one system does a certain amount of work on a second system, then the first system loses that amount of energy and the second system gains that amount of energy.

It is useful to state this idea another way. Let's think of the two systems referred to in the work-energy principle as a single, more complex system composed of two parts. When the first part does work on the second part, according to the work-energy principle the second part gains as much energy as the first part loses. Thus the total system neither gains nor loses energy, regardless of what might have happened to the individual parts. When stated in this form we call this idea the

Principle of Conservation of Energy. The total energy of an isolated system always stays the same, regardless of any processes occurring within the system.

That is, energy cannot be created or destroyed; energy can be transformed from one form to another, but the total amount of energy stays the same.

Many physical processes are best understood in terms of *energy transformations.* For example, as an object slides faster and faster downhill, it loses gravitational energy while gaining kinetic energy—energy is transformed from gravitational to kinetic. In the ideal case, when friction and air resistance are negligible so that no energy is transferred to the hill or to the air, conservation of energy tells us that the amount of gravitational energy lost equals the amount of kinetic energy gained.

Consider a skier skiing down a mountain; air resistance and friction are no longer negligible. Our skier starts from the top of the mountain with a certain amount of gravitational energy. Some of this energy is transferred into the skier's kinetic energy. It can't all become kinetic energy, however, because air resistance and friction slow the skier down, reducing the final kinetic energy. Into which of the eight types of energy was the rest of the energy transformed? By experiment we know that the remaining energy becomes thermal energy. Careful measurements of the temperature of the air and snow would show that they had been warmed a little by the run downhill. Quantitatively, the principle of conservation of energy says that the kinetic energy the skier picks up, plus the thermal energy created in the air and snow during the run, precisely equals the skier's gravitational energy at the top of the mountain.

You use electric energy when you mix orange juice in a blender. Where does this energy go? It must go into warming the juice. A lot of it goes into warming the motor, also, as you can discover by touching the blender near the motor. A careful measurement would show that, in the absence of other energy transfers (such as heating or cooling the surrounding air), the thermal energy gained by the juice and motor equals the electric energy purchased from the power company.

You can demonstrate this effect for yourself. Measure the temperature of two bowls of cold water, then mix one of them for a few minutes in an electric blender. Compare their temperatures after mixing.

Energy transformations are all around you. When you walk up a hill, you increase your gravitational energy. The principle of conservation of energy tells us that this energy must come from somewhere. Where did it come from? The answer: from chemical energy in the molecular structure of your muscles. The chemical energy your muscles lose should equal the gravitational energy you gain plus whatever heat energy you produce during the climb.

A *hydroelectric power plant* illustrates several energy transformations. As its name indicates, this device transforms the energy in a body of water (a lake, usually) into electromagnetic energy. The gravitational energy of water in a lake behind a dam is converted to the kinetic energy of spinning turbines (waterwheels, essentially). This spinning motion can be used to produce an electric current if the right combination of wires and magnets is attached. The resulting device, called an *electric generator,* converts the kinetic energy of the turbines to electromagnetic energy.

Exercise 12
Where did the gravitational energy in the lake come from? Trace this energy as far back as possible, listing the types of energy before and after each energy transformation.

Exercise 13
Suppose that the electric energy produced in a power plant travels through wires to

your house, where it operates a blender, a toaster, and a light bulb. What energy transformations (initial and final types of energy) are performed by each of these devices?

Exercise 14
What would have become of the energy that powers a hydroelectric plant if the plant and dam had never been constructed?

The principle of conservation of energy seems to apply everywhere we've looked. It is not subject to the limitations of Newtonian physics (Chapter Eight): Energy is conserved even for systems moving at nearly the speed of light, for microscopic systems, and for systems subject to strong gravitational forces. The principle of conservation of energy is one of the most all-embracing physical principles known and one of the most important.

Thought Question 2
Can we say, then, that here we have finally found a scientific principle that is known for certain to be true, a principle that will never be found to be in error?

As always, the answer is: No, we can't say that. In science, nothing is absolute.

The First Principle of Thermodynamics

The work-energy principle and the principle of conservation of energy are two ways of expressing the same idea. In this section, this idea is stated in yet another form, a form that is useful in understanding processes involving thermal energy.

A few examples suffice to show that the energy of a system is higher when its temperature is higher. Hot steam can rattle a kettle top or drive a locomotive and hot gases in the cylinder of a car can push a piston and drive the car. In these examples, a system is doing work that it could not do if the system were at a lower temperature. The work that a system can do as a consequence of its temperature is called *thermal energy.*

Any device that measures temperature is called a *thermometer.* The expansion of mercury when thermal energy is added to it is the basis of the familiar mercury-in-glass thermometer.

In the United States the *degree Fahrenheit* (°F) is the most widely used unit of temperature; nearly everywhere else the *degree Celsius* (°C) is used. The standard reference points for both Fahrenheit and Celsius scales are the freezing and boiling points of water. The freezing point of water is 32°F or 0°C; the boiling point of water is 212°F or 100°C. Figure 9.1 shows the wide range of temperatures in nature.

It is a simple, familiar, and very important property of temperature that *heat flows from hot to cold.* More precisely:

The Principle of Thermal Energy Transfer. If a high-temperature object is connected to a low-temperature object in such a way that thermal energy can move, or flow, from one object to the other, then thermal energy always flows from the hotter object to the colder one.

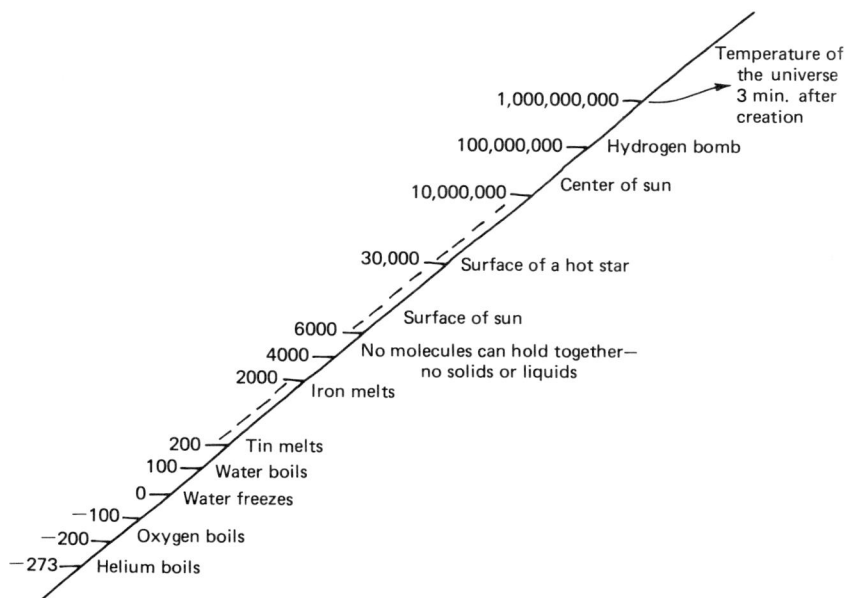

FIGURE 9.1 A few Celcius temperatures.

It follows that, as long as the two systems remain connected, thermal energy will continue to flow until the hotter system has cooled and the cooler system has warmed to identical temperatures. The two systems are then said to be in ***thermal equilibrium,*** by which we mean that no further thermal energy transfer will occur.

This property of temperature has been observed in countless experiments. Examples: Ice melts in a drink, and the drink cools, as thermal energy flows from the hotter liquid to the colder ice. When you grasp a cold doorknob, thermal energy flows from your hand to the knob, warming the knob and cooling your hand. The 6000°C radiation from the sun warms the much cooler earth.

Any such transfer of thermal energy is called ***heating.***

It is possible to explain the principle of thermal energy transfer in terms of atoms. Imagine heating a pan of water. What is the effect, at the atomic level, of the increased thermal energy? For roughly the past century, scientists have answered that heating increases the kinetic energy of the molecules of the system. While the water is heating, the kinetic energy of the H_2O molecules increases—the molecules move faster.

According to our microscopic description in Chapter Three, the molecules of a liquid constantly move around throughout the liquid. Heating just makes them move faster. This increased random, or disorganized, kinetic energy would not be apparent to an observer unless a microscope capable of following the motion of individual molecules were available, which is why you can't ordinarily see anything going on in a pan of water as you heat it—at least not until the water is close to boiling.

We'll call this idea the

Kinetic Principle. From the molecular point of view, thermal energy is random (or disorganized) kinetic energy of the molecules of the system.

Scientists believe this idea because it is so useful in explaining so many things. A few examples:

Exercise 15
Why does the pressure increase in a tire on a hot day? Why do balloons expand when the air inside them is heated.

Exercise 16
Liquids evaporate faster at higher temperatures. Explain. (*Hint:* See Chapter Three on ''evaporation.'')

Why are solid surfaces heated as they slide across each other? Recall (Chapter Three) that solids consist of atoms vibrating around certain fixed positions. Sliding sets the atoms of both surfaces into faster vibration, increasing the thermal energy of both surfaces. From the atomic point of view, frictional forces represent a transfer of the organized kinetic energy of the large-scale motion of the solid into the disorganized kinetic energy of vibrating atoms. This transfer increases the thermal energy, and hence the temperature, of the surfaces.

Finally, why can thermal energy flow from hot objects to cold objects, but not from cold objects to hot objects? Consider a hot block of iron and a cold block of iron put in contact with each other. Since the blocks are in contact, the vibrating atoms of one block frequently bump the vibrating atoms of the other block. Now, when two moving objects collide, energy is usually transferred from the faster object to the slower object. Our kinetic hypothesis tells us that the atoms of the hot block have more kinetic energy than the atoms of the cold block. Thus when atoms from the two blocks bump each other, energy is usually transferred from the atom in the hot block to the atom in the cold block. The atoms of the hot block slow down a little, and the atoms of the cold block speed up a little; the hot block cools and the cold block warms.

With the kinetic principle we have explained hot and cold in terms of atoms and empty space. Apparently, the only difference between an object when it is cold and the same object when it is hot is that the random kinetic energy of its atoms is greater when it is hot. We have reduced the intuitive, human notions of hot and cold to the motion of atoms, just as Democritus (Chapter Three) said we could.

It's a remarkable idea. For instance, when your hand touches a cold doorknob, your hand feels cold. But from the atomic viewpoint it would be more accurate to say that the atoms of your hand have slowed down a little. Your brain interprets this slower atomic motion as coldness.

Again from the atomic point of view, heating is simply a small-scale version of work: a force acting through a distance. In the case of heating, though, the forces and distances are individual atomic forces and distances, so that, from our human-scale point of view, we can't observe heating as ordinary work. Throughout this book the word **work** will mean ordinary, human-scale work. Similarly, **kinetic energy** will mean ordinary, human-scale kinetic energy. Atomic-level kinetic energy will be called **thermal energy.**

Heating, like ordinary work, can be measured quantitatively. Its units of measure are the same as the units of work or energy: joules, calories, foot-pounds, and so on.

When the work-energy principle is put into a form that explicitly takes heating into account, it is called the

First Principle of Thermodynamics. Whenever work or heating or both are done on a system, the system's total energy increases by an amount equal to the sum of the heating and the work done.

For historical reasons, we usually measure heating in calories and work in joules or foot-pounds. They must, of course, be put into the same units before they can be added.

Exercise 17
You place a pot on an electric stove, where 20 cal of heating are done on it. You then remove the pot from the stove and stir its contents, doing 40 J of work. How much has the total energy of the pot increased? What type (or types) of energy have increased? Where did this energy come from?

The Human Energy Machine

In terms of energy, what happens when you eat a slice of bread? Dieticians tell us that a slice of bread contains about 60 dietician's calories, or 60 kcal. This means that 60 kcal, in the form of chemical energy stored in adenosine triphosphate (ATP) molecules in your body, are produced when you eat one slice of bread. This chemical energy is generated by the reaction of oxygen with the organic molecules of the bread, that is, by the respiration reaction (Chapter Three). The energy stored in ATP remains in that form until you move a muscle, at which time it is transformed into other forms of energy.

Exercise 18
60 kcal is about 240,000 J. If all of this chemical energy is converted to gravitational energy, how high will it lift a person whose weight is 800 N? (Answer below.)

Because gravitational energy = weight × height, 240,000 J will lift an 800-N person 300 m (nearly 1000 ft). This sounds a little farfetched—surely one slice of bread doesn't provide enough energy to climb a 300-m mountain! We should have taken into account the inefficiencies of the human body. Only 25 percent of the chemical energy in ATP actually goes into muscular contractions; the remaining 75 percent goes into the production of thermal energy. Part of this thermal energy is used to keep your body at its required 98.6°F, and any excess thermal energy is released to the atmosphere by evaporation when you sweat. Only 25 percent, or 15 kcal, of the energy in the slice of bread is available for doing work. Furthermore, part of the remaining 15 kcal is spent in such internal activities as breathing and pumping blood. Only a small fraction of the original 60 kcal is actually available for such activities as climbing mountains.

The "human energy machine" needs about 2500 kcal per day to run smoothly. That's 100 kcal/hr, or 400,000 J/hr, or slightly more than 100 J/sec. So, if human energy production were 100 percent efficient, your power output would be 100 J/sec, or 100 W. Your actual power output averages about 25 W, the output of a dim light bulb; the other 75 W goes into thermal energy. This is why a roomful of people gets so hot: two dozen people have the thermal energy output of two dozen 75-W heaters!

The Second Principle of Thermodynamics

Experience tells us that most systems eventually run down. Unless energy is provided from outside the system, clocks run down, automobiles run out of gas, and so on. Such examples are not violations of the principle of conservation of energy. In these examples, energy has not been destroyed, it has simply changed forms, usually into thermal energy. For example, the automobile's chemical energy transforms into thermal energy of the atmosphere, road, engine, and tires.

Systems that run down in this fashion are examples of the second great restriction that nature places on energy transformations. We call this restriction the

Second Principle of Thermodynamics. It is impossible, within an isolated system, to transform all the system's thermal energy into nonthermal forms of energy. That is, thermal energy can be only partially transformed into other forms. However, it is possible to transform the system's nonthermal energy entirely into thermal energy.

This principle says there is a "one-wayness" about energy transformations involving thermal energy. Without outside assistance, a system's thermal energy can't be entirely converted to kinetic energy, gravitational energy, or another form of energy. But it is possible to go the other way, to convert nonthermal energy entirely into thermal energy. For example, it is easy for a wound-up watch to convert all its elastic energy into thermal energy, but you will never see a watch spontaneously wind itself by converting its own thermal energy into elastic energy! An outside agent is needed to wind the watch.

Whenever energy is used, it is transformed: when you wind a watch, you transform some of your chemical energy into the elastic energy of the spring; when the watch runs, it transforms elastic energy into the kinetic energy of the moving parts and into thermal energy. The second principle of thermodynamics puts a restriction on our ability to transform, and hence to use, thermal energy. It says that, whereas the other forms of energy can be entirely transformed, thermal energy can be at best only partially transformed. So thermal energy is less useful than the other forms.

In many situations, thermal energy is partially converted to nonthermal forms: the thermal energy in a pot of boiling water is partially converted to kinetic energy of the rattling lid; the thermal energy in steam is partially converted to the kinetic energy of a moving steam engine. In such situations, there is always an exhaust of the unused thermal energy. The exhaust is not just a minor detail, not something that can be removed by improved design; it is a direct consequence of the second principle of thermodynamics, which states that thermal energy can be at best only partially transformed into other forms.

Any device that partially converts thermal energy into kinetic energy is called a **heat engine.** Typically, the energy conversion occurs when the high pressure of a hot gas is used to push a piston (Fig. 9.2). According to the second principle of thermodynamics, only part of the gas's thermal energy can be converted into the piston's kinetic energy, thus there must be an exhaust to get rid of the remaining thermal energy. Briefly, no heat engine can be 100 percent efficient.

Entropy

It is intersting to look at the second principle of thermodynamics from the atomic point of view. The kinetic hypothesis says that thermal energy is the disorganized kinetic energy of individual atoms and molecules. So, from the atomic point of view, the second principle of thermodynamics says that other, more organized forms of energy can always be turned into disorganized energy, but that disorganized energy can be only partially turned into organized energy. In other words, it's easier to disorganize things than it is to organize them.

For example, a bullet moving north at 1000 mph represents a fairly organized form of kinetic energy. Every molecule in the bullet is doing the same thing, namely, heading north at 1000 mph. Now imagine that the bullet strikes a wall and becomes embedded in the wall. All of the initial energy is still there, but it has turned into thermal energy of the bullet and the wall. Microscopically, this energy has none of the organization of the energy in the moving bullet; all the molecules of the bullet and of the heated part of the wall are now doing different things (Fig. 9.3).

Using the concepts of thermal energy and temperature, it is possible to give a precise quantitative meaning to the concept of disorganization. The resulting, precisely defined quantity is called the **entropy** of the system. I won't present the quantitative definition of entropy here, because we'll have no need for it. Suffice it to say that the entropy of a system is the amount of disorganization the system has when viewed microscopically.

One nice thing about entropy is that the second principle of thermodynamics has an elegant and meaningful form when stated in terms of entropy:

Second Principle of Thermodynamics (entropy form). The entropy of an isolated system can increase, but it can never decrease.

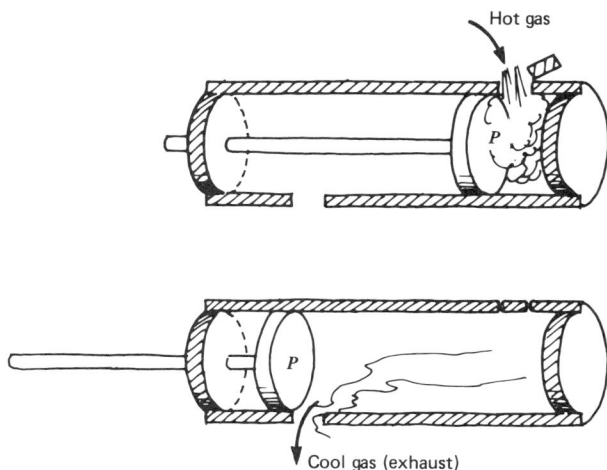

FIGURE 9.2 Hot gas can do work by expanding against a piston.

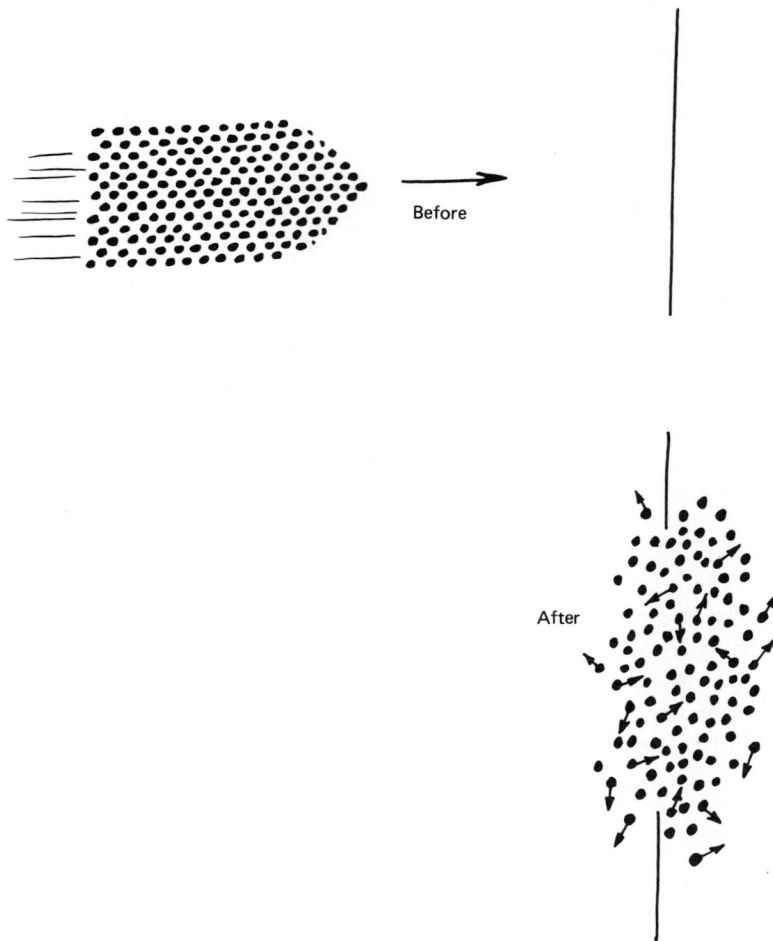

FIGURE 9.3 One way to transform organized energy into disorganized energy.

This definition says quite directly that isolated systems can easily become more disorganized, but that systems can become more organized only with outside assistance.

Entropy and Life

At first glance, the growth of a biological system seems to violate the second principle of thermodynamics. After all, a growing leaf puts together, or organizes, sugar molecules from the CO_2 and H_2O molecules in its environment. The sugar molecules are a more organized form of matter than the CO_2 and H_2O molecules from which they came, because the carbon, oxygen, and hydrogen atoms must link together in just the right way to make a sugar molecule.

Exercise 19
Does the growing leaf represent a violation of the second principle of thermodynamics?
Explain. (Answer below.)

A growing leaf is not an isolated system. The system consisting of the leaf plus the surrounding CO_2 and H_2O molecules has a crucial interaction with the outside world: it uses energy from the sun! Without this outside help, the sugar molecules won't form and the leaf won't grow. According to the second principle of thermodynamics, isolated systems cannot become more organized. The growing leaf does not violate this principle. In fact, the complete system, consisting of sun plus leaf, does have an increase in entropy.

If the principles of thermodynamics were violated by every growing leaf, we wouldn't consider them to be principles of physics. In fact, we have never observed a violation of these principles.

Your brain represents an extremely organized state of matter. The human brain is undoubtedly the most highly organized form of matter within a distance of some 4×10^{13} km (the distance to the nearest star) from Earth. Depending on your estimate of the likelihood of intelligent life elsewhere, the human brain might be the most highly organized form of matter in our galaxy or in the entire universe.

The molecules of the human body are so well organized that they have the ability to be aware of a portion of the universe and even to be aware of themselves. That's pretty special.

So take good care of yourself.

You can maintain yourself in this highly organized, low-entropy state only by being in constant interaction with your environment. Take away the air, the water, and the food in your environment and you will soon run down. The molecules of your brain will lose the organization that enables them to be conscious. Your body will eventually return to the disorganized dust of Earth.

Planet Earth itself has become more organized during the past few billion years, the period during which life evolved. According to the second principle of thermodynamics, outside help was needed to do the job. The outside help came, obviously, from the sun. Turn off the sun, and Earth will soon become highly disorganized.

You may have run across the well-worn, and totally wrong, argument that Earth could not have arrived at its present, highly organized state without some sort of divine intervention to circumvent the second principle of thermodynamics. Divine intervention may or may not have occurred during Earth's evolution to its present state—that's a religious question which each person should answer individually (you might, for example, think of sunlight as divine intervention). But the question of whether the second principle of thermodynamics must be circumvented to permit Earth to become increasingly organized is a scientific question, and you can see from the preceding paragraphs that the answer is no.

Checklist

work

units for work

power

units for power

energy transformations

hydroelectric power plant

thermal energy

thermometer

energy

units for energy

the eight types of energy

gravitational energy

kinetic energy

work-energy principle

principle of conservation of energy

exhaust

units for temperature

principle of thermal energy transfer

heating

kinetic principle

first principle of thermodynamics

second principle of thermodynamics

heat engine

entropy

Further Thought Question

3. Does the kinetic principle really explain away the human feelings of hotness and coldness? Are these feelings really *nothing but* atomic motions? Be careful: Biochemists and physiologists might someday explain more subtle human feelings, such as pain, or even love, in terms of atomic motions in the brain. Would love, then, be *nothing but* certain motions of atoms?

Further Exercises

20. Compare the kinetic energy of a moving object on Earth with the kinetic energy of the same object moving at the same speed on the moon. Compare the gravitational energy of an object held above the surface of Earth with the gravitational energy of the same object held the same distance above the surface of the moon.

21. A metal spring is compressed tightly and clamped in that position. The clamped spring is then tossed into a bottle of acid where the spring completely dissolves. What type of energy was in the clamped spring before it was tossed into the acid? What became of this energy, that is, what type of energy did it turn into?

22. What type of energy is consumed and what type is produced by each of the following: (a) a burning log; (b) a nuclear plant for generating electricity; (c) a coal-burning electrical generating plant; (d) a photovoltaic cell (Chapter Eight); (e) a slingshot.

23. A 2-kg book slides across a table, starting with a speed of 3 m/sec. (a) How much kinetic energy did the book start with? (b) By how much will the thermal energy of the book and table increase by the time the book slides to rest? (c) At the atomic level, what happens to the book and the table?

24. The escape velocity from Earth is about four times as large as the escape velocity from the moon. In terms of energy, how much more energy is required to boost an object off Earth, compared to the amount needed to boost the same object off the moon? The answer explains much of the economic justification for the proposed space colonization project (Chapter Eight). (Answer: 16 times as much.)

25. An inventor proposes to manufacture a clock that will run forever once it is wound. The idea is to use a fraction (say 10 percent) of the clock's energy to continually rewind the clock, with the remaining 90 percent of the energy to be used in actually turning the hands of the clock. What is your response to this idea?

26. Our inventor became more subtle since you pointed out the difficulties with the device described in the previous exercise. The new plan is to capture all the thermal energy produced in the air and in the clock mechanism as the hands turn. The inventor proposes to continually recycle this thermal energy to keep the watch moving forever without outside assistance. What is your response to this idea?

Recommended Reading

1. Erwin Schroedinger, *What is Life?* and *Mind and Matter,* Cambridge University Press, London and New York, 1967. In the first of these two essays, Schroedinger examines life in terms of quantum mechanics and entropy and asks whether life is based on the laws of physics. The author is one of the founders of quantum mechanics.
2. Louise B. Young, ed., *The Mystery of Matter,* Oxford University Press, New York, 1965. An anthology of essays on a wide range of physics-related topics. Parts 6, "The Origin of Living Matter"; 7, "Is Living Matter Immortal?"; 8 "Does Order Arise from Disorder?"; and 9, "What is Life?" are relevant to this chapter. See especially "The Origin of Life," by I. Oparin, and "The Chemistry of Life," by I. Asimov, in Part 6. Schroedinger's "What is Life?" (Reference 1) is reprinted in Part 8.

10

ENERGY RESOURCES

During the 1950s and 1960s, when oil and gas were cheap, very few people discussed energy; we all took energy for granted. Today everybody is talking about energy (Fig. 10.1). The industrialized countries are in trouble because easy-to-get and convenient-to-use energy resources (oil and gas) are (1) running out, in the sense that they are increasingly difficult to obtain; (2) being consumed so rapidly that supply doesn't always keep up with demand; and (3) so widely used that we can no longer ignore their harmful environmental effects. The problem is not going to go away, certainly not in this century, so we had best learn a little about it. Chapters Ten, Eleven and Twenty-three are devoted to this topic.

The energy problem is controversial, as it is bound to be because it affects all of us in so many ways. In those areas in which informed people disagree, I will try to present the options impartially.

Thought Question 1
In your experience, are many people really concerned about the energy problem? Why or why not? Are people as concerned about energy as they are about inflation? Unemployment? Their favorite football team? Given a choice between a television special about the energy crisis and a typical situation comedy, which would most people choose? Which would you choose?

Energy and Society: An Outline

Energy and society is a complex, multidisciplinary topic that requires at least an entire course for an adequate introduction. The following outline indicates one way to organize our thoughts about energy.

A. Energy *resources* such as coal and direct solar energy.
B. Energy *conversion* schemes such as the electric generating plant and the automobile.
C. Energy *transmission* methods such as electric power lines and natural gas pipelines.
D. Energy *storage* methods such as the storage battery and the gas tank.
E. Energy *efficiency* in the home, in industry, and in transportation.

I won't try to touch on all these items in this book. This chapter is an introduction to the energy resources. The next chapter presents just a few of the many energy conversion schemes of today and tomorrow. We will have to skip the important topics of energy transmission and energy storage.

We should take at least a passing glance at energy efficiency, because many

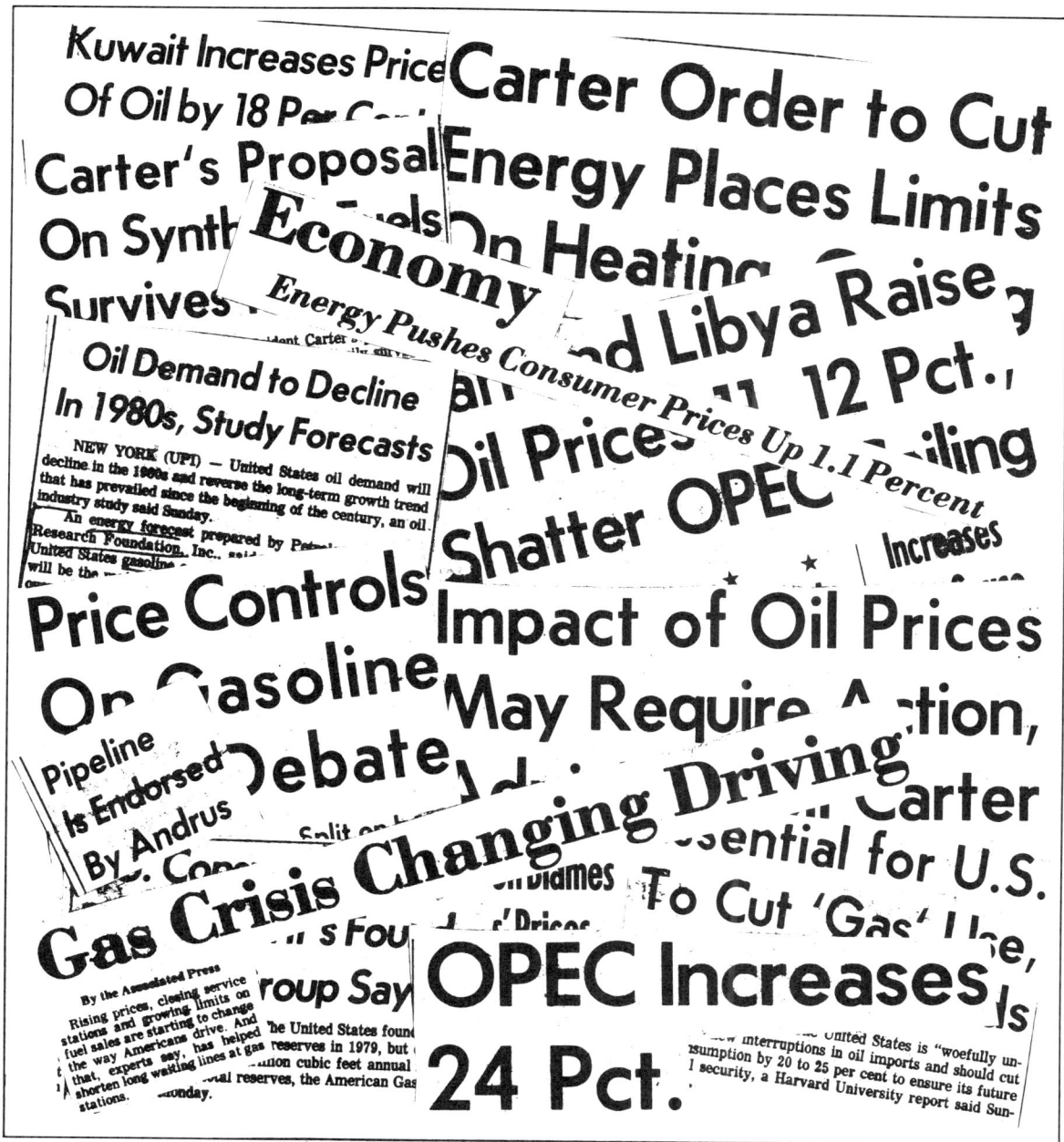

FIGURE 10.1 Energy in the news.

analysts consider it the key to solving the energy riddle. Energy, the ability to do work, is important to us because of the services that it can provide. It is not the energy itself, but rather the services, that we value. For example, we can heat a drafty room with large amounts of energy, or we can insulate the room and heat it with smaller amounts of energy. In either instance, the room gets heated. One method, however, is more *energy efficient* than the other, because it performs the same service with less energy. A room heated with electricity from a steam-electric generating plant is much less energy efficient than the same room heated by a natural gas heater or by solar energy, because of the inefficiency of steam-electric generating plants. For every three units of energy put into such a plant, only about one unit of electric energy is produced (see the next chapter).

Energy efficiency, often called energy conservation, does not necessarily mean giving up valuable services. Rather, it can mean performing the same services with less energy.

As we will see in this chapter and the next, the United States is rapidly consuming scarce energy resources. It is clearly desirable to (a) increase the availability of the scarcer resources, the **production solution;** (b) convert to the abundant resources, the **alternative energy solution,** and (c) use energy more efficiently, the **conservation solution.** A wise energy policy will rely to some extent on all three approaches. Most recent studies have emphasized the central importance of energy efficiency. For example, the Energy Project at the Harvard Business School (Ref. 4) calls conservation the "key energy source." The study of the National Academy of Sciences Committee on Nuclear and Alternative Energy Systems, probably the most complete energy assessment ever conducted in the United States, states that "All in all, conservation deserves the highest immediate priority in energy planning."

Turning now to energy resources: An **energy resource** is a raw material, obtained directly from the natural world, that can be used to do useful work. Such resources are usually classified as either **renewable** or **nonrenewable.** A renewable energy resource is one that renews itself within a normal human life span. Other energy resources are nonrenewable.

Exercise 1
Which of the following resources are renewable? (a) Coal; (b) firewood; (c) solar radiation; (d) uranium; (e) wind; (f) water behind a dam; (g) oil.

Exercise 2
Which of the eight basic types of energy (Chapter Nine) is each of the resources in Exercise 1?

First we'll study the nonrenewable resources—fossil, nuclear, geothermal—then we'll move on to the renewable resources—direct solar, hydraulic, tidal, biomass, wind.

Fossil Fuels

Fossil energy, by far the most widely used energy resource in the United States today, is the chemical energy available in the gaseous, liquid, and solid remains of plants that grew on Earth hundreds of millions of years ago. All fossil fuels are **hydrocarbons,** that is, their molecules are primarily combinations of hydrogen atoms and carbon atoms. Do not confuse *hydrocarbon* with *carbohydrate.* The sugars and starches are carbohydrates. Their molecules are made of hydrogen, carbon, and oxygen, whereas the hydrocarbon molecules are made of only hydrogen and carbon.

Table 10.1 lists the major fossil fuels, their approximate chemical compositions, and their heating values. For comparison, the list includes one nonfossil gaseous fuel, hydrogen, and one nonfossil solid fuel, wood.

The gaseous hydrocarbons are called **natural gas,** and the liquid hydrocarbons are called **oil.** Heating values of gases are given as energy per cubic meter rather than as energy per kilogram, because the user is more interested in the volume needed to contain them than in their weight.

The heating value of these fuels is closely related to their carbon content. It is interesting that some wood has a higher heating value than peat and lignite. Peat and lignite are subcoals, which, although not widely used today, may become more popular in the future.

Exercise 3

Why would you expect the heating value of a fuel to be related to its carbon content rather than, say, to its hydrogen or sulfur content?

Nearly all technology in the United States today, from light bulbs to automobiles to giant industries, is driven by the fossil fuels. The nation relies on this resource for over 90 percent of its total energy. We have consumed this resource intensely because it has been plentiful, easy to obtain, and convenient to use and because our energy consumption has been small enough that we have been able to neglect the environmental impact.

But the energy joyride is over.

The energy joyride is over because the fossil fuel joyride is over. To grasp the nature and dimensions of the energy crisis, then, we need to understand the fossil fuel supply-and-demand picture.

First of all, the situation cannot be summarized simply by saying that we are running out of fossil fuel. We are close to running out of some fossil fuels, which is

Table 10.1 Some major fuels, mostly fossil

Fuel	State	Composition	Heating Value
Methane	Gas	CH_4	35×10^6 J/m³
Ethane	Gas	C_2H_6	60×10^6 J/m³
Propane	Gas	C_3H_8	87×10^6 J/m³
Hydrogen	Gas	H_2	10×10^6 J/m³
Gasoline	Liquid	5—12 C atoms/ molecule	44×10^6 J/kg
Kerosene	Liquid	12–16 C atoms/ molecule	43×10^6 J/kg
Lubricating oil	Liquid	16–20 C atoms/ molecule	43×10^6 J/kg
Peat	Solid	30% C	4×10^6 J/kg
Lignite	Solid	35% C	12×10^6 J/kg
Coal	Solid	70% C	28×10^6 J/kg
Wood	Solid	25–50% C	14×10^6 J/kg

Table 10.2 U.S. fossil fuel resources

	Coal	Oil	Natural Gas
Resources remaining in the United States	1560×10^9 ton	100×10^9 barrel	540×10^{12} ft³
Total energy content of remaining resources	36×10^{21} J	0.52×10^{21} J	0.55×10^{21} J
Cumulative production of U.S. resources through 1980	42×10^9 ton	140×10^9 barrel	560×10^{12} ft³
Fraction of U.S. resources produced through 1980	2.6%	59%	51%
Current yearly production rate of U.S. resources	0.8×10^9 ton	3×10^9 barrel	18×10^{12} ft³
Time remaining until resource would be gone *at current production rate*[a]	1900 yr[a]	33 yr[a]	33yr[a]

[a]These figures may be misleading, especially in the case of coal, because production rates are not actually expected to remain constant. Coal production may increase dramatically because oil and gas are being phased out and because total energy use is increasing. Thus U.S. coal resources could be exhausted in 2 or 3 centuries. On the other hand, oil and gas will last longer than 33 years as a result of decreasing production rates and more sophisticated recovery methods.

a problem, but it is not the main problem. The main problem is that the U.S. and world *consumption rate* of fossil fuel has gotten so large that the supply rate has not kept pace; the price of fossil fuels has increased and will continue to increase, producing inflation in the countries that use these fuels; and oil imports are creating an untenable international *balance-of-payments deficit* in the industrialized countries.

Many informed observers are saying that survival of the world's industrialized democracies depends on citizens (you and me) understanding this situation. So let's look at it more closely.

Table 10.2 shows the extent of U.S. fossil energy resources. The figures quoted for resources remaining are estimates of eventual total production. Major technological or economic changes could alter these figures. Nevertheless, they do provide a rough picture of the situation.

As you can see, the United States has hundreds of times more energy available in coal than in oil or natural gas. Furthermore, oil and natural gas are actually running out, whereas we have only begun to dip into our enormous coal resources. One would think that the country would reduce its oil and gas consumption and begin to use coal instead. However, our nation, especially the enormous transportation sector, is organized to run primarily on liquid and gaseous fuels. It is possible to liquify and gasify coal, and the nation has a synfuels program designed to do just that, but it is much cheaper to use oil and natural gas. In addition, the mining and burning of coal presents a host of environmental problems.

The resource problem outlined in Table 10.2 is bad enough, but the *supply problem* is worse and even more immediate. The United States consumes oil and natural gas at a much greater rate than can be supplied from domestic sources. There is a limiting rate at which these resources can be drawn out of the ground, and we are exceeding this rate. The United States, along with all the other industrial countries, has turned to foreign sources of oil and gas.

But international oil and gas resources will run out in a century or two, even at present rates of consumption. There are also limits on the rate at which oil from the Middle East and from other oil-rich areas can be drawn out of the ground, and the world is close to this limit. Demand for foreign oil is beginning to exceed supply, which naturally means that energy prices are going up, which in turn means that all prices are going up.

Figure 10.2 captures the grim essentials of the fossil fuel problem. The graph shows U.S. production of oil and U.S. consumption of oil, both in millions of barrels per day. The difference between the two graphs must be made up by imports. During 1960–1970, U.S. production and consumption were both rising, and imports accounted for about 20 percent of total consumption. Around 1970, U.S. production leveled off and then began to decline while consumption continued its previous rise. Hence the fossil fuel crisis.

The situation could be improved only somewhat by new U.S. discoveries or by the development of synthetic fuels. At present, the United States depends on imports for one third of its oil, and efforts to increase production cannot appreciably change this figure because the United States is simply running out of its easily obtainable oil reserves.

Although the problem became severe in 1970 and had been predicted long before 1970, Americans first became really aware of the oil crisis in 1973, when Arab oil producers embargoed the United States. The embargo caused a sudden dip in U.S. consumption, as shown in Figure 10.2. Despite this warning, U.S. consumption resumed its upward course in 1975. Consumption began a slow decline, finally, in 1980.

It is not difficult to see that chaos would ensue if oil imports were cut off for any reason. The situation is chaotic even though foreign supplies have not been cut off, because of the money that must flow out of the country to keep the oil flowing in.

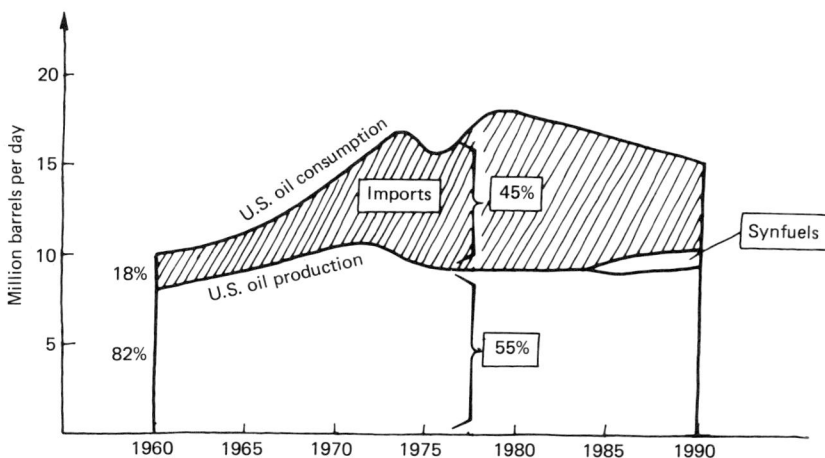

FIGURE 10.2 The energy crisis in a nutshell.

The resulting balance-of-payments deficit depletes our financial resources, leading to financial instability and unemployment.

The culprit in all this is obvious: our insatiable appetite for oil. Many energy analysts argue that conservation plus greater reliance on nonfossil energy resources could bring this appetite under control. But the fact is that we continue to consume fossil resources as though there were no tomorrow.

Thought Question 2
Some analysts have argued that democratic government is inadequate to deal with the energy crisis, because the crisis can only be averted by long-range planning, whereas democratic governments are elected for short-range periods. For instance, American presidents are more disposed to think about the coming 4 years than about the coming half-century. Can democracies deal successfully with the energy problem? If so, how?

Nuclear Fuels

The energy of the nucleus has been heralded as the solution to the energy problem and damned as the destroyer of human health. In the 1950s and 1960s it was widely thought that nuclear energy was the most promising future energy resource, but a variety of problems have conspired to give this resource a smaller role than had been expected. Nuclear energy currently accounts for about 3 percent of the nation's total energy budget and for 11 percent of our electric energy.

Nuclear energy will be studied in detail in Chapter 23 in connection with nuclear physics. Here, I'll briefly list the nuclear resources.

Whereas fossil fuels are a form of chemical energy, obtainable by chemical reaction (combustion), **nuclear energy** is obtainable by *nuclear* reaction. A **nuclear reactor** is any device that obtains useful energy from nuclear reactions. Two distinct types of nuclear reactions yield useful energy transformations: fission and fusion. Briefly, fission means splitting the nucleus and fusion means combining two or more nuclei.

There are three types of reactors: the uranium-based **fission reactor,** the plutonium- or uranium-based **fission breeder reactor,** and the hydrogen-based **fusion reactor.** Almost every reactor now in operation is a uranium-based fission reactor. Breeder reactors are under intense technological development and could make a major contribution to the energy budget before long, but it will be at least the year 2025 before fusion reactors have any significant impact.

Corresponding to the three types of reactors, there are three distinct nuclear resources.

1. **Uranium 235,** one of the forms, or isotopes, of the element uranium, is the fuel for the uranium-based fission reactor. Estimates of U.S. and world uranium deposits vary widely, but there is general agreement that uranium 235 resources are limited and that the uranium reactor can make a major contribution to the energy picture for only 30 to 100 years.

2. **Plutonium** is the main fuel for the breeder reactor. It cannot be considered a natural resource because it isn't found in natural deposits on Earth. Plutonium is manufactured from Uranium 238, which is found in natural deposits on Earth, so the

basic natural resource for the breeder reactor is uranium 238. There is over 100 times as much uranium 238 as there is uranium 235. Thus resources for the breeder reactor are nearly unlimited. On the other hand, breeder reactors carry with them a number of undesirable side effects, and it is not known whether they will ever be widely used.

3. *Hydrogen* is the fuel for the fusion reactor. Supplies are as unlimited as anything can be in this world, as the universe is made primarily of hydrogen. However, despite an intense attack on this problem over many years, no one has yet gotten the fusion reactor to work. Researchers have made steady progress, but it will still be many years, at best, before the fusion reactor emerges from the research and development stage.

Nuclear energy is one of the few energy resources that doesn't come directly or indirectly from the sun: uranium comes from the rocks of Earth, and hydrogen has been around literally since the beginning of time.

Geothermal Energy

Energy resources are all around us. Solar is above, wind and biomass surround us, fossil and nuclear lie below, and buried more deeply beneath the surface of the earth lies the enormous reserve of thermal energy associated with Earth's hot interior. This resource is known as *geothermal energy.*

In a few areas of the world, *steam* arises from water that is heated at some depth and then flows close to the surface. This steam can be easily and inexpensively used to run electric generators or for direct heating. The Geysers generating plant in California uses steam to produce 600 mW of electric power, with 600 more mW planned. Steam is a small resource, however. In the United States, only about 10^{20} J are available—one-tenth the energy available in oil or natural gas and far less than that available in coal.

Much more energy than that available in steam exists in the form of *hot water* that is heated at some depth and then flows close to Earth's surface. The total energy available in this form in the United States is perhaps 10^{21}–10^{22} J—as much as ten times the energy available in oil or gas. But hot water is not as convenient as steam, especially if the purpose is to provide electricity. Most hot water resources are uneconomical and have not yet been exploited. It is estimated that California's Salton Sea area contains enough underground hot water to produce 20,000 mW of electric power for southern California, the equivalent capacity of 20 large coal or nuclear plants.

Present world output from geothermal energy is about 1500 mW of electric power, equivalent to the output of one or two large coal or nuclear plants. All of this geothermal energy comes from hot steam and hot water close to Earth's surface.

Farther below the surface, at a depth of 3 to 10 km, are much larger geothermal resources in the form of *high-pressure hot water, hot rock,* and *molten rock.* The energy available in the United States from each of these three forms, at less than 10 km (6 miles) below the surface, is some 10^{23} J. This is far more energy than is available in all the fossil fuels combined. A small fraction of this would fill our energy needs far into the future.

Extraction of these deep-lying geothermal resources is in the research and development stage. Figure 10.3 shows how thermal energy is tapped in hot rock. The process is simple in principle: (*a*) Drill 3 to 10 km into hot rock; (*b*) fracture the rock

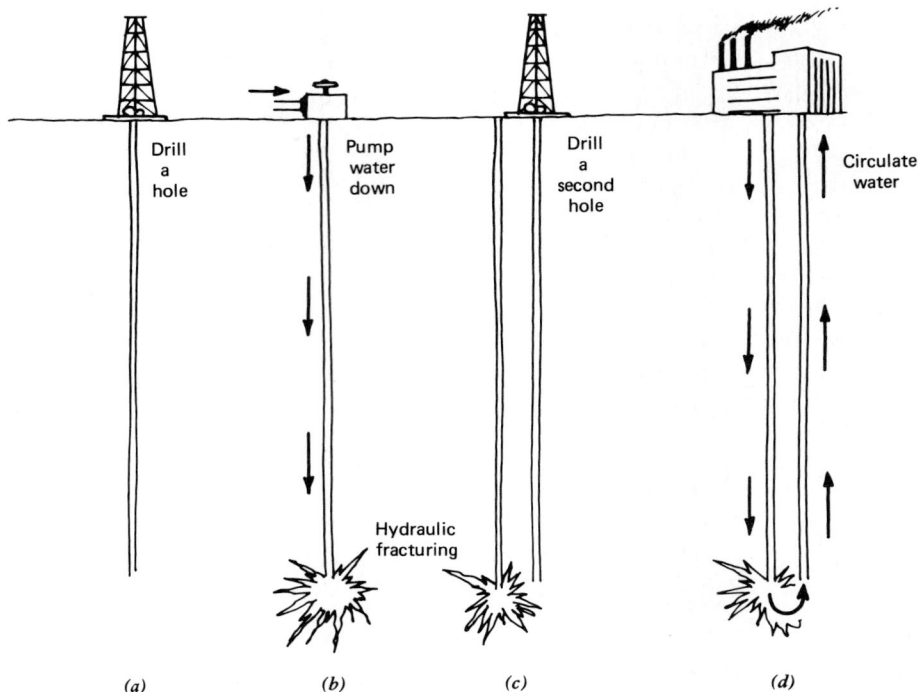

FIGURE 10.3 (a-d) How to build a geothermal power plant. The holes are 3 to 10 km deep.

forcing water down the hole; (c) drill a second hole into the same rock; and (d) circulate a closed loop of water down one hole, through the hot rock, and up out the second hole. Above ground the hot water provides thermal energy for direct heating or for steam-electric power generation. The world's first hot-rock geothermal electric generating plant started up in New Mexico in 1980. It is a tiny, 60-kW affair, but a 5 to 10 MW plant is planned. The plant uses water pumped through fractured granite 3 km underground.

Use of geothermal energy may have significant environmental impacts: some geothermal fluids contain gases, salts, and metals that can pollute air and water; geothermal energy production involves the commitment of large land areas; the removal of large quantities of underground fluid can result in serious land subsidence (sinking of land); and exploitation of hot-rock geothermal energy could produce slippage along geologic faults, possibly causing earthquakes.

Unfortunately, it is difficult and expensive to drill the necessary 3 to 10 km into the earth. As drilling technology improves and as the other resources become more expensive, geothermal energy might begin to play an important role.

Direct Solar Energy

The sun, giver of light and life, is one of humankind's important long-term energy options. Many observers believe that, given proper encouragement, direct and indirect renewable solar energy will provide us, in the short term, with a large fraction of our total energy, perhaps 25 percent by the year 2000 and more than 50 percent by the year 2020.*

Nearly every energy resource can be traced back to the sun.

Exercise 4
Which of the following resources can be traced back to the sun? (a) Coal; (b) firewood; (c) uranium; (d) water running downhill; (e) water behind a dam; (f) oil.

The term **solar energy** has come to mean all the renewable resources stemming directly or indirectly from the sun; **direct solar energy** means only the energy directly available in sunlight. For instance, firewood, wind, and the energy behind a dam are solar energy resources, but they are not direct solar energy resources. Fossil fuels are not renewable, so they are not considered solar resources even though they can be eventually traced back to the sun.

Direct solar energy may be received above the atmosphere, on the surface of the sea, or on the ground. We've already discussed (Chapter Eight) a proposed solar-satellite project to capture solar energy above the atmosphere and convert it to electricity. In the next chapter we will discuss several ways of using the solar energy that strikes the sea and the ground. For now, let's just look at the size of this resource.

In the United States, the average rate at which solar energy strikes a single square meter at ground level is 200 W. This rate is averaged over day and night (about 400 W is received during the day, and none at night), over the year (more than 200 W is received during summer, less during winter), and over the entire United States (more than 200 W is received in the southwest, less in the north). The radiation striking a single square meter could continuously light a 200-W bulb if it could be converted to electricity with 100 percent efficiency and if the radiation could be smoothed out so that 200 W were received day and night in all seasons and in all weather.

Here are some exercises to give you a feeling for the meaning of this radiation rate and to check your understanding of energy, work, and power.

Exercise 5
Football fields are about 100 m long and 30 m wide. (a) At what rate, on the average, does solar radiation strike a football field? (b) Americans consume total energy at the average rate of 10 kW (10,000 W) per person. Roughly how many U.S. citizens would be supported by the solar radiation hitting a football field, assuming that the solar energy is used with 100 percent efficiency?

Exercise 6
(a) How much solar energy (not power) strikes 1 square meter during 1 second? (b) During 1 minute? (c) How far could the energy in part (b) lift a 600-N (about 150-lb) person?

Exercise 7
The rate of U.S. total energy consumption is about 2×10^{12} W. If solar energy could be used at 100 percent efficiency, what land area would provide energy at this average

*These figures come from a 1979 report issued jointly by the Department of Energy, the President's Council on Environmental Quality, and other federal agencies.

rate? Convert your answer from m² to km², using the fact that a million square meters equals 1 square kilometer.

The radiation hitting a football field could, in principle, support 60 U.S. citizens. Radiation striking a land area of 10,000 km², a square 100 km on a side, could, in principle, support the total energy needs of the entire U.S. population. For comparison, 10,000 km² is about the size of Connecticut or 4 percent of the area of New Mexico.

The sun certainly provides impressively large amounts of energy. But do not suppose that our energy problems can be solved by blocking off 4 percent of New Mexico, or even all of New Mexico, for solar energy. We consume energy in a variety of forms, in different places, under different conditions, at all times of the day and night, and in all kinds of weather. It is no simple matter to fit solar energy to these consumption patterns. Furthermore, realistic devices use solar energy at efficiencies of 20 percent or less, far below the 100 percent efficiency assumed above.

Direct solar energy provides very little of the energy presently consumed in the United States. Nevertheless, most observers predict that in the future this resource will play a much bigger role in space heating, water heating, production of thermal energy for industrial processes, and electricity generation.

Hydraulic Energy

Although direct solar energy provides practically none of the present U.S. energy budget, indirect solar energy provides a significant fraction of that budget, in the form of **hydraulic energy**. Hydraulic energy is the gravitational energy in water that has risen into the atmosphere by evaporation (courtesy of the sun), fallen to Earth as rain, and become trapped behind dams.

Hydraulic resources provide about 4 percent of the energy used in the United States. This fraction is small but significant, because all of it is used to generate electric power and because the hydroelectric method is one of the most efficient and nonpolluting ways of producing electricity today. Some 13 percent of our nation's electrical energy is generated hydroelectrically. On the other hand, dam failures are a significant hazard, and new dams often destroy the desirable natural qualities of free-flowing rivers.

Hydraulic energy is, of course, a renewable resource. It is, however, a fairly limited renewable resource, because in order to use it we must impound water behind a dam and there are only a limited number of places where such dams are feasible. We've already dammed most of the larger rivers that can deliver significant hydraulic energy, but small rivers offer significant new potential. Some analysts predict that maximum utilization of undeveloped or abandoned small dams could double the contribution of hydraulic energy, without requiring any new dams.

Exercise 8
At Hoover Dam on the Colorado River the difference in water levels above and below the dam is 170 m. The average rate of flow of water at this point on the Colorado River is about 6 million N (1.5 million lb) per sec. Assuming that hydroelectric power generation is 100 percent efficient (the actual efficiency is 85 percent), find the electric power output of this plant.

The actual output of the Hoover Dam power plant is about 1 billion W. Electric power plants are usually rated in megawatts (MW), which means millions of watts. Thus the output of Hoover Dam is 1000 MW.

Exercise 9
The electric generating capacity of the United States is about 600,000 MW. How many Hoover Dams would we need to provide this?

Tidal Energy

The oceans are raised and lowered twice a day by the gravitational pull of the moon and, to a lesser extent, by the gravitational pull of the sun. The gravitational energy available in the raised water is called *tidal energy.* It is one of the few energy resources that cannot be traced to the sun's radiation.

Tides can be used to generate hydroelectric power along a coastline by building a dam across the mouth of a bay. The bay is closed off at every high tide. Six hours later, the ocean outside has dropped to low tide and the water behind the dam is released to generate hydroelectric power.

It's a simple and environmentally clean idea that, unfortunately, will never amount to much. There aren't many places in the world where the tides are high enough to justify the expense of building a dam.

Only one large-scale tidal plant is operating in the world. It is at the mouth of the Rance River, on the coast of France, at a point where the tides reach 35 ft. This plant's output is 240 MW, one-fourth of the output of Hoover Dam. Canada and the United States are studying the possibility of tidal plants along the Bay of Fundy between Nova Scotia and Maine, where tides often exceed 50 ft.

Biomass Energy

Wood burning captures *biomass energy:* the chemical energy stored in biological materials. Most biomass energy is stored in plants by photosynthesis (Chapter Three). This renewable resource comes indirectly from the sun and thus is one of the forms of solar energy.

Surprisingly large amounts of biomass energy are in use today, mainly as firewood. The nonindustrial countries of the world obtain nearly half their energy by burning agricultural wastes, wood, and dung. The United States currently gets 2 percent of its energy from biomass—half the amount obtained from hydraulic or nuclear energy. The industrial burning of wood wastes from the pulp, paper, and forest industries accounts for most U.S. biomass usage, and the half-million wood-burning stoves sold in the United States in recent years account for the remainder.

A recent study estimates that 5 percent of U.S. energy needs could be provided by burning the wood that could be grown on just 10 percent of the now-idle forestland and pastureland.

Here's another way of looking at the size of this resource: If trees were grown as a fuel-crop on a 200-km² area, they could be burned to provide energy for a 150-MW power plant, which could provide electricity for a city of 75,000. The area required is a square 14 km (10 miles) on a side.

There seem to be one thousand and one schemes afoot for capturing biomass energy. Some observers regard the use of wood for small-scale space heating as the most important opportunity now available in the United States. Many cities burn their trash to produce thermal energy, which is then sold to local industries. Biomass subjected to heating in the absence of oxygen results in a high-energy tarlike material that burns at high temperatures. This material could provide high-temperature steam for steam-electric power plants.

There are many schemes for producing gaseous and liquid fuels by the fermentation of biomass. **Fermentation** is the metabolism, or breakdown, of sugars into simpler molecules by the action of bacteria. This process produces such fuels as methane and alcohol. Many cities are considering generating methane from garbage. Alcohol produced from crops or from garbage might eventually compete with gasoline as an automobile fuel in the United States; gasahol, a gasoline-alcohol mixture, is in widespread use already. In Brazil, officials hope to completely replace gasoline and diesel fuel with alcohol generated by sugarcane. Brazil will soon produce a full 20 percent of its liquid fuels this way.

Researchers are studying a bushy plant whose sap contains a high percentage of hydrocarbons. Maybe we can grow gasoline! One nice feature of this idea is that the plant grows well in the arid regions of the southwestern United States, where it wouldn't compete with food crops.

Floating "rafts" of kelp might be grown in the oceans and fermented as a source of methane.

Biomass schemes, and solar energy schemes in general, are for the most part uneconomical today. In other words, other energy sources are cheaper. But the prices of the other sources continue to rise. This is especially true of the fossil fuels, primarily as a result of the large demand for them. As the other energy sources become more expensive, solar resources will become more competitive and will play a larger role in the energy budget.

Wind Energy

Wind energy, the kinetic energy available in moving air, is a renewable, indirect form of solar energy. The winds arise from the combined effects of solar heating of the earth's atmosphere and the rotation of the earth. Wind energy contributes practically nothing to the national energy budget. As the United States turns to currently unused resources, however, wind is likely to assume a larger role.

The energy of the wind is surprisingly large, rivaling direct solar energy in many regions. The average wind power on the Great Plains (from North Dakota through Texas) is somewhat greater than the 200 W/m^2 received from the sun.

Wind energy might contribute 1 or 2 percent of the nation's energy budget by the end of the century. As in the case of hydraulic energy, this percentage sounds small, but it would be significant because it would be entirely electric and reasonably clean. This contribution might represent 10 percent of our electric energy budget, which is comparable to the amount provided by either hydraulic or nuclear sources.

The principle behind the use of wind power is simple: Make the wind turn a fan blade, and make the fan blade turn an electric generator to produce electricity. It's just like the hydroelectric plant, with wind rather than falling water doing the work. These direct conversion schemes, from falling water to electricity and from moving air to electricity, are much more efficient than indirect schemes that convert fuel to thermal energy and then to electricity. As we'll see in the next chapter, this difference in efficiency is a consequence of the second principle of thermodynamics.

Today's wind machines are often called **wind turbines,** a more accurate term than *windmill* because they are used to generate electricity rather than to grind grains. The large machines, such as the 1-to-2-MW devices being planned, could supply the electrical needs of 500 to 1000 people. Several smaller wind turbines, in the 2-to-10 kW range, are already on the market (Fig. 10.4).

At Tvind College, in wind-rich Denmark, students, teachers, and environmental buffs have banded together to design, finance, and assemble the world's largest wind turbine. The device, with a capacity of 2000 kW, or 2 MW, supplies electricity as well as heat and hot water for the dormitories, classrooms, and other buildings of this campus of 500 students and 80 teachers. Because the school's needs average only about half of the windmill's output, excess power is fed into the regional electric grid and sold to raise funds to pay for the project. Thanks to the self-help efforts of the participants, the entire project cost only $720,000.* The project earned the skeptical chuckles of local residents and Danish government officials when it was proposed in 1975. Three years later, when the giant blades began producing electric power, nobody was chuckling except perhaps the young people and teachers who financed and carried out the entire project.

Checklist

energy efficiency	the three nuclear fuels
energy resource	geothermal resources
renewable resources	direct solar energy
nonrenewable resources	megawatt
hydrocarbons	hydraulic energy
fossil fuels	tidal energy
natural gas	biomass energy
oil	fermentation
coal	wind energy
balance of payments problem	

Further Thought Questions

3. The oil-producing nations want to keep their oil in the ground in order to sell it in the future when it is worth more and when those nations need the money. They don't want to use their oil up rapidly and leave nothing for the future. The oil-consuming nations would like the oil producers to increase production rates in order to supply the oil consumers' needs and in order to hold oil prices down. How would you propose to resolve these conflicting interests?

4. The 1973 Arab oil embargo created many problems for the United States and stirred up resentment toward the Arab countries. In view of subsequent history (see Fig. 10.2), can you think of any ways in which the oil embargo was actually helpful

---

*This is $720,000/2000 = $360 per installed kW, which compares very favorably with, for instance, the $1,000 per installed kW cost of nuclear power plants (1978 dollars). And the fuel is free!

FIGURE 10.4 Block Island, Rhode Island wind turbine generates electricity for the community on Block Island. Here a resident turns on the light with the wind turbine in the background. (Department of Energy photo by Malcolm Greenway.)

to the United States, perhaps in the way that unpleasant medicine is sometimes helpful?

5. How can colleges and universities make a significant contribution to solving the energy problem? Are they making a significant contribution at present? Have you taken any college courses that deal with the energy problem or with any other significant contemporary social problems? Which courses and which problems?

6. Is it possible for nonexperts to form intelligent opinions about the energy crisis? Would it be better to leave it to the experts?

Further Exercises

10. A typical coal-fired electric generating plant uses about 10,000 tons of coal every day. If this plant were fueled with wood instead of coal, how many tons per day would be needed? (See Table 10.1.)

11. The hydrocarbon fuels produce thermal energy by combustion in air (Chapter Three). The chemical reaction is $C + O_2 \rightarrow CO_2$. Make a guess as to the chemical reaction by which hydrogen gas (H_2) produces thermal energy.

12. Fossil fuels produce large amounts of carbon dioxide. What does gaseous hydrogen fuel produce? Note: Hydrogen fuel is not widely used but it might become more popular in the future. (Answer: Water!)

13. Suppose we doubled our rate of coal production. How long would U.S. coal resources last? Suppose instead that the production rate doubled during the next 10 years, then doubled again during the following 10 years, and then leveled off. In this case, how long would U.S. coal resources last?

Recommended Reading

1. John M. Fowler, *Energy and the Environment*, McGraw-Hill, New York, 1975. A good treatment of most of the socially important aspects of energy. Designed as a textbook and for general reading.
2. Robert H. Romer, *Energy: An Introduction to Physics*, W. H. Freeman and Company, San Francisco, 1976. An excellent nontechnical physics text, emphasizing energy concepts and the energy problem.
3. Alvin M. Saperstein, *Physics: Energy in the Environment*, Little Brown and Company, Boston, 1975. Another good nontechnical text based on the energy-and-society theme.
4. Robert Stobaugh and Daniel Yergin, editors. *Energy Future: Report of the Energy Project at the Harvard Business School*, Random House, New York, 1979. Maps out a realistic energy path for the United States.

11

ENERGY CONVERSION

Look around at the technology in your environment: light bulbs, air conditioners, television sets, automobiles. Each of these devices is an **energy converter,** transforming energy from one form to one or more other forms. The light bulb, for example, converts electrical energy into radiant energy and thermal energy. In this chapter we'll study several important energy conversion methods.

Exercise 1
What type of energy conversion occurs in (a) an automobile engine; (b) a spring-wound wristwatch; (c) an electric fan; (d) a gas heater?

Much of this chapter is devoted to one of the world's most important energy converters, the heat engine, and in particular to the internal combustion engine and the steam-electric power plant. This chapter also looks at several solar energy conversion schemes.

Heat Engines: General Principles

The heat engine might be the most important piece of technology since the wheel, or at any rate, since the invention of printing. The heat engine was the basis for the Industrial Revolution, the change from a farm-based society to an industry-based society that occurred in Europe and the United States during 1750–1900. Heat engines first pumped water from the coal mines, later they ran the railroads that hauled the coal, and finally, by 1860, they powered the coal-burning electric generators that lit the world with a new form of energy.

A **heat engine** (Chapter Nine) is any device that converts thermal energy to kinetic energy. An automobile engine, for instance, converts the thermal energy of burning gas into the kinetic energy of a moving piston.

Exercise 2
What restriction does the second principle of thermodynamics (Chapter Nine) put on the conversion of thermal energy to kinetic energy? (Answer below.)

According to the second principle of thermodynamics, a heat engine cannot convert a given quantity of thermal energy entirely into kinetic energy. That is, heat engines are prohibited from operating in the manner shown schematically in Figure 11.1; there must be an exhaust. The engine must operate as shown in Figure 11.2.

The second principle of thermodynamics also tells us that the exhaust must differ from the input in some essential respect. Otherwise, the exhaust could simply be routed back into the heat engine and reused as shown in Figure 11.3. The net effect

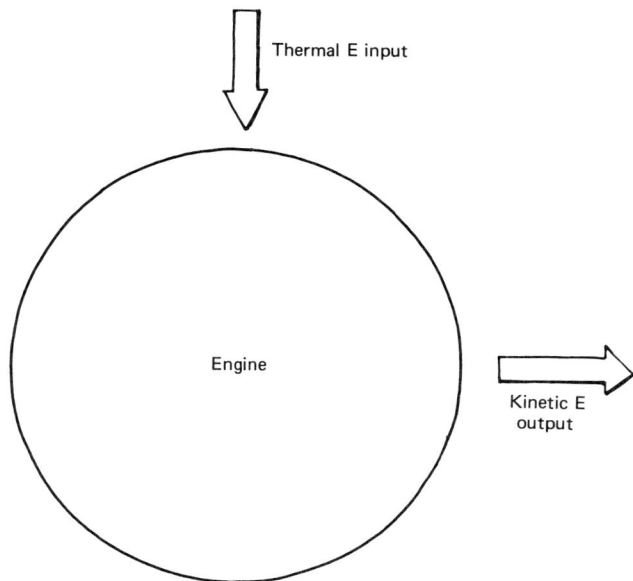

FIGURE 11.1 An impossible heat engine. The arrows represent heat flows.

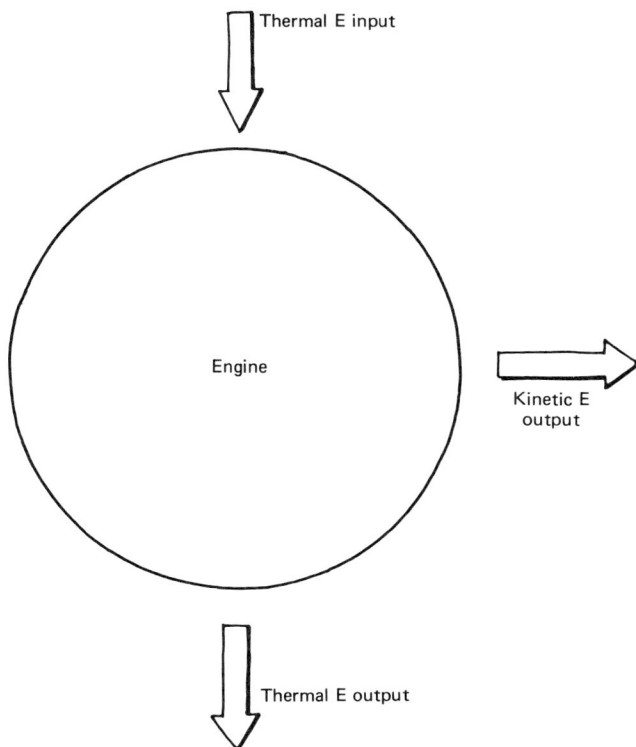

FIGURE 11.2 A possible heat engine.

of this routine would be the same as the scheme shown in Figure 11.1, which the second principle of thermodynamics forbids.

A more complete, quantitative statement of the second principle of thermodynamics would explain that the exhaust energy is at a lower temperature than the input energy. An automobile's exhaust must be cooler than the burning gases in the engine. Figure 11.4 shows the general situation, the input and output thermal energies along with their temperatures.

What does the first principle of thermodynamics have to say about the heat engine? It says that energy is conserved. If the engine operates in a repetitive cycle, as most engines do, then after one complete cycle, the engine has returned to its original condition. Thus the engine's energy is unchanged, and, because of conservation of energy, we know that the energy input during one cycle must equal the energy output:

Thermal energy input = kinetic energy output + thermal energy output

Efficiency is a useful quantitative indicator of the quality of any energy converter. It is defined as the fraction of the input energy that becomes useful output energy.

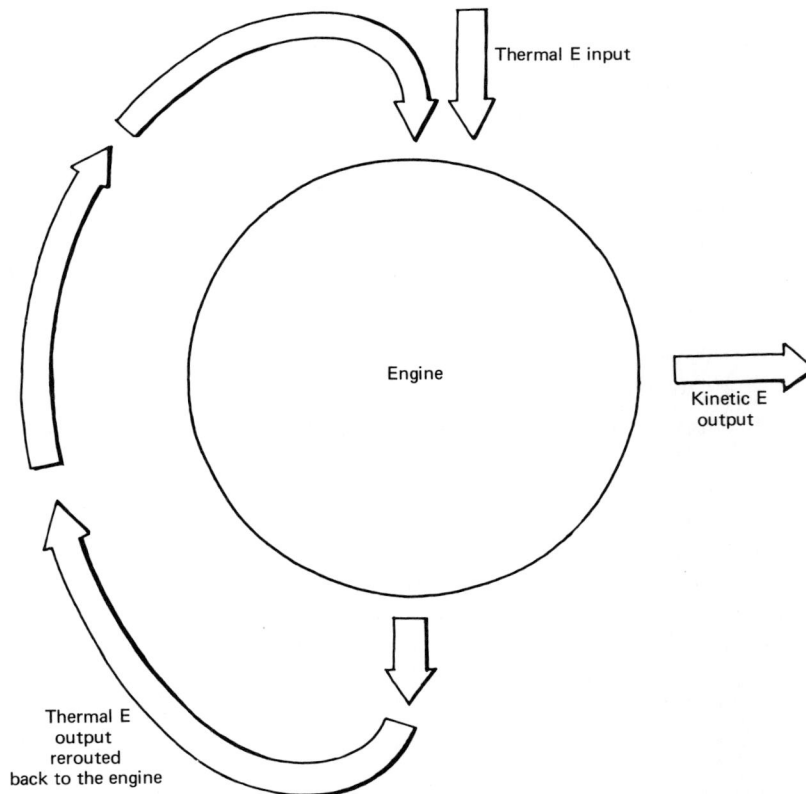

FIGURE 11.3 Another impossible heat engine.

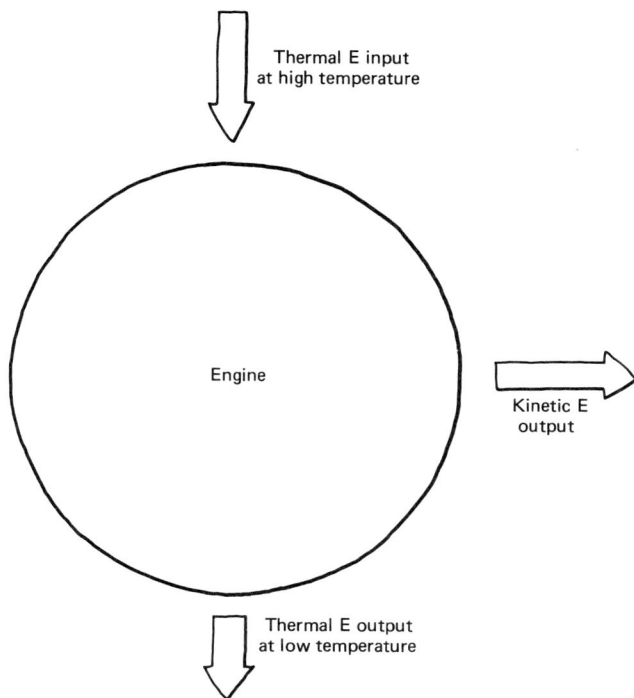

FIGURE 11.4 A possible heat engine, showing input and output temperatures.

For heat engines, the kinetic energy output is considered useful and the exhaust is usually considered not useful:

$$\text{Efficiency} = \frac{\text{kinetic energy output}}{\text{thermal energy input}}$$

To summarize: According to the first principle of thermodynamics, the energy output from an engine must just equal the energy input—you can't get ahead! However, according to the second principle of thermodynamics, the *useful* energy output must be *less* than the energy input—you can't even get even!

The quantitative form of the second principle of thermodynamics tells us how large the exhaust must be. More precisely, it tells us how much of the input energy must be wasted as exhaust. Thus the second principle of thermodynamics puts a limit on the efficiency of heat engines. Furthermore, the quantitative analysis shows that this limit depends only on the temperatures of the input and exhaust thermal energy. This is rather amazing. So far as the limits on efficiency are concerned, the details of the engine's construction and the form of the thermal energy input (hot air, hot steam, etc.), make no difference. All that matters is the temperature of the input and the temperature of the exhaust.

Again according to the quantitative form of the second principle of thermodynamics, hotter sources of thermal energy produce more efficient engines and colder exhausts produce greater efficiency as well. What we want, for high efficiency, is as great a difference as possible between input and output temperatures. Engines that "burn hottest" and "exhaust coolest" are the most efficient.

Table 11.1 lists several examples of input and exhaust temperatures and the corresponding efficiency limits allowed by the quantitative form of the second principle of thermodynamics.

Thermal energy is fundamentally different from other types of energy because the second principle of thermodynamics limits our ability to use this particular type. We can convert other types of energy from one to another with no fundamental restrictions on efficiency, but the efficiency of thermal energy conversion is restricted. Thermal energy is less useful than other types of energy in the sense that the principles of physics place limits on our ability to convert thermal energy to other forms of energy.

Furthermore, lower-temperature thermal energy is less useful than higher-temperature thermal energy. For a given exhaust temperature, high-temperature thermal energy can be converted with greater efficiency than low-temperature thermal energy (see Table 11.1). For example, 50 J of thermal energy from hot coffee is not as useful as 50 J of thermal energy from burning oil, even though the energy content of the two is the same. The 50 J from the burning oil can be used with greater efficiency because the burning oil is so much hotter than the coffee.

Whenever a heat engine is run, high-temperature thermal energy (the input) is converted into lower-temperature exhaust. The thermal energy is degraded in this process because lower-temperature thermal energy is less useful. By the same token, if gasoline is converted to high-temperature thermal energy by combustion in a heat engine, the chemical energy of the gasoline is degraded because thermal energy is less useful than chemical energy.

All this has important social consequences. A million joules of gravitational energy in the form of water behind a dam is a high-grade form of energy because it is nonthermal. A million joules of thermal energy in the form of 500° C steam is a low-grade form of energy. Even in the absence of friction and all other such "imperfections," the steam can produce useful kinetic energy with an efficiency of only 60 percent (assuming that it exhausts to the atmosphere), whereas the water behind the dam could, in principle, produce useful kinetic energy with an efficiency of 100 percent. This difference between thermal and gravitational energy is the primary reason for the actual 85 percent efficiency of hydroelectric generating plants as compared with the 40 percent efficiency of steam-electric generating plants.

Nuclear fission can produce thermal energy at temperatures of millions of degrees.

Table 11.1 Maximum permitted efficiencies for heat engines

Input temperature, °C	Exhaust temperature, °C	Maximum efficiency, %
25 (warm seawater)	5 (cold seawater)	7
120 (burning gas in water-cooled car engine)	25 (atmosphere)	25
500 (power plant steam)	25	60
6000 (surface of sun)	25	95

This is an extremely high-grade form of thermal energy because at this temperature it could be converted to other forms with an efficiency of nearly 100 percent. But in a nuclear power plant this thermal energy is immediately degraded to much lower-grade thermal energy by using it to heat water to a mere 500° C. If a way could be found to utilize the higher temperatures potentially available from nuclear fission (for instance, by heating some substance to thousands of degrees), nuclear power plants would be much more efficient.

Exercise 3
Steam-electric generating plants produce only 40 J of electric energy for each 100 J input. How much exhaust does this 100 J produce? What is the efficiency of such a plant?

Exercise 4
A hydroelectric plant operates at 85 percent efficiency. How much energy is wasted for each 100 J of gravitational energy input?

The Automobile Engine

Transportation consumes 38 percent of the U.S. energy budget* and produces 42 percent of the nation's pollution. Most of this consumption and pollution is the result of the internal combustion engines that power our automobiles, trucks, and buses. In addition, the internal combustion engine consumes more than 40 percent of our oil, the resource behind the present energy crisis. Most observers agree that out addiction to the automobile is the most important single cause of the energy crisis. But such undoubted benefits as mobility, comfort, and independence make it difficult for us to reduce our use of the internal combustion engine.

We had best try to understand this device.

Figure 11.5 shows the main steps in the operation of a single cylinder in an internal combustion engine. Your automobile, if you own an automobile, contains four, six, or eight such cylinders. Naturally, the four-cylinder engines usually consume less energy and produce less pollution; also, they are used in the smaller cars, which consume less iron, take up less space, produce less wear and tear on highways, and so on.

A *cylinder* is simply a hollow space in the engine, shaped like the inside of an empty tin can. It is fitted with a tight-fitting, movable "lid" called a *piston,* which moves up and down inside the cylinder. The piston delivers power to the car's drive shaft and thence to the wheels. The drive shaft is turned by the piston moving back and forth in the cylinder. Figure 11.5 shows a side view of one complete cycle of operation of one cylinder.

There are four steps in one cycle: (a) A value opens to allow a gaseous mixture of gasoline and air into the cylinder. (b) The piston, temporarily driven by the drive-shaft power from the other cylinders, compresses the gas-air mixture until, at an appropriate time, a spark ignites the mixture. The resulting combustion converts chemical energy into thermal energy at about 120° C (250° F). This energy conversion

---

*Twenty-nine percent for fuels, plus 9 percent for related activities, primarily highway construction.

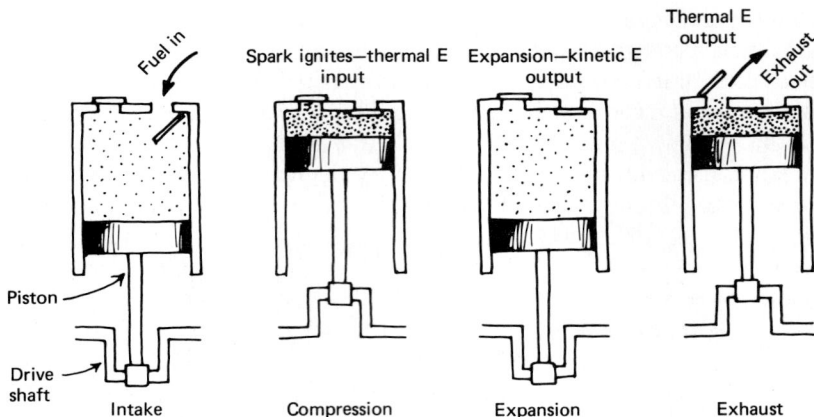

FIGURE 11.5 One cycle of one cylinder in the internal combustion engine.

is the thermal energy input (compare Fig. 11.4). (c) The gas now has a high pressure because it is hot, thus it expands against the piston and turns the drive shaft. This expansion is the kinetic energy output. The gas cools during this process, as it must according to the first principle of thermodynamics: the gas does some work, so it must lose some thermal energy. (d) The piston pushes the cooled and burned gas out the cylinder through a valve. This is the thermal energy output or exhaust.

This device is called an **internal combustion engine** because combustion occurs inside the engine.

Since this device consumes our most precious, and rapidly vanishing, form of energy, we would expect it to perform at high efficiency. Unfortunately, it doesn't.

The main problem is that the internal combustion engine is a heat engine, subject to the second principle of thermodynamics. High efficiencies are impossible without high input temperatures, low exhaust temperatures, or both. The lowest conceivable exhaust temperature is atmospheric temperature, about 25° C. Any lower temperature would necessitate having a refrigerator to cool the gas, and carrying a refrigerator around in your car wouldn't be very efficient. The input temperature, the temperature of the burning gas, is limited to around 120° C, because the engine must be small, and movable, and safe. According to Table 11.1, these temperatures allow the automobile engine to have an efficiency of, at most, 25 percent.

The 25 percent figure is bad enough, but actual automobile efficiencies turn out to be only 12 percent, or about ⅛. That is, for every 8 J of fossil energy put in, 1 J goes into moving the car down the road.

Energy flow diagrams such as that shown in Figure 11.6 are very nice for understanding a wide variety of energy-consuming devices. Figure 11.6 shows the rate of energy use (the *power*) in a 2-ton car moving at 40 mph. In this example, 72 kW (72,000 J/sec) enters the engine as chemical energy in the form of gasoline. The engine has an actual efficiency of 20 percent, somewhat less than the thermodynamically allowed limit of 25 percent. Thus 14 kW, rather than the thermodynamically allowed 18 kW, is delivered to the pistons. Evaporation of the fuel accounts for 1 of the extra 4 kW lost by the engine. **Incomplete combustion** accounts for the remainder: Many of the hydrocarbons don't react with oxygen, or react incompletely to produce CO (carbon monoxide) instead of the desired CO_2. The resulting soup of hydrocarbons plus CO plus CO_2 plus an assortment of other things is then pushed out of the cylinder and into the air, where people can breathe it. Carbon monoxide is a deadly poison, and carbon monoxide pollution is largely the result of the internal combustion engine. An engine with improved combustion would be a major energy saver *and* pollution solution.

Some of the energy from the pistons goes to the lights, fan, power steering (if there is power steering), and the like. Friction in the transmission and drive train converts some of the pistons' energy to thermal energy. Because these internal consumers of energy don't really move the car down the road, they are inefficiencies in the car's overall operation.

Finally, 9 kW gets to the drive wheels. Thus our automobile is $9/72 = 1/8 = 12.5$ percent efficient. For a car moving at constant speed at 40 mph, this 9 kW goes about equally into air resistance and road friction, that is, into producing wind and thermal energy of the air and road.

Exercise 5

Suppose you accelerate the car by pressing on the accelerator. Into what forms of energy is the drive-wheel energy going?

Exercise 6

Continuing the previous exercise, suppose that a drive wheel power of 36 kW goes into accelerating the car in addition to the 9 KW needed to overcome air resistance and friction. What total power must now be delivered to the drive wheels? Assuming that the car's efficiency is still $1/8$, what is the rate of consumption of fossil energy? How many 100-W light bulbs could this power light up, assuming that all of it appeared as electric energy? Moral: Don't accelerate unless you have to!

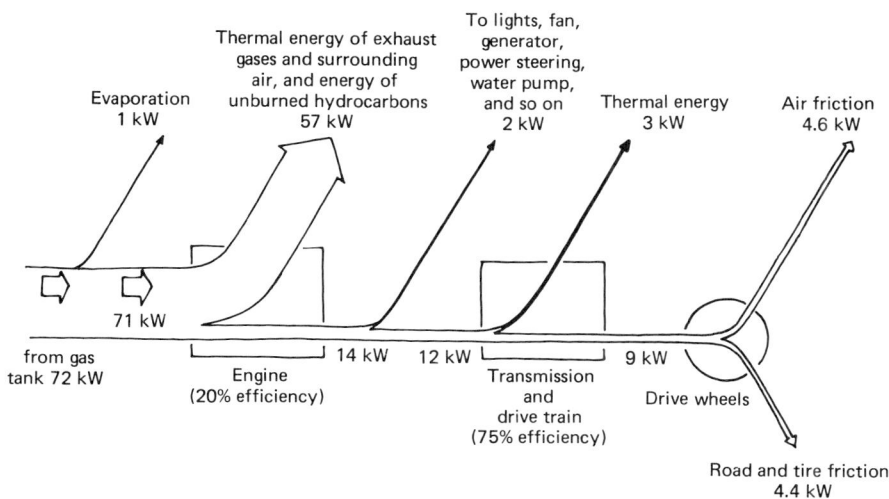

FIGURE 11.6 Energy flow in a 2-ton car traveling at 40 mph. (From *Energy: An Introduction to Physics* by Robert H. Romer, W. H. Freeman and Company, copyright © 1976.)

Exercise 7

Smaller cars expend less energy against air resistance, roughly in proportion to their size,† because they don't have to push aside as much air. They expend less energy against road friction, roughly in proportion to their weight, because they don't press as hard against the pavement. Roughly how much power must be delivered to the drive wheels of a car that has half the weight and half the size of our 2-ton car to keep it moving at a constant 40 mph? Assuming an efficiency of one eighth, what is the rate of consumption of fossil energy? How would pollution from this car compare with pollution from the 2-ton car?

Thought Question 1

List at least 20 ways in which we could reduce our reliance on the automobile. It might be helpful to group your items into categories, such as alternate transportation, redesign of cities, changes in life-style, and economic and tax incentives. Which of these items would reduce the average person's quality of life (or satisfaction with life)? Which items would enhance it?

Thought Question 2

List the pros and cons of the automobile. In light of these pros and cons, should this country try to depend less on the automobile? What measures, if any, should we take to reduce automobile use?

The Steam-Electric Power Plant

Electric generating plants powered by thermal energy consume about 25 percent of the nation's fossil energy and much of its financial capital. Let's see how they work.

Figure 11.7 shows the essentials of the operation. Thermal energy boils water to produce hot steam. The high-pressure steam forces its way through the blades of a waterwheellike device called a **turbine**, thereby turning the turbine. The turbine then turns the **electric generator**, which produces electricity. The cooled steam condenses to water, and the water circulates back to the boiler.

Let's run through this tale again, more slowly. A furnace or nuclear reactor produces the thermal energy that heats the water in the boiler. Because the furnace is located outside the boiler, the generating plant is an **external combustion engine.**

Exercise 8

List several conceivable sources of the thermal energy used in the operation described above. (Answer below.)

The thermal energy source can be anything that is hot enough. Over 80 percent of today's thermally generated electricity comes from fossil fuels, and the remainder comes from nuclear fuels. Most of the fossil fuel plants burn coal; the rest burn oil and natural gas. The law now requires that all new fossil fuel plants burn coal, lignite, or peat. Other nonfossil sources are possible: wood, agricultural wastes, trash, geothermal, direct sunlight, and so on.

†More precisely, size means cross-sectional area as seen in a front view of the car.

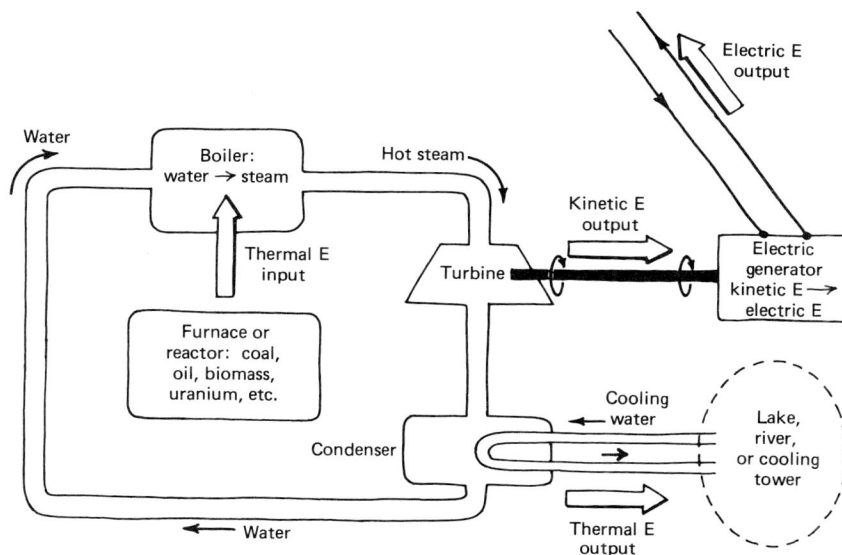

FIGURE 11.7 Schematic diagram of the steam-electric generating plant.

High-temperature (hence high-pressure) steam from the boiler is circulated to the front end of the turbine. The back end of the turbine is cooled by the **condenser** (we'll see how this works in a moment). The turbine feels a high pressure on the front end and a low pressure on the back, so naturally the steam forces its way through the turbine from front to back, thus turning the turbine.

For maximum efficiency, the front end of the turbine should be as hot as possible and the back end should be as cold as possible (remember the second principle of thermodynamics). The condenser cools the back end, using something in the environment that is already cool, like a lake. In the process, the lake is heated, an effect called **thermal pollution.** As an alternative, the atmosphere can be used to cool the back end of the turbine. The waste thermal energy is transferred to the air in a large, curved tower called a **cooling tower** (Fig. 11.8). In the process of cooling the back end of the turbine, the condenser converts the steam into water, that is, the steam is condensed.

Modern power plants operate on steam heated to about 500°C, so according to the second principle of thermodynamics, their efficiency could be as high as 60 percent. However, because of friction, thermal energy losses from pipes, and so on, the actual electrical energy output of modern plants is only about 40 percent of the input thermal energy.

Plants based on either nuclear or fossil fuels produce various waste products of the nuclear or chemical reactions. Fossil plants filter out some of the wastes as solids or liquids for later use or disposal and emit the remainder into the atmosphere as gases or small dust particles. These stack emissions from fossil plants are a significant source of atmospheric pollution. The main gaseous emission is, as you might expect, CO_2, which isn't exactly a pollutant because it is a natural component of air and because it is nontoxic in the human body. The emitted CO_2 does add appreciably to the total amount of CO_2 in the atmosphere, however, and the resulting worldwide increase in CO_2 might have an important effect on climate.

Natural gas is the cleanest of the fossil fuels, and coal the dirtiest. The world is running out of natural gas and oil, and coal is undoubtedly the fossil resource of the future. The most important pollutant in coal is the sulfur found in all coal deposits. When coal burns, this sulfur combines with atmospheric oxygen to produce various undesirable gases, mainly SO_2 (sulfur dioxide).

Several methods are in use or under development to alleviate the sulfur problem; (1) Mine low-sulfur coal to begin with. Unfortunately, only a small fraction of all deposits are clean enough to be burned directly. (2) Remove some of the sulfur from the coal before burning. This is difficult because most of the sulfur is chemically united with the coal. (3) Use combustion techniques that minimize the reaction of sulfur with atmospheric oxygen. Such techniques can require expensive changes in power plant boilers. (4) Increase the height of the smokestack to disperse the pollution over a wider area. This prevents high concentrations near the plant but can obviously create problems downwind. (5) Remove the sulfur oxides from the stack gases by **scrubbing.** The stack gases are scrubbed by passing through a limestone-in-water mixture. The SO_2 combines with the small limestone particles, reducing sulfur emissions by over 90 percent. Nearly all future plants will have scrubbers. (6) Convert the coal into a

FIGURE 11.8 The Three Mile Island nuclear power plant, under construction. Visible in this view are two cooling towers, a reactor containment building under construction, and a turbine building under construction. (Courtesy Metropolitan Edison Company.)

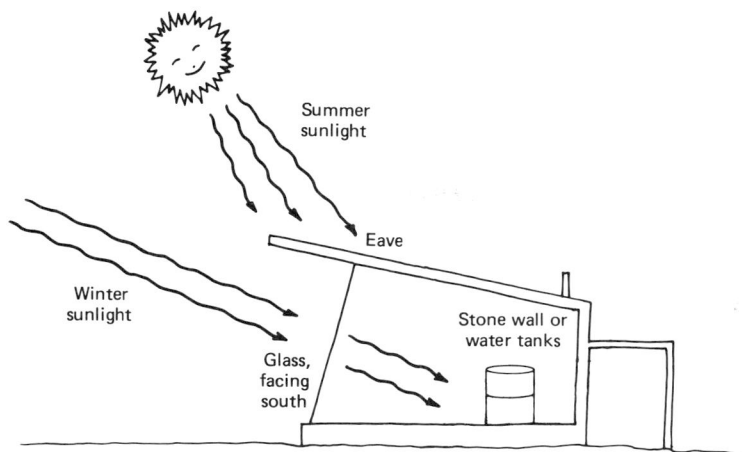

FIGURE 11.9 Cross section of a house designed for passive solar heating.

liquid or gaseous hydrocarbon fuel. Such **synthetic fuels** are similar to petroleum or natural gas and much cleaner than coal. Essentially all of the sulfur is removed in the process. In 1980, the United States embarked on a large synfuel research and development project.

Catch the Sun

There is a fascinating variety of proposals for converting solar energy into other energy forms. We'll look briefly at six of the most widely discussed schemes for converting direct solar energy. The previous chapter outlined several methods for using the indirect solar resources: hydraulic, biomass, and wind.

You can heat your house by just letting the sun shine in. This **passive solar heating** must be humanity's simplest and oldest energy-conversion scheme, but it is often overlooked. Just put some south-facing windows in your house and build an eave over them to keep out light from the high summer sun while admitting light from the low winter sun (Fig. 11.9). To collect and store thermal energy for release at night, place a large tank of water, a bin of pebbles, or a massive stone wall inside the house, facing the sun. To absorb the maximum amount of solar energy, paint the collector black.

You can let the sun heat water for your hot water tank and for home heating. A few countries make significant use of this scheme, and the United States will undoubtedly use this form of solar heating more widely in the future. The idea is to circulate water through **flat-plate collectors** placed in the sun (Fig. 11.10). Radiant energy heats a black metal surface; a plate of glass covers this collecting surface so that air currents will not cool it; and water is heated by circulating underneath the collecting surface. The water then circulates inside the house and deposits its thermal energy in a storage reservoir, such as a tank of water or a bin of pebbles.

Sophisticated flat-plate collectors can heat water up to 300°C. With the help of certain techniques to focus sunlight on the collecting surface, 500°C is attainable. Temperatures this high can run a steam-electric generator. To provide enough energy

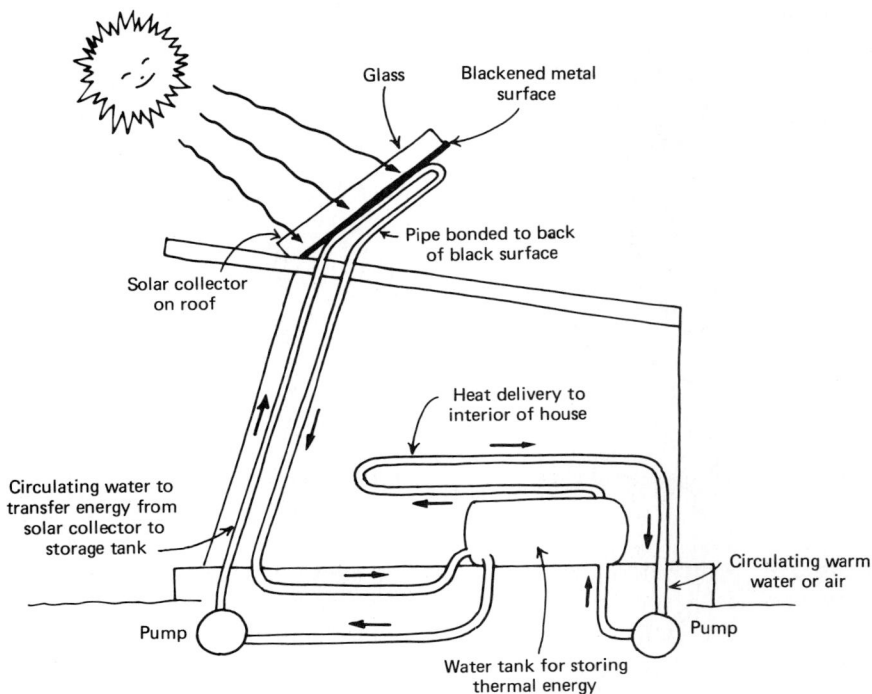

FIGURE 11.10 Flat plate collector design. (From *Energy: An Introduction to Physics* by Robert H. Romer, W. H. Freeman and Company, copyright © 1976.)

to run a large power plant, the collectors would have to cover a sizable area; a collecting area, or **solar farm,** of about 1 square mile would run a 120-MW power plant.

Sunlight reflected from a large array of mirrors and focused on a boiler can heat steam to the high temperatures needed for efficient power plant operation. This is the **power tower** concept. The 10-MW plant developed for the Department of Energy at a site near Barstow, California, uses 1760 slightly curved mirrors to track the sun daily. The mirrors focus sunlight onto a central solar-energy absorber 250 ft above the ground.

The sun has been heating ocean water for some time now. We can tap this energy by using the thermal energy in the warm water near the ocean's surface to run a heat engine. Several research organizations are studying this **solar-sea** idea as a method of operating large electric plants (Fig. 11.11). Because warm seawater is not hot enough to produce steam, some other "working fluid" would have to turn the turbine. Ammonia is one possible choice because seawater is hot enough to turn ammonia from a liquid into a gas. Surface seawater has a temperature of 25°C, which would thus be the temperature of the gaseous ammonia at the front end of the turbine. Nature has kindly provided a coolant for the back end of the turbine—the 5°C water a few thousand feet below the surface. This cool water could condense the ammonia back to liquid form. The maximum efficiency of this device is only 7 percent (see Table 11.1), but the input thermal energy is free, courtesy of the sun, so the low efficiency is not a drawback.

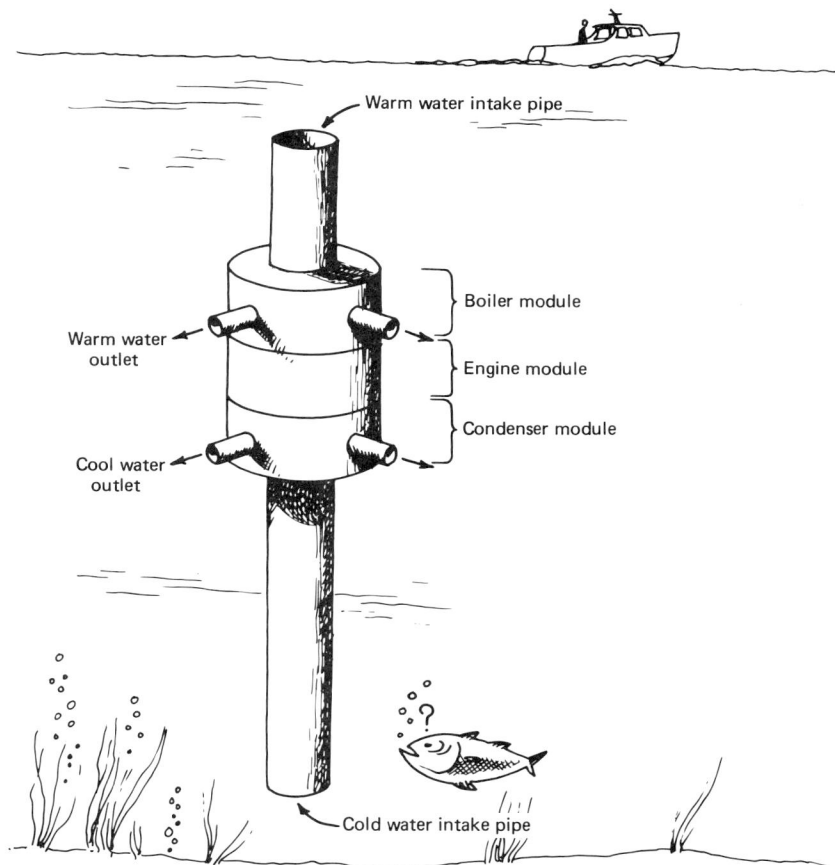

FIGURE 11.11 Solar-sea power plant.

Exercise 9

Would a solar-sea power plant warm the ocean, cool the ocean, or neither? Why?

Radiant energy hitting certain materials can free electrons from their places in an atom, causing a flow of electrons, or an electric current, in the material. Such **photovoltaic materials** are the basis for the photographer's light meter and the electric eye. Photovoltaic materials that are arranged in an array to capture solar energy and convert it to electric current to operate electrical devices are called **solar cells.** These cells provide power for space vehicles and may someday capture solar energy in the giant solar-satellite scheme described in Chapter Eight.

The most common photovoltaic material is the element silicon. Large silicon crystals are produced in the laboratory and then sliced into thin "wafers" that are distributed over the surface of the cell. This process is very expensive, especially at present, when production is so low that mass-production methods cannot be justified. So, photovoltaic energy is not economical today. Other photovoltaic materials, such as amorphous (i.e., not arranged into a regular crystal pattern at the atomic level) silicon or thin layers of cadmium sulfide, might turn out to be much cheaper to make.

Solar cells are inefficient producers of electric energy. Typical photovoltaic materials convert only 10 percent of the solar energy they receive into electricity; 80 percent goes into thermal energy, and the remaining 10 percent is reflected. This breakdown suggests a much more economical way of using solar cells, as you can see in the following exercise.

Exercise 10

(a) A solar cell costing $20,000 powers a demonstration solar home electrically. For heating, 1.5 kW are needed; for appliances and lighting, 0.5 kW. The cell converts 10 percent of the incoming solar energy into electricity and 80 percent into thermal energy. At what rate must the cell receive solar power? (b) Suppose that a $3000 collector is installed to collect the thermal energy produced in the cell. At what rate is thermal energy collected by this method? Could this energy heat the house? (c) The home owner, having worked through parts (a) and (b) of this exercise, cuts the solar cell down to 30 percent of its original size and attaches a thermal collector. How much electrical power does the cell provide now? Could this power run the appliances and lights? (d) At what rate is thermal energy provided by the reduced cell with thermal collector attached? Could this energy heat the house? (e) How much money did the owner save by altering the original set up?

Solar energy, direct and indirect, is a great field for people who like to tinker, either with gadgets or with ideas. Every issue of such magazines as *Science News*, *The Futurist*, *Science*, *Technology Review*, and *Bulletin of the Atomic Scientists* contains some new scheme for manufacturing solar cells, converting biomass to automobile fuel, capturing solar energy in storage batteries, or the like. Newspapers also contain many reports of new energy developments. I hope that you will seek out such magazines and newspaper reports. If enough of us get interested in the energy problem, we can solve it.

Checklist

heat engine, and its operation

efficiency of heat engines and of other devices

maximum permitted efficiency

high-grade energy

low-grade energy

internal combustion engine and its operation

automobile efficiency

incomplete combustion

steam-electric power plant and its operation

external combustion engine

thermal pollution

scrubbers

passive solar heating

flat-plate collector

solar farm

power tower

solar-sea project

solar cells

3. List 20 ways in which we could reduce our reliance on fossil-fuel generated electricity. Again, it might be helpful to group the items into categories, such as alternative power plants, use of nonelectric devices, changes in life-style, and so on. Which of these items would reduce the average person's quality of life? Which items would enhance it?

4. Because of the increasing economic significance of energy, some analysts have suggested that our present monetary system, based on paper certificates of gold reserves, be replaced with an "energy certificate" system. As a variation of this idea, we might employ both our present gold-based system and a new energy-based system. Do you think this is a sensible idea? Would gasoline rationing based on transferable gasoline coupons actually be a system of this sort?

5. Some two-thirds of the energy that goes into a steam-electric power plant comes out as waste heat. When this waste heat enters a lake, a stream, or the air, we call it thermal pollution. Discuss the possibilities of using this waste heat constructively. Would we still call it thermal pollution?

Further Exercises

11. Do higher-efficiency power plants produce less pollution for a given power output? Explain.

12. In each cycle of operation, a certain engine burns 80 cal of chemical energy and exhausts 60 cal of thermal energy. What is the engine's efficiency? How many joules of useful work does the engine do in each cycle?

13. A particular engine operating at 5 cycles per sec exhausts 500 cal of thermal energy and does 1000 J of useful work in every cycle. (a) What is this engine's efficiency? (Recall that 1 cal = 4 J.) (b) What is its useful power output, in watts? (c) What is its required power input, in watts?

14. How many square meters of solar cells would be required to supply the electrical needs of a typical U.S. home? Assume that solar cells convert solar energy to electricity with an efficiency of 10 percent, that the sun delivers energy to Earth's surface at a rate of 200 W/m², and that a home consumes 1000 W of electric power. Note that some method for the storage of the energy received from the sun would have to accompany this scheme, because energy from the sun is not provided at the rates and times that it is needed.

Recommended Reading

1. The references listed in Chapter Ten contain information about energy conversion schemes.

2. You will find excellent articles about all facets of energy and society in such general interest science magazines as *Science, Science News, Technology Review, Environment, The Futurist,* and *Bulletin of the Atomic Scientists.*

3. E. F. Schumacher, *Small is Beautiful,* Harper & Row, New York, 1973. This book's theme is indicated by its subtitle: "Economics as if People Mattered." The author, who died in 1977, was an economist and British governmental adviser of wide experience. His call for a smaller-scale and more humane technology is regarded with near reverence by some and is passionately criticized by others.

PART 3

Transition to the New Physics

12

WHAT IS LIGHT?

What is light? This is one of the oldest questions in physics, and one of the most important. Historically, it led nineteenth-century scientists beyond Newton's physics to the physics of the twentieth century. Socially, it is important because of the human impacts of light and other forms of electromagnetic radiation.

To understand light we must study two new phenomena: waves and the electromagnetic force. These concepts will lead us to the description of light as it was understood during the nineteenth century. This older, electromagnetic wave theory of light is sufficient for many purposes (see the introduction to Chapter Five). Chapter Sixteen will present the modern quantum theory of light.

What Is a Wave?

Suppose you fasten one end of a long rope to a wall and hold the other end in your hand. You shake the end you are holding, one time, up and back down. A disturbance called a *pulse* will travel along the rope to the wall, reflect from the wall (provided the pulse hasn't died out by then), and travel back to your hand (Fig. 12.1).

QUESTION. What is moving down the rope toward the wall?
(pause)

There are several correct answers, but perhaps the most direct is, simply, that a *bump* (a raised place on the rope) is moving along the rope. A particular shape or configuration (namely the bump) of a portion of the rope is being transmitted along the rope.

You can also say that energy is moving down the rope. This answer is true, but it isn't specific enough because energy is transmitted whenever *anything* moves. For example, a baseball moving toward the wall would represent energy transmission, but the moving baseball is quite different from the moving bump.

One thing that is not moving toward the wall is the rope itself. Any specific point on the rope, such as the part of the rope marked X in the figure, just moves straight up and straight back down as the bump passes through that point. Points on the front side of the bump (the right side) are moving up, and points on the rear side of the bump are moving down, as indicated by the small arrows in the figure. The bump moves to the right even though no portion of the rope moves this direction.

The shape that is moving down the rope doesn't have to be a single bump. Depending on how you choose to shake your end of the rope, you can send all sorts of shapes (Fig. 12.2).

Our example illustrates the meaning of the word **wave.** A **wave** is a shape or configuration that is transferred from one place to another without a corresponding transfer of matter.

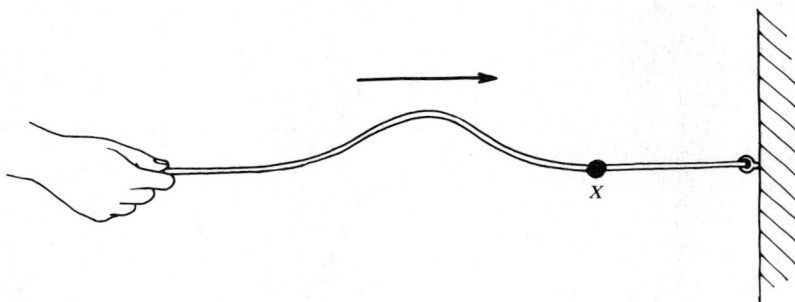

FIGURE 12.1 What is moving toward the wall?

There is a fundamental difference between **wave motion,** such as the pulse moving down the rope, and **projectile motion,** such as a moving baseball. In wave motion no single object actually moves from source to receiver (from your hand to the wall in Fig. 12.1), whereas in projectile motion some object actually moves through the entire distance.

This feature of waves makes them very nice for communication. For example, suppose you are on one side of a pond and a friend is on the other side. You want to communicate with your friend, but for some reason you can't just shout your message; maybe it's a secret or maybe someone is using a jackhammer nearby. How are you going to get the word across?

Exercise 1
Any ideas? (Answer below.)

Well, you could find a piece of paper and a pencil, write down your message, give it to your pet dog Archibald, and let Archibald carry the message around to your friend. This would be communication by transfer of matter. It requires that you have paper, pencil, and a pet dog.

Or you could communicate via water waves. If you and your friend worked out a prearranged water-wave language, in which, perhaps, a single set of waves means yes and two separate sets of waves mean no, you could send your message by dropping either one or two rocks (one a few seconds after the other) into the water. Your friend would get the message when the waves reached the other side. This would be communication by waves.

One obvious means of communication is the human voice. Try this: Place your fingers on your throat while you hum. Feel the vibrations. Your vocal cords are shaking back and forth probably 300 to 400 times per second (women) or 150 to 200 times per second (men). This motion disturbs the air near the cords, much as the moving hand in Figure 12.1 disturbs the rope. The disturbance in the air then moves outward as a wave—a **sound wave.**

Sound waves are similar to the wave along the Slinky toy in Figure 12.3. If you hang a Slinky from the ceiling and push the bottom coil up then back down one time, you will send a disturbance, a compressed place in the Slinky, along the Slinky from your hand to the ceiling. The motion of this disturbance is a wave, as defined above, because no portion of the Slinky actually moves from your hand up to the ceiling. It is the shape, the compression, that is transmitted.

Exercise 2
Suppose you had jerked your hand down and back up instead of up and back down. How would this affect the transmitted shape?

Each vibration of your vocal cords moves the neighboring air molecules back and forth, just as a single vibration of your hand can move a Slinky coil back and forth. This motion of the air is then transmitted to neighboring air molecules, just as the motion of a Slinky coil is transmitted to neighboring coils. The vibrating vocal cords send a series of such disturbances outward through the air. One type of disturbance that is transmitted is a compression, similar to the compression in the Slinky in Figure 12.3. Sound, then, is a compression wave in air. Note that sound actually is a wave because air molecules are not transferred from one place to another along with it; individual air molecules simply vibrate back and forth as the compression wave passes by. Figure 12.4 shows how a vibrating object, such as a musician's tuning fork, produces sound.

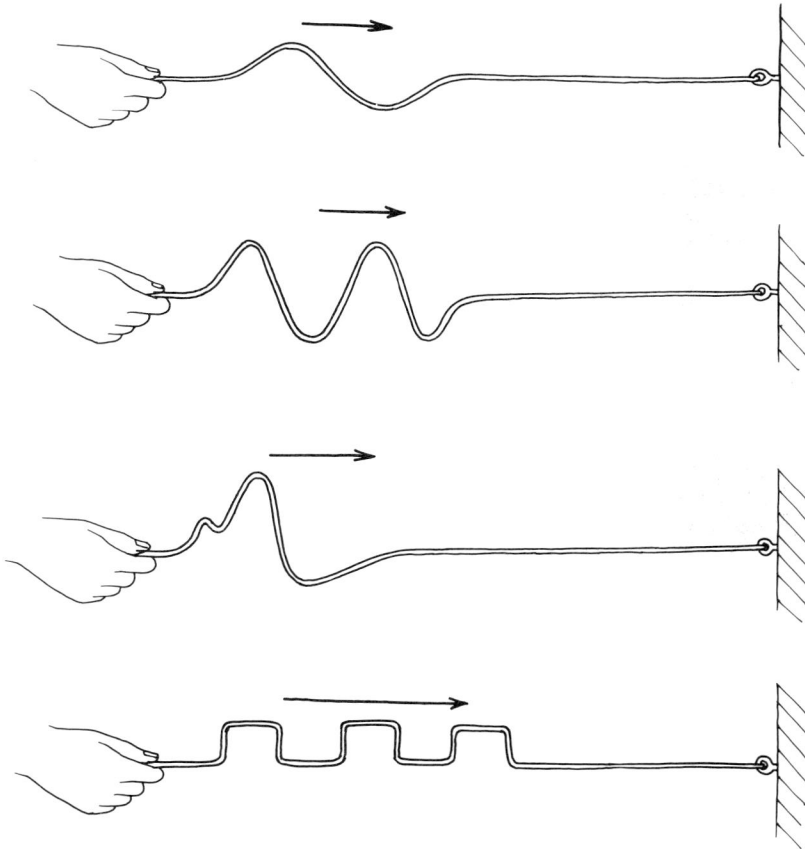

FIGURE 12.2 Some other wave shapes.

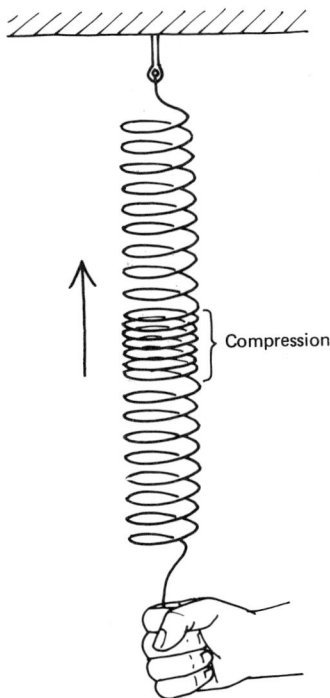

FIGURE 12.3 A wave along a Slinky.

Describing Waves

Physicists have several useful concepts for describing long, repeated waves such as that shown in Figure 12.5. The **wavelength** of such a wave is the distance from any given point in the wave to the next similar point in the wave, for instance from one crest (high point) to the next crest (Fig. 12.5). Thus the wavelength of a sound wave is the distance from the center of one compression to the center of the next compression. Typical sound waves, corresponding to notes in the middle of the piano keyboard, have wavelengths of 1 or 2 ft.

The height of a wave, measured from its midpoint to the top of a crest, is called the **amplitude** (Fig. 12.5).

The **frequency** of a wave is the number of complete wavelengths that pass a given fixed point in 1 sec. For example, the frequency of middle C on the piano is 270 vibrations per second. This means that 270 complete waves strike your eardrum every second and that the piano string vibrates 270 times per second to produce this tone. According to its definition, frequency is measured in "vibrations per second," also called **hertz** (Hz). Thus middle C has a frequency of 270 Hz.

The **speed** of a wave means, naturally, the speed at which it is transmitted. The speed of a sound wave is the speed at which a given compression moves.

Most waves have a definite fixed speed in a given medium, regardless of the frequency or wavelength of the wave.* For example the speed of any sound wave in air is about 340 m/sec, or 750 mph, regardless of frequency or wavelength.

*There are exceptions to this statement. Different wavelengths often travel at slightly different speeds. Furthermore, the wave speed often depends on the temperature and the pressure of the medium.

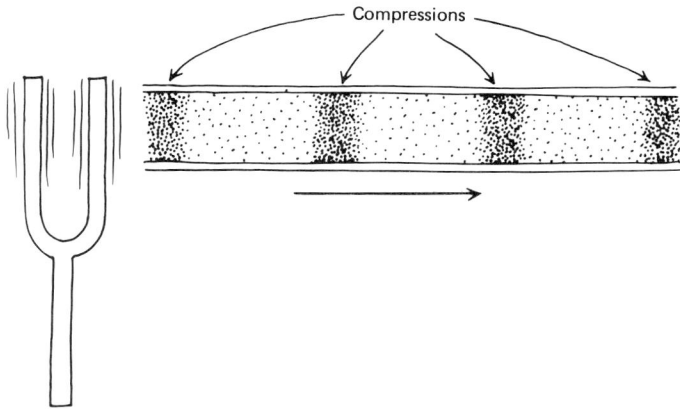

FIGURE 12.4 Tuning fork producing sound inside a long pipe. A cross-section of the pipe is shown. The dots inside the pipe represent air molecules. The sound wave is moving to the right.

As all waves of a given type in a given medium have the same speed, waves with shorter wavelengths must have higher frequencies, because, if the wavelength is short, a large number of waves must pass a given point in a second in order to maintain the required wave speed. Briefly, long wavelength means low frequency, and short wavelength means high frequency (Fig. 12.6).

Interference

The phenomena that occur when two or more waves pass simultaneously through the same point are called *interference effects.*

Consider two single pulses, two single bumps, started from opposite ends of the same rope (Fig. 12.7*a*).

Exercise 3

Assuming that both pulses are upward, as in Figure 12.7*a*, and both are the same size, make a guess about how the rope will look as the pulses pass through each other. How will the rope look if one pulse is upward and one is downward as in Figure 12.7*b*? (Answers below.)

Each pulse in Figure 12.7*a* raises the rope a little. Suppose that each pulse is 4

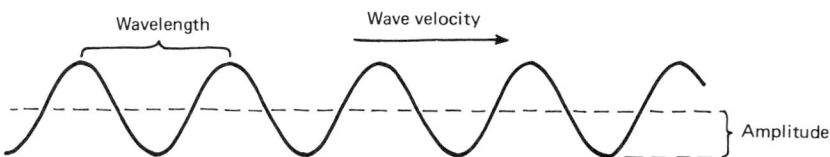

FIGURE 12.5 The meaning of wavelength and amplitude.

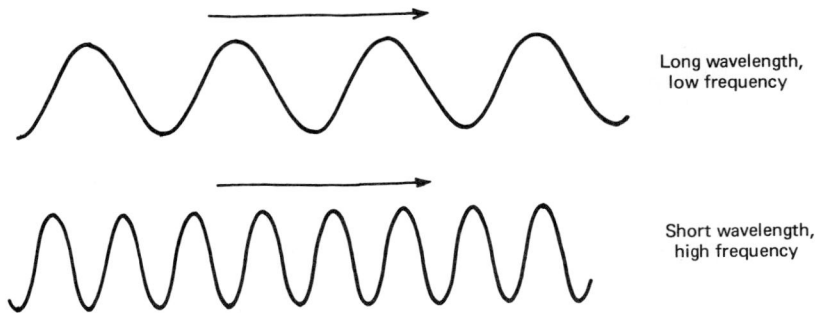

FIGURE 12.6 Long wavelength means low frequency, and short wavelength means high frequency. Both waves are assumed to have the same speed.

cm high. Then each part of the rope is raised 4 cm as each pulse passes. As both pulses pass simultaneously through the center of the rope, the center is raised 8 cm, because it is raised 4 cm by the first pulse and an additional 4 cm by the second pulse (Fig. 12.8a). The two pulses then pass through each other and continue on their way (Fig. 12.9a).

This situation, in which interference produces an effect larger than that produced by either of the individual waves, is called **constructive interference.**

The rope in Figure 12.7b is raised by the upward pulse and lowered by the downward pulse. As the two pulses pass through the center of the rope, the center is neither raised nor lowered because it is raised by one pulse but lowered an equal amount by the other. The two waves momentarily cancel each other (Fig. 12.8b). However, it would be a mistake to think that nothing is going on at the instant pictured in Figure 12.8b. Actually, the center portion of the rope is moving: part of it is moving upward (this part will soon become the upward pulse that will reemerge from the center of the rope), and part of it is moving downward (this part will soon become the downward pulse). Thus the two pulses pass through each other (Fig. 12.9b).

This situation, in which the interference effect is smaller than the effect of either of the individual waves, is called **destructive interference.**

Waves along a rope are one-dimensional waves because a rope has only one

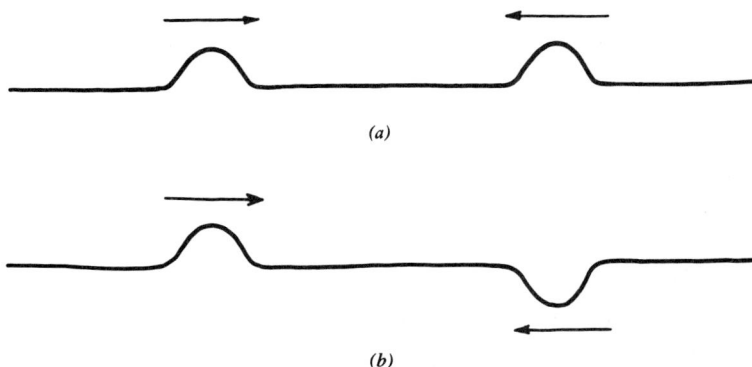

(a)

(b)

FIGURE 12.7 Before two pulses meet.

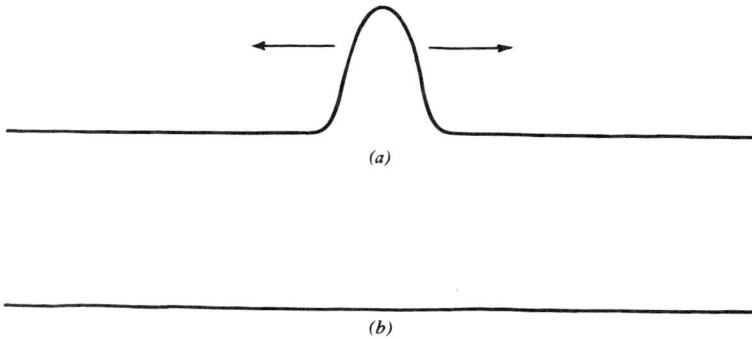

(a)

(b)

FIGURE 12.8 Two pulses meet.

significant dimension, length. Water waves are two-dimensional waves because they occur on the surface of the water, which has two dimensions, length and width.

Figure 12.10 illustrates interference between two-dimensional waves, such as water waves. Figure 12.10a is a diagram showing each of the two waves separately, viewed from above. The circles in this diagram represent crests emanating from two vibrating wave sources at S_1 and S_2. Because these waves are continuous waves that occupy the entire surface of the water, interference will occur everywhere on the surface. Constructive interference occurs wherever a crest from one source meets a crest from the other source and wherever a trough (a low point) meets a trough. (When trough meets trough, there is an especially low point, which is constructive interference because it is a larger-than-usual disturbance). Where a crest from one source meets a trough from the other source, destructive interference occurs.

Try this: On Figure 12.10a mark an x at each point where crest meets crest. Noting that the troughs occur halfway between the crests (between the circles), also mark an x wherever trough meets trough. These x's are points of constructive interference. You should be able to see a pattern developing out of these x's. If so, draw lines connecting the x's to show the pattern. Now mark an o at each point where crest meets trough. These are points of destructive interference. Draw lines connecting adjacent o's.

If you carried out the suggested activity, you discovered the **interference pattern**

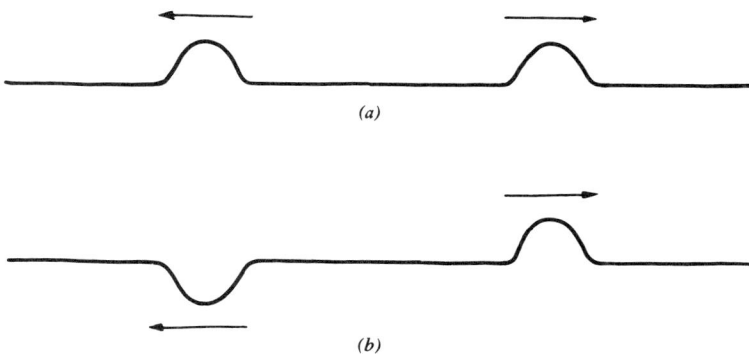

(a)

(b)

FIGURE 12.9 After two pulses meet.

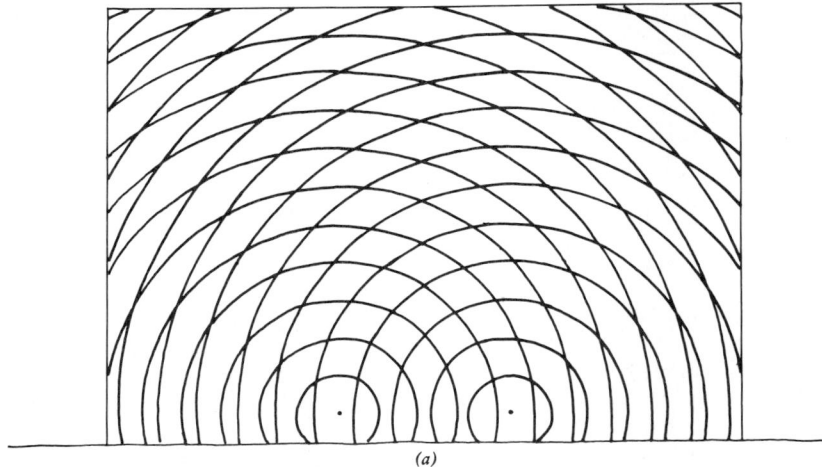

(a)

(b)

FIGURE 12.10 (a–b) Water waves from two sources, demonstrating interference. Viewed from above. (Photograph from *PSSC Physics,* 5th edition, 1981).

that occurs when continuous two-dimensional waves interfere. You should have discovered lines of constructive interference radiating outward (toward the top of the diagram) from roughly the midpoint between the two sources. These lines of constructive interference are interspersed with similar lines of destructive interference.

Figure 12.10b is experimental confirmation of all this. It is an actual photograph of the water waves produced by the oscillation of two objects, at S_1 and S_2, moving up and down in the water. The photograph, taken from above, shows alternating lines, radiating from the midpoint of the two sources, of constructive interference and destructive interference. Large waves move outward along each line of constructive interference, and there are no waves at all along lines of destructive interference.

The Waviness of Light

During the nineteenth century considerable evidence accumulated in support of the view that light is a wave emitted by, or reflected from, the seen object. The most convincing evidence came from interference experiments.

For example, if light passes through two closely spaced thin slits onto a screen, the resulting pattern on the screen is precisely analogous to the water-wave interference pattern of Figure 12.10. Alternate bright strips and dark strips appear on the screen (Fig. 12.11). Apparently, light waves are coming through each of the two slits on the left side of the figure. Thus each slit acts as a source of light waves, similar to the sources, S_1 and S_2, of water waves in Figure 12.10. Light waves move outward (to the right) from each slit, just as water waves move outward from S_1 and S_2. These light waves interfere throughout the three-dimensional region between the slit and the screen, just as the water waves interfere throughout the two-dimensional surface of the water. The bright strips on the screen are regions where the waves from the two slits interfere constructively, and the dark strips are regions where the waves interfere destructively.

Figure 12.12 is a photograph showing the outcome of this double-slit experiment. The photograph was made by placing a photographic plate at the screen position in Figure 12.11.

An interference experiment that you can do involves only a single slit, rather than the two slits described above. Hold your thumb and your forefinger close together, about a foot in front of one eye. Close the other eye. Gaze through the space between your fingers and focus on a distant, well-lit object. Gradually bring

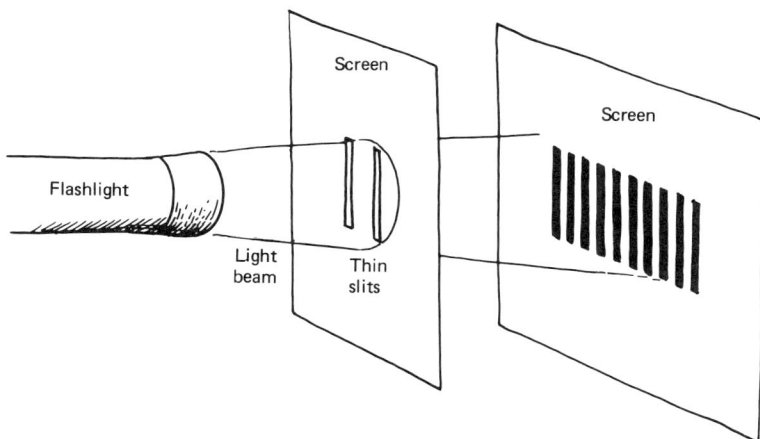

FIGURE 12.11 Light from two sources, demonstrating interference. Light from the flashlight on the left is passing through two thin slits. These slits act as two sources for the light that strikes the screen at the right.

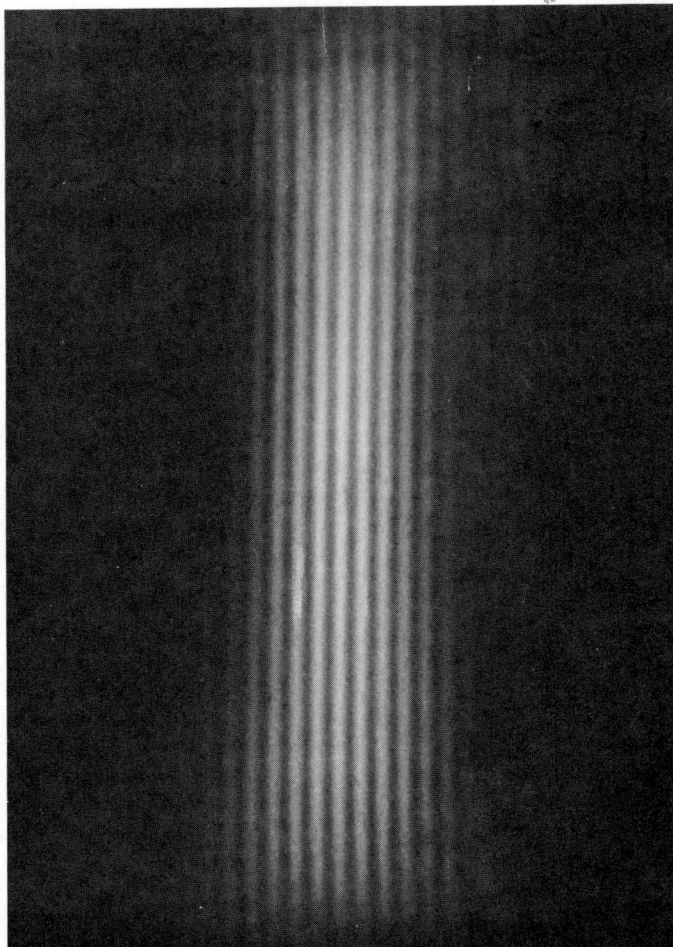

FIGURE 12.12 The double slit interference pattern made by light.

thumb and forefinger close to each other until the gap, or slit, between them is nearly closed. You should see narrow light and dark lines between your fingers, running parallel to your fingers. These bands are caused by interference between the light coming through different portions of the single slit. You have just made a direct confirmation of the wave nature of light. As an improved version of this experiment, cut a thin slit (perhaps 1 millimeter wide) into the edge of an index card. You can narrow the slit by turning the card partially sideways as you look through the slit. When you perform the above experiment using this improved slit, you should be able to see distinct dark and light lines running parallel to the slit.

As a result of interference experiments and other phenomena explicable only in terms of waves, scientists today accept the view that light is a wave. However, you should be forewarned of one of the perplexities of twentieth-century physics: Scientists today also accept the view that light consists of small particles shot from the seen object. This sound like nonsense: How can light be both a moving wave and a beam of moving particles? It's strange, but it appears to be true. For now we'll focus on the older wave theory of light.

The Electromagnetic Force

If light is a wave, then what kind of wave is it? What is doing the waving? What is it a wave in? These questions require a venture into the phenomenon of *electromagnetism*.

If you have ever washed, dried, and then combed your hair on a cold, dry day, you may have noticed that your hair is unusually "difficult" under these conditions. Many individual hairs stand out from your head, and hairs tend to stick to the comb. The phenomenon is often accompanied by mysterious crackling noises as you run the comb through your hair. "Electric hair," it is sometimes called. An invisible force is acting among some of the hairs and between the hair and the comb. Different hairs seem to be trying to get away from one another, indicating that a repulsive force is acting among them, whereas an attractive force acts between the hairs and the comb. Figure 12.13 shows an extreme example of electric hair, produced in this case by the electrical device, known as a Van de Graaff generator, that the student is touching.

These effects are caused by the *electric force* or, better, the **electromagnetic force.** Like gravity, this is one of a small handful of fundamental forces in nature (Chapter Eight). The electromagnetic force is responsible not only for electric hair, but also for electric power, magnetism, and light and other forms of electromagnetic radiation. Furthermore, as we have seen in Chapter Eight, electromagnetic forces

FIGURE 12.13 A severe case of electric hair, produced by the charged object that the student is touching. Combing your hair on a cold, dry day can also produce electric hair. (From Faughn and Kuhn, *Physics for People who Think They Don't Like Physics,* W. B. Saunders, 1976).

acting among atoms are responsible for such common forces as your hand pushing down on a desk, a bat hitting a ball, friction, and air resistance.

Any object that is, like one of the electric hairs mentioned above, in a condition to exert or feel the electric force is said to be **electrically charged.** The simplified theory of the atom presented in Chapter Three will help us understand how objects become charged. It happens (nobody knows why—it just happens) that electrons and protons exert and feel the electric force. In other words, they are permanently charged objects. Neutrons, on the other hand, are not charged. Because electrons and protons produce electric forces of equal magnitude, we say that they both have the same amount of electric charge. Furthermore, whereas the force between two protons or between two electrons pushes the two particles away from each other, the force between a proton and an electron pulls the two particles toward each other. Thus we say that protons possess one type of charge, called **positive charge,** and electrons possess a different type of charge, called **negative charge.** Briefly, like charges repel each other, and unlike charges attract each other.

Now we can see how a hair becomes "electric." When the comb rubs a hair, it may rub some electrons off the atoms of the hair and onto the comb. Because an atom's electrons are fairly loosely bound to the atom, one or more electrons can relatively easily be removed. An atom from which one or more electrons have been removed is called a **positive ion.** An **ion** is any atom that has an excess or deficiency of electrons. The electrons removed from the hair become attached to the atoms of the comb, so these atoms become **negative ions** (they now have an excess negative charge), while the atoms of the hair become positive ions (because they have lost some electrons). Each hair acquires a net positive charge (a deficiency of electrons) while the comb acquires a negative charge (an excess of electrons), thus the comb and hair attract each other while individual hairs repel each other.

The phenomenon of **magnetism,** as exhibited by compass needles or by dime-store magnets, was once viewed as an independent force in nature caused by magnets, just as the force between electric hairs is caused by charge and just as the gravitational force is caused by mass. Then, during the nineteenth century, magnetism came to be viewed as one consequence of the force between moving charges, so that the independent concept of magnets was no longer needed; only the concept of charge was needed. The two concepts, electric and magnetic force, were unified into a single, electromagnetic force, providing another example of the scientific tendency to unify, or reduce, phenomena to the smallest possible number of basic principles.

As an introduction to magnetism, imagine a pair of charged particles such as the two positively charged particles in Figure 12.14. Suppose they are both moving in the directions indicated. Because these objects are charged, they exert a repulsive force on each other. At the instant pictured, this electric force acts toward the left on particle 1 and toward the right on particle 2. This force is present whether the objects are moving or at rest and is unaffected by their motion. If the particles are at rest, this electric force is the only force between them (except for an extremely small gravitational force). But experiments show that if both charges are moving, an additional force is exerted by each particle on the other. This additional force between charged particles, felt only when both particles are moving, is called the **magnetic force.** Experiments show that the magnetic force on a moving charged object is always directed sideways to the object's motion, that is, perpendicular to the direction of motion of the object. Thus the total electromagnetic force between two charged particles consists of an

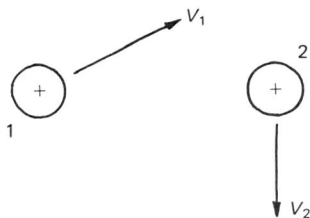

FIGURE 12.14 A pair of charged particles in motion.

electric part, which is felt even when the charges are at rest, and a magnetic part, which is felt only when both charges are moving.

As an example of the magnetic force, let's consider the forces between permanent magnets, such as the small magnets available in toy shops. These forces arise at the atomic level from the motion of the electrons around nuclei in the atoms of each magnet. That is, the moving electrons in one magnet exert magnetic forces on the moving electrons in the other magnet. Individually, these atomic forces are small. In most materials, these tiny forces are randomly directed (i.e., the individual atomic forces point in different directions) and produce no overall large-scale effect. But the iron of which each magnet is made is prepared in such a way that all of these tiny forces act in roughly the same direction, so that the overall effect between all the atoms of one magnet and all the atoms of the other magnet is sizable.

As another example of magnetic force, Earth's magnetism arises from the motions of charged particles in Earth's molten interior. These motions, probably related to the spin of the planet, produce magnetic forces that can be felt by permanent magnets, such as compass needles, on the surface of the earth.

As you can see, the electromagnetic force between charged objects is complex, more complex, for instance, than the gravitational force described in Chapter Seven. The electromagnetic force can be attractive or repulsive or it can act sideways to the motion of the particle that feels it. It acts on moving particles differently than on stationary particles, because the electric part is present whether the particles are moving or at rest but the magnetic part is present only if the particles are moving.

In this section we have briefly outlined the *theory of electromagnetism,* which was developed during the eighteenth and nineteenth centuries. This theory, modified somewhat in our century by the principles of quantum mechanics, is still the scientifically accepted way of looking at electric and magnetic phenomena. Like all other theories, its key concepts, such as charge and electromagnetic force, were invented in order to account for the behavior of compass needles, electric hair, and the like.

The Electromagnetic Field

Scientists necessarily invent new concepts. In fact, the history of science is the history of our observations of the natural world and of concepts invented to help make sense out of these observations. Copernicus invented (or reinvented) the concept that Earth goes around the sun in order to make a more beautiful organization of astronomical observations. Newton invented universal gravitation in order to unify the observed motions of apples and the moon in one grand scheme. The electromagnetic field is another such invention. Like Earth orbiting the sun, and like Newton's theory of gravity, this invention has turned out to be so useful that it has taken on a certain amount of reality.

Any charged particle is said to produce an ***electromagnetic field*** everywhere in the particle's vicinity. This electromagnetic field is said to exist at any point in space where a second charged object would feel a force if such an object were placed at that point. This field fills the space in the particle's vicinity even though there might not be anything (any other objects) actually in that space.

Let's consider an example: the negatively charged comb discussed previously. Suppose we put the comb in an empty room or, better yet, in empty space far from the earth and far from all other objects. Then the comb is not actually exerting any significant electromagnetic force on any object, because there are no other objects around. Nevertheless, physicists like to think of the comb as the source of an electromagnetic field that surrounds the comb and fills up space out to some distance from the comb.* This electromagnetic field represents the comb's ability to exert a force on charged objects that could be placed in the comb's vicinity.

Notice that there isn't really any matter, any *thing*, in the vicinity of the comb in empty space. Nevertheless, we agree to imagine that there is an electromagnetic field there. This field must be thought of as something that the comb does to the space around it.

For example, permanent magnets produce electromagnetic fields in their vicinity. These are often called magnetic fields because they arise from the magnetic part of the electromagnetic force law. The force exerted on other objects by a permanent magnet can be thought of as arising from the magnetic field produced by the permanent magnet.

As another example, lightning strokes (Chapter Eighteen) are the result of the large electromagnetic fields produced in the atmosphere when excess electrons build up on clouds and a corresponding deficiency of electrons builds up on the ground. These fields are often called electric fields because they arise from the electric part of the electromagnetic force law.

As a third example, the force on a compass needle arises from the electromagnetic field produced everywhere in the vicinity of the earth by the motions of charged particles in the earth's interior.

Light: An Electromagnetic Wave

We've seen that charges produce electromagnetic fields all around them. Now suppose you caused a charged object to vibrate in some manner. For instance, you could electrically charge a comb, by running it through your hair, and then shake the comb up and down. This would produce a *vibrating* electromagnetic field.

Now we must introduce a fundamental principle about the electromagnetic force. Like the other basic principles of physics, nobody knows why it is true, but it happens that the electromagnetic force by one charged particle on a second charged particle is not established instantaneously; instead, the force is communicated from the first charge to the second charge at 300,000 km/sec.*

*According to the detailed quantitative definition of the electromagnetic field, the field is weaker at larger distances from the comb, so that, for all practical purposes, the field is nonexistent at large distances from the comb.

*More precisely, 298,000 km/sec.

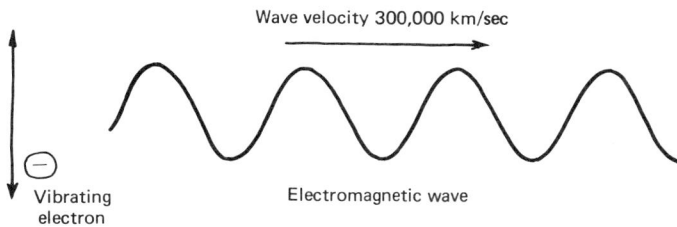

Wave velocity 300,000 km/sec

Vibrating electron

Electromagnetic wave

FIGURE 12.15 Electromagnetic waves are produced by vibrating charged particles.

Imagine that two charged objects are 300,000 km apart, and that the first object starts moving. This motion will result in a magnetic force on the second object (provided the second object happens to also be moving). However, according to the foregoing principle, the second object will not feel this magnetic force right away, but will feel it only after a delay of 1 sec, that is, it takes 1 sec for the force to be communicated from the first object to the second object.

In terms of the field concept, the new principle says that changes in the electromagnetic field produced by a charged particle are not communicated instantaneously. Such changes are communicated outward at 300,000 km/sec, much as disturbances in the water move outward from a pebble dropped into a pond.

Now let's return to the vibrating electromagnetic field produced by a vibrating electron. The new principle tells us that the vibrations of the electromagnetic field are communicated outward from the electron at 300,000 km/sec. This is precisely the sort of thing that has been defined as a wave: a certain shape, or pattern, of the electromagnetic field is moving outward from the vibrating charge. But this wave is quite unlike the rope waves, water waves, and sound waves that we studied earlier. Because electromagnetic fields can exist even in vacuum, the "stuff" into which this pattern is woven is not made of atoms, not made of ordinary matter. Because the electromagnetic field is the "stuff" doing the waving, we call this phenomenon an *electromagnetic wave.* (Fig. 12.15).

And that is what light is!

The Decline of the Newtonian Universe

So light turns out to be a traveling pattern of vibrations in the electromagnetic field, moving at a speed of 300,000 km/sec (186,000 mi/sec). But wait—the electromagnetic field was introduced earlier as a purely imaginary, invented concept, useful in helping us think about the electromagnetic force. How can light be made of an imaginary concept? Is light just a figment of our imagination?

The answer, surely, is that light is a real phenomenon. After all, many of our direct and reproducible sense impressions are visual, making light about as real as anything can be in the natural world. But then it must also be true that our invented concept, the electromagnetic field, is real, because light is an electromagnetic wave.

A little thought will reveal that the principles of the electromagnetic force developed in this chapter force us to accept the reality of the electromagnetic field concept. Let's return to our negatively charged comb. Suppose that at noon someday you begin shaking this comb up and down. This motion causes the comb to emit an electromagnetic wave, which travels outward until it encounters other charged particles, which will then begin to vibrate in response to the wave. But some time must elapse before these receiving charges can begin to vibrate because the electromagnetic force is not transmitted instantaneously. Suppose that this delay time is 1 sec. Then at 1 second past noon the receiving charges begin to vibrate. Look at it from the energy point of view: At 1 second past noon the receiving charges acquired a certain amount of kinetic energy; however, this energy must have been sent out by the comb at noon, because that is when you started shaking the comb. Where was this energy between noon and 1 second after noon?

Exercise 4
Where, indeed? (Answer below.)

Do you see? If we want to preserve the idea of conservation of energy then we have to assume that during the 1-sec delay time the "missing" energy resides somewhere—somewhere other than the comb or the receiving charge. Therefore, it must reside in the electromagnetic field. And we must regard this field as more than just an imaginary concept.

The energy carried by an electromagnetic wave is called **radiant energy.** As we have seen, radiant energy travels at the speed of light from the charged object that emits it to the charged object that receives it. Unlike a sound wave, no air is required for the transmission of an electromagnetic wave because the electromagnetic force can be exerted by one charge on another across empty space. You demonstrate the ability of electromagnetic waves to move through a vacuum every time you glance at the sun or stars.

When the electromagnetic field concept was invented in the early nineteenth century, it seemed to be only a convenient fiction, and it received little attention. But when the delay time for the electromagnetic force and the **electromagnetic wave theory of light** were discovered later in the century, this concept became more than a convenient fiction. The electromagnetic field became real.

It was a revolutionary idea. For 2 centuries physicists had regarded the world as a kind of clockwork device built of parts, built of atoms moving in empty space and interacting in definite, predictable ways by means of specific forces. Newton's laws ruled supreme in this mechanistic universe. And now comes the electromagnetic field, filling up entire regions of so-called "empty" space, not composed of parts, not composed of atoms, but nevertheless laying claim to reality! Apparently, the world is not like a clock.

Thus does the electromagnetic field concept, and the phenomenon of light, forecast the decline of the Newtonian world view and the rise of the nonmechanistic physics of our own century.

CHECKLIST

wave	electromagnetic force
sound wave	electric charge
wavelength	ion
amplitude	electric force

frequency	magnetic force
hertz	electromagnetic field
wave speed	electromagnetic wave
interference	speed of light
evidence for light waves	radiant energy

Thought Question

What is reality? Discuss the ways in which each of the following items is real or unreal or neither. Are some items more real than others? *Objects,* such as (a) a baseball; (b) the sun; (c) the planet Neptune; (d) a visible dust particle; (e) an invisible dust particle; (f) a molecule; (g) an atom; (h) a proton. *Concepts,* such as (a) the gravitational force; (b) electric charge; (c) the electromagnetic field; (d) electromagnetic waves; (e) the circular motion of the sun around Earth (Ptolemy); (f) the circular motion of Earth around the sun (Copernicus); (g) the elliptical motion of Earth around the sun (Kepler); (h) the wobbly, nearly elliptical motion of Earth around the sun as a result of the gravitational effects of the sun, moon, and other planets; (i) capitalism; (j) communism. *Experiences,* such as (a) the sight of light streaming through a window; (b) the sound of a radio; (c) warmth from a stove; (d) the feel of satin; (e) the impact of a baseball landing in your hand; (f) your own conscious thoughts; (g) someone else's conscious thoughts; (h) a dream you had last night; (i) a mirage in the desert; (j) love.

Further Exercises

5. A pebble falls in the water; ripples spread over the surface and eventually die out. What has become of the energy in these waves?

6. Could we talk normally to each other on the moon? Explain.

7. Suppose that sound waves of different frequencies traveled at different speeds. How would distant music sound?

8. Explain why you sometimes get a shock when you walk across a rug on a cold, dry day and then touch a piece of metal.

9. A bell rings inside a jar. If the air is pumped out we can no longer hear the bell but we can still see it. What does this indicate about the differences between sound and light?

10. Suppose that the light from the sun were suddenly shut off all at the same time. Would you be immediately aware of this catastrophe, or would there be a delay? Viewed from the earth, would the sun appear to turn off all at once? How would it appear? How would it appear from Earth if all the stars (except the sun) suddenly stopped shining, all at the same time? (*Hint:* a small black dot would first appear at the center of the sun, as seen from Earth. The dot would then spread. Why?)

Recommended Reading

Robert H. March, *Physics for Poets,* McGraw-Hill Book Company, New York, 1978. Chapters 6 and 7 deal with electromagnetism and light waves.

13

THE COMPLETE SPECTRUM

Light is an electromagnetic wave, but it is not the only type of electromagnetic wave. You may have heard of the other types: radio, infrared, ultraviolet, X ray, gamma ray. Each one is similar to light; each is produced by vibrating charged particles; each is a vibrating electromagnetic field that travels at the speed of light, 300,000 km/sec. What distinguishes the various types of electromagnetic waves is their wavelengths. Figure 13.1 shows the wavelength range corresponding to each type. These waves are known collectively as **electromagnetic radiation** because they radiate outward from the charged particles that produce them. The complete range of electromagnetic wavelengths is known as the **electromagnetic spectrum.**

Electromagnetic radiation affects your life in many ways. In this chapter we will study just a few of the human implications of the three regions of the spectrum that predominate in the radiation from the sun: visible, infrared, and ultraviolet.

Visible

We'll begin our tour through the varieties of radiation in the middle of the spectrum, with visible light, the most familiar type of radiation.

Visible light is electromagnetic radiation that can be detected by the light-sensitive part of your eye known as the **retina.** The retina plays a role similar to that of the film in a camera; it contains light-sensitive cells, just as photographic film contains light-sensitive chemicals. Charged particles in these cells are the receivers for visible light. The retina passes the received pattern of light on to the optic nerve which transmits the pattern to the visual center of the brain. The transmission along the optic nerve is a complicated process involving electrical forces and the chemical composition of the nerve.

The retina is sensitive to electromagnetic radiation only within a certain narrow range of wavelength, about 4×10^{-7} to 7×10^{-7} m. Radiation in this range is produced by the motion of electrons within atoms or (see the following section) by the thermal vibrations of molecules. That is, atoms and molecules are the senders of visible light. An explanation of the details of the emission of light by atoms and molecules will be found in Chapter Eighteen.

Whether you are looking at Michelangelo's ceiling of the Sistine Chapel or at a pile of garbage, the mechanism of seeing is the same: atoms and molecules in the seen object emit electromagnetic waves that travel at the speed of light to your eye, where they are received by electrons in the retina and transmitted to your brain by the optic nerve.

Once again the scientist has reduced a human experience to a purely physical process, to the working of atoms, to mechanism. Once again we can ask the deeper question: Is that all there is to seeing? Is the scientist's description of the physical processes that occur when you view the Sistine chapel ceiling really equivalent to

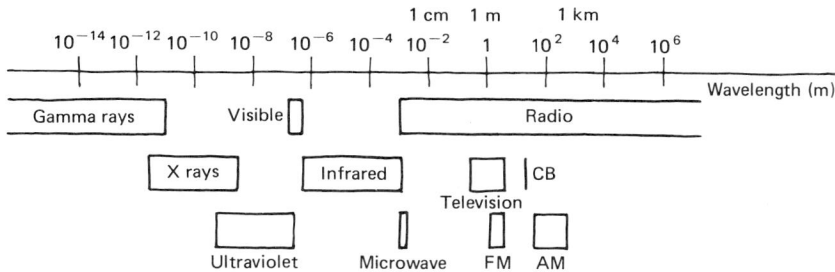

FIGURE 13.1 The electromagnetic spectrum.

your experience of viewing that ceiling, even assuming that the scientist's description included all the details about the frequencies coming from every part of the painting, all the processes going on in your optic nerve, and all the physical processes (mostly unknown at present) going on in your brain?

Or is it perhaps the case that the scientist's description is only one way of viewing this situation; that the scientist's description, although correct, is not complete? Are there other, equally valid and perhaps more meaningful ways of describing what happens when you view Michelangelo's conception of the creation of humanity on the ceiling of the Sistine Chapel (Fig. 13.2)?

Thought Question 1
Think about that.

Thought Question 2
If we used the word *see* to mean not only the purely physical process described above but also the subjective experience (or feeling, or whatever you want to call it) of seeing, then do two people really "see" the same thing when they look at a painting?

Thought Question 3
For the sake of argument, let's restrict the definition of *see* to include only the purely physical processes described above. Using this definition, do two people "see" the same thing when they look at a painting? Are all retinas physically identical? Are all optic nerves identical? Are all brains identical?

Your eye responds differently to different wavelengths of visible radiation. Your brain interprets the different wavelengths as different colors. If the incoming light has a wavelength of about 0.00004 cm (4×10^{-7} m), you see violet. At wavelengths around 0.00007 cm (7×10^{-7} m), you see red. Between these wavelength extremes you see the other colors of the rainbow.

Scientists have observed an enormous range of electromagnetic wavelengths (Fig. 13.1)—from gamma rays at 10^{-12} m and shorter to radio waves at 100 m and longer. In the middle of this span is the tiny region called visible. Nearly all our direct information about the world around us comes from this narrow range of wavelengths. But most objects also emit a wide variety of electromagnetic wavelengths outside the visible region. The world would look quite different if our eyes responded to a broader, or different, set of wavelengths.

FIGURE 13.2 Section of the Sistine Chapel ceiling. (Alinari/Editorial Photocolor Archives.)

Infrared and Global Heating

Imagine a beam of visible radiation whose wavelength can be varied. If the wavelength is set at 5×10^{-7} m and the beam strikes your retina, you will see yellow. If the wavelength is gradually increased to 6×10^{-7} m and then to 7×10^{-7} m, you will see red and then a darker and darker shade of red. Now, 7×10^{-7} m is about the longest wavelength to which your retina can respond. Beyond this wavelength, although the beam of radiation still enters your eye, your retina does not respond by sending signals to your brain. The source of the beam is now black. The radiation that enters your eye is beyond red, it is so red that you can't see it, *infrared*. Although you can't see it, this radiation has great human significance.

Wavelengths in the infrared region are produced primarily by the ***thermal vibrations*** of molecules. As you know, the molecules of any substance are in continual thermal motion, due to their thermal energy. This continual, random molecular motion, whether it be in a solid, a liquid, or a gas, amounts to a kind of disorganized vibration of the molecules. These vibrations are the source of infrared radiation.

Entire molecules, for example the H_2O molecules in a glass of water, are usually uncharged, so how can they produce electromagnetic waves? The answer is that, whereas most molecules are electrically neutral overall, some regions of the molecule have an excess of protons over electrons and other regions have an excess of electrons over protons. Figure 13.3 shows these regions of charge concentration for the H_2O molecules. The motion of these charge concentrations, caused by thermal motion of the molecule, is the source of infrared radiation.

Because there is more thermal motion in a hotter object than in a cooler object, hotter objects must emit more infrared than cooler objects, and in turn, any object that receives infrared is warmed by it because this radiation has just the right frequency to set molecules into thermal motion.

The solar radiation received on Earth comes from the sun's 6000° C surface. It is not surprising that the energetic thermal motion in the hot surface of the sun produces large amounts of infrared. Although we receive solar radiation throughout the entire electromagnetic spectrum, most of this radiation is concentrated near the center of the spectrum, in the ultraviolet, visible, and infrared regions.

Although you can't see infrared, you can feel it warm you on a sunny day because infrared produces thermal motion in the receiver, your skin. Because of their ability to warm the body, infrared lamps are used for the treatment of muscle strains and bruises. On the other hand, it is the higher frequency ultraviolet radiation from the sun that burns your skin.

Photographic film contains chemicals sensitive to the visible region of the spectrum. These chemicals undergo a reaction when they are hit by light. Some films, however, are sensitive to infrared rather than to visible radiation, and the resulting *infrared photograph* is a pattern of the infrared radiation emitted by the photographed object. Such a photograph is really a temperature map of the object. Infrared photographs taken from airplanes or satellites provide useful information, based on temperature differences, about forest growth, crop health, pollution, and ocean temperatures.

There has been much discussion lately of a possible worldwide atmospheric warming trend caused by carbon dioxide and infrared radiation. This warming trend will be discussed in detail here, because it may turn out to be the most significant environmental problem of our century.

Earth's temperature results from the balance between the energy received from the sun and the energy emitted back into space. Energy from the sun is received in the form of radiant energy, most of it in the ultraviolet, visible, and infrared regions. About 35 percent of this incoming radiation is immediately reflected back into space;

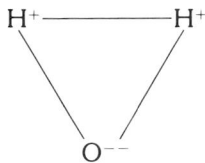

FIGURE 13.3 Structure of the water molecule, showing regions of charge concentration. The oxygen atom has two excess electrons in its vicinity, and each hydrogen atom has a deficiency of one electron.

the remainder is absorbed by the atmosphere, the seas, and the land masses of Earth, thereby warming the earth. The warmed earth, in turn, reradiates energy back into space. Most of this reradiated energy is in the form of infrared electromagnetic waves.

Certain trace molecules in the atmosphere are excellent absorbers of infrared. Most notable among these is carbon dioxide, CO_2. Carbon dioxide constitutes less than 1 percent of Earth's atmosphere, hence the name **trace molecule.** Prior to about 1900, CO_2 formed only 0.029 percent of Earth's atmosphere—about three-hundredths of 1 percent, or 3 parts in 10,000. But this small part is important because it accounts for much of the infrared absorption by the atmosphere.

Humans can change the percentage of CO_2 rather easily, precisely because CO_2 is a trace molecule. It would be very difficult, for instance, for humans to tamper with the percentage of O_2 in the atmosphere. This percentage is so large (about 20 percent) that it represents an enormous total amount of O_2, and a huge addition would be needed to increase this 20 percent significantly. Since there isn't much CO_2 in the atmosphere, its percentage can be changed significantly by adding or removing relatively small amounts.

There is a lesson here. Human activities can significantly alter such trace chemicals as carbon dioxide, argon, ozone, and lead, so we should be cautious with any activities that could upset the balance of these molecules.

Atmospheric absorption of infrared helps warm our planet. The continents and oceans absorb solar radiation, which warms them, causing them to emit infrared. The atmosphere absorbs the reradiated infrared, which warms the atmosphere, causing the atmosphere to reradiate infrared. All the infrared from the atmosphere doesn't go into space; it's just as likely to go downward, back to the continents and oceans. In effect, some radiation is trapped by the atmosphere, making the earth warmer than it would be without atmospheric absorption of infrared.

As you know (Chapter Three), combustion of carbon-based fuels produces carbon dioxide.

Exercise 1
Suppose you burn 1 ton of coal. Assuming that coal is 100 percent carbon (the true figure is 70 percent) and assuming complete combustion (i.e., all the carbon is actually burned), roughly how many pounds of CO_2 will be produced? *Hint:* The mass of an oxygen atom is about 4/3 the mass of a carbon atom. (To simplify the problem, you might want to assume that oxygen and carbon atoms have the same mass.)

Exercise 2
With all the CO_2 produced by burning a fossil fuel, why isn't there a tremendous mass of leftover CO_2 at fossil fuel power plants?

The lesson of these exercises is that fossil fuel plants, not to mention automobiles and all other fossil-fuel burners, produce tremendous amounts of CO_2—enough to significantly alter the worldwide balance of this trace molecule. Nobody worried about this when fossil fuel plants or automobiles were introduced. After all, CO_2 is a natural constituent of the atmosphere, is nontoxic to animals, and is beneficial for plant growth. We are learning late in the game that CO_2 may have more dangerous implications than we had thought.

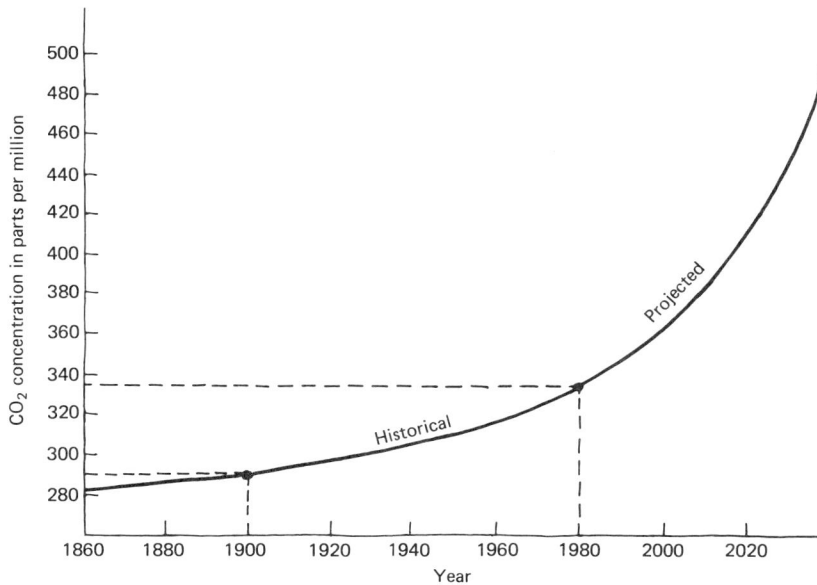

FIGURE 13.4 Historical and projected concentration of CO_2 in the atmosphere.

Carbon dioxide formed 0.029 percent of the atmosphere in 1900; it forms nearly .034 percent today—a 17 percent increase. And the increase continues. The increase since 1880, along with estimated future increases, is shown in Figure 13.4. Note that the graph reads 290 parts per million (0.029 percent) in 1900, and 335 parts per million in 1980.

The world's use of fossil fuel is not leveling off. It has been increasing each year since 1860 at a remarkably consistent 4.3 percent. The projection shown for 1980 and beyond in Figure 13.4 is based on a continued 4.3 percent increase.

Exercise 3

Which of the following energy policies would help decrease the rate of injection of CO_2 into the atmosphere? (a) Switch from oil and natural gas to coal; (b) switch to nuclear energy; (c) conservation of energy resources; (d) switch to solar energy; (e) switch to biomass fuels.

So far, the increase in world temperature resulting from increased CO_2 has not been noticeable, and it might not be noticeable for several years to come.

The most comprehensive study of this problem to date was carried out in 1979 by the National Academy of Sciences. The study group found that atmospheric CO_2 will double by about 2030 if use of fossil fuel continues to grow at about 4 percent per year. Doubling will be delayed until 2050 if the fossil fuel growth rate is only 2 percent and until 2100 if fossil fuel use remains at today's level. The report estimates that a doubling of atmospheric CO_2 will produce a global warming of probably 3° C, give or take 1.5° C.

Now, a 3° C (5.4° F) increase in temperature might not sound like a lot, but remember that this is a worldwide increase and it is permanent. To make matters worse, the increase is predicted to be several times larger than this in the region of the polar ice caps. There would also be local "hot spots" over all the heavily industrialized regions because there is more CO_2 over these regions. In fact, this effect today makes our cities warmer than the surrounding countryside.

Everyone knows that it is difficult to predict the weather, but a 2° or 3° C warming would have vast effects on ocean waters, polar caps, agriculture, plant photosynthesis, and climate zones. Some of the repercussions would be harmful, some possibly beneficial. But the world would have to change.

Examples: The greatest temperature increases would occur in the polar regions. Thus the temperature difference between polar and equatorial regions would decrease. Because this temperature difference drives the atmospheric heat engine, we could expect alterations of global airflow patterns. The large temperature increases in the polar regions could partially melt the polar ice caps, perhaps eventually raising ocean levels by 1 to 5 m. Portions of the Antarctic ice cap could slip into the ocean, causing an even larger rise in global sea levels. If a lot of ice slid into the sea, it could spread to cover much of the southern oceans. The brilliant white surface of this vast ice shelf would reflect enough solar energy back into space to cool the global climate and just possibly initiate an ice age.

These are possible effects that could occur if the CO_2 level keeps rising. The human race can continue using fossil fuels at an increasing rate and wait some 20 years for nature to perform the experiment that tells us whether these possible effects will actually occur. If the effects appear to be harmful, we would then no doubt embark on a crash program to reduce fossil fuel use, a program that might reverse the undesirable effects. Or we can decide now that we will not experiment with the whole Earth, that we will reduce our use of fossil fuel until we know more about its effects. Of course, the latter decision has significant effects in itself: for example, any large-scale switch from the internal combustion engine would displace workers, alter many personal habits, and increase the use of such nonfossil resources as nuclear and solar or reduce our total energy consumption.

There are no easy solutions.

Exercise 4
The U.S. government has recently embarked on a crash program to develop synthetic fuels, liquid or gaseous fuels developed from coal and from an oily rock known as shale. This program could help solve our fossil fuel shortage, but will it help solve the global heating problem? (Answer below.)

The so-called synfuels cannot help us with the CO_2 dilemma because they are based on carbon just like the conventional fuels. In fact, synfuels make the problem worse, because CO_2 is released not only during combustion of synfuels but also during their production from coal or shale. On an energy-equivalent basis, the total CO_2 released from synfuels is 1.4 times greater than that from coal and 1.7 to 2.3 times greater than from oil and natural gas, respectively.

Thought Question 4
What should we do?

Ultraviolet and Ozone Depletion

The preceding section opened with a beam of visible radiation whose wavelength could be varied. Let's now start this beam at a wavelength of 5×10^{-7} m (yellow) and gradually decrease the wavelength. The source will appear to change from yellow to green, blue, violet, dark violet, and finally, around 4×10^{-7} m, black. The wavelength is now so short that it is beyond violet, or **ultraviolet**.

Ultraviolet radiation originates in the motion of electrons in atoms. This process will be described in detail in Chapter Eighteen.

Whereas the infrared rays from the sun warm you, the ultraviolet rays burn you. Ultraviolet radiation penetrates the top few layers of your skin and initiates a chemical reaction that darkens, or "burns," the cells and that may initiate skin cancer. Ultraviolet radiation from the sun can penetrate the clouds even on overcast days, when little direct sunlight gets through, so you can get a sunburn even when the sun isn't shining.

Above Earth's atmosphere, solar radiation consists of roughly equal parts of infrared, visible, and ultraviolet. Most of the ultraviolet is absorbed in the upper atmosphere, at a height of 20 to 50 miles, by a dilute band of gas known as the **ozone layer.** Scientists are beginning to realize that the ozone layer is fragile and easily altered by human activities.

The ozone molecule, O_3, consists of three oxygen atoms. Ozone is created in the upper atmosphere by the action of solar radiation on ordinary oxygen molecules (O_2). Ozone molecules are excellent absorbers of ultraviolet radiation. In fact, the ozone layer absorbs some 98 percent of ultraviolet from the sun, so the radiation reaching the ground is mainly visible and infrared. Sunburns are a consequence of the 2 percent of ultraviolet that does get through. Imagine what life would be like if a much larger fraction of this high-energy radiation reached the ground. We would have to stay out of the sun; farm animals would have to stay out of the sun; wildlife as we know it could not exist; crops and other plants could not exist, at least not out in the open—life as we know it could not exist.

Exercise 5

Make a rough estimate of the number of minutes in the sun that would be required for a Caucasian to get a sunburn if the atmosphere absorbed no ultraviolet radiation. (Caucasians ordinarily sunburn in about an hour.)

Several human activities appear to threaten the ozone layer. These activities could reduce the amount of ozone in the upper atmosphere, leading to an increase in the ultraviolet radiation reaching Earth.

Our knowledge about this problem has been changing rapidly in recent years. At the latest count, there appeared to be three major threats to the ozone layer. (For the most recent information, read a good science newsmagazine, such as *Science* or *Science News.*)

1. Aerosol Sprays

In 1979 the U.S. government banned the sale of most ozone-destroying sprays. Prior to the ban, nearly all aerosol sprays, such as deodorants, hair sprays, and insect sprays, were powered by **chlorofluorocarbon** gas. (CFC for short). As its name indicates, the molecules of CFC contain chlorine, fluorine, and carbon. One of its primary advantages as a spray propellant is that it is **inert,** that is, it doesn't react chemically with much of anything, and thus it can't harm you, your food, your pet dog, and so on.

But its inertness poses a new problem. Being a gas, CFC molecules rise in the atmosphere, and being inert, they rise high without reacting with the air. They eventually reach altitudes of 20 to 50 miles, where there is a lot of ultraviolet radiation from the sun. This high-energy radiation splits the molecules into their component elements, one of which is chlorine. Unfortunately, chlorine reacts strongly with ozone. Worse yet, at the end of this reaction the chlorine atom is freed to react with yet another ozone molecule. The net effect is to turn O_3 into O_2; we say that chlorine **catalyzes** (promotes) the conversion of O_3 to O_2.

Prior to the phaseout in 1979, the United States was responsible for about 50 percent of the world's CFC sprays. However, the sprays account for only 50 percent of the world's use of CFC. Refrigerants (discussed below) account for the other 50 percent and the refrigerants are not being phased out. So the U.S. phaseout of sprays solves perhaps 25 percent of the world's CFC problem.

The spray-can industry is now using various other propellants. We will presumably discover in the future whether these propellants have any harmful side effects. One obvious solution to the whole mess is the old-fashioned, hand-operated pump spray.

Thought Question 5
The research leading to the discovery of the chlorofluorocarbon problem was conducted many years after the introduction of chlorofluorocarbon refrigerants and propellants. Should the research have been done earlier? What can we learn from this?

2. Refrigerants

A refrigerator works like a heat engine (Chapter Eleven) in reverse as shown schematically in Figure 13.5. Figure 13.5a shows the energy flows for a heat engine. By turning all the arrows around, we get Figure 13.5b, the schematic diagram for a refrigerator. The purpose of a refrigerator is to remove some low-temperature thermal energy from the inside of a container. Some work is needed to do this. The electric company provides this work, which runs the compressor in the cooling apparatus. The refrigerator then exhausts a larger amount of thermal energy at a higher temperature, outside the cooled container.

If the container is a small box with food inside, the device is an ordinary refrigerator; if the container is a room, the device is an airconditioner.

Every heat engine operates by allowing some working fluid, such as steam, ammonia, or a gas-air mixture, to expand and contract as its temperature changes. Refrigerators operate along similar lines. One of the favorite working fluids for refrigerators and air conditioners is **freon.** Unfortunately, freon is a chlorofluorocarbon.

Refrigerants represent about 50 percent of worldwide CFC production. Phaseout of this part of CFC production is difficult, however, because there seems to be no convenient replacement for freon, because refrigerators are fairly essential items, and because many jobs are related to freon production. Thus the U.S. government has banned CFC sprays but has not banned freon.

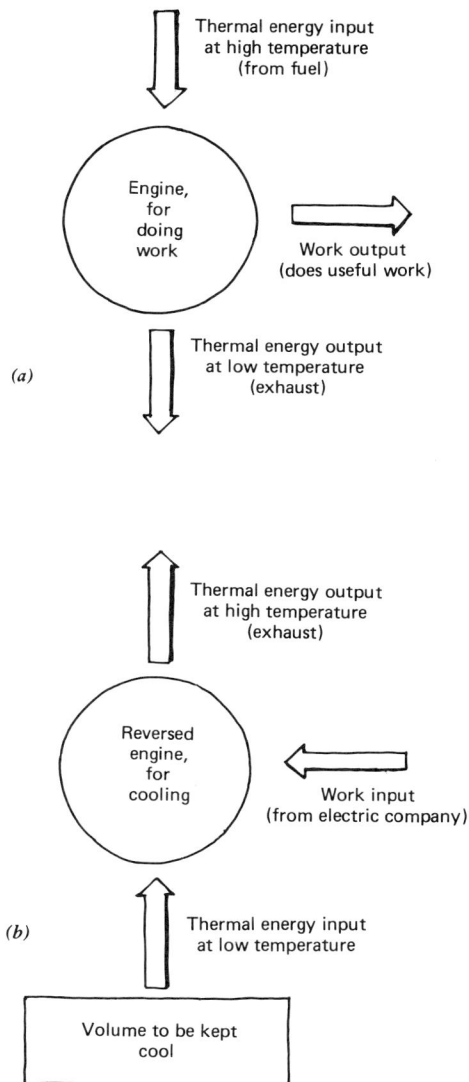

Thermal energy input
at high temperature
(from fuel)

Engine,
for
doing
work

Work output
(does useful work)

(a)

Thermal energy output
at low temperature
(exhaust)

Thermal energy output
at high temperature
(exhaust)

Reversed
engine,
for
cooling

Work input
(from electric company)

(b)

Thermal energy input
at low temperature

Volume to be kept
cool

FIGURE 13.5 (a) The heat engine. (b) The
refrigerator.

3. Methyl Chloroform

This chemical is used as a degreasing agent. Like the chlorofluorocarbons, it rises to
the stratosphere, where it breaks down under the impact of high-energy radiation.
This breakdown releases chlorine and thus poses a threat to the ozone layer. The
production of methyl chloroform is currently doubling every 5 years. The Environmental Protection Agency is considering regulating this chemical.

The hypothesis that CFCs may harm the stratosphere was first suggested in 1974
by F. Sherwood Rowland and Mario Molino of the University of California. Although
Rowland and Molino's idea was initially scoffed at, a great deal of laboratory work,
measurement of chemical concentrations in the stratosphere, and computer modeling
of the atmosphere have failed to find any significant holes in it.

The National Academy of Sciences carried out studies of the CFC problem in 1976 and 1979. The 1979 study confirmed and extended most of the results of the 1976 study. The academy estimated that if CFC emissions continue at the 1977 rate, the loss of stratospheric ozone will total 16 percent, plus or minus 11 percent. That is, the loss will total between 5 and 27 percent. A 16 percent reduction in ozone will produce a 44 percent increase in ultraviolet radiation reaching Earth's surface, and a 27 percent decrease in ozone will nearly double the ultraviolet. In addition, CFC emissions may increase despite the 1979 ban on CFC sprays in the United States because production outside the United States has been increasing. Furthermore, even if all CFC emissions suddenly stopped, ozone levels would continue to drop for many years, the result of the long lifetime of CFCs in the atmosphere.

Confirmation of the Rowland and Molino hypothesis by direct measurement of ozone depletion in the stratosphere is still far in the future. Because of natural variations in stratospheric ozone concentrations, a 5 percent reduction caused by CFCs must occur before it can be detected reliably. In 1980 the depletion was estimated to be about 2 percent. A 5 percent depletion is not expected until the year 2000. If *all* production of CFCs were halted after such a depletion was observed, the eventual total depletion would still rise to about 7 percent.

According to the academy report, the estimated 16 percent reduction in stratospheric ozone would lead to "very many thousands of additional cases every year" of the nonfatal form of skin cancer and "a likely probability of thousands of new cases every year" of the fatal form, in the United States alone. Furthermore, the 16 percent reduction in ozone "might cause an appreciable reduction in yield for at least a few crops." The effects on sea life could also be severe, especially on crabs, shrimps, anchovies, and many aquatic microorganisms, as these species are already near their tolerance limit for ultraviolet radiation.

The estimate of a 16 percent reduction in ozone does not take the recent methyl chloroform emissions into account. Methyl chloroform now contributes about 50 percent as many chlorine atoms to the stratosphere as contributed by the CFCs. If its production continues to grow, methyl chloroform may become the largest single source of stratospheric chlorine.

Thought Question 6
What should we do?

To help solve this problem, governments throughout the world could ban inessential uses of CFCs, such as aerosol propellants; require the use of the best available technology for reducing emissions; tax the production and use of CFCs; put quotas on production or use of CFCs; require purchasers of refrigerators and air conditioners to make a monetary deposit, refundable when the unit is turned in at a recycling center at the end of its useful life; and require and subsidize collection centers for refrigeration equipment.

Radio, X Ray, Gamma Ray

Turning to the extremes of the electromagnetic spectrum, we will briefly look at the long-wavelength end (radio) and the short-wavelength end (X rays and gamma rays).

Humans can build electrical circuits containing electrons that vibrate as fast as 10^{10} Hz. These oscillations produce radiation with a wavelength of about 1 mm (10^{-3} m) or longer. Electromagnetic waves in this range, whether produced in electronic devices or in natural processes, are called **radio waves**.

The shortest waves in the radio region are about 1 mm long and are called **microwaves.** They have some characteristics in common with their shorter-wavelength neighbors, the infrared waves. For example, they can warm food and eventually cook it in a microwave oven. If we build satellites to capture solar energy (Chapter Eight), microwaves will beam the energy down to Earth.

Because we can produce radio waves in electronic circuits, we can control many of their properties. By varying such properties as frequency and amplitude we can use these waves to transmit information. Thus the detailed information needed to construct a television picture or the sound of a human voice is transmitted via waves in the radio region. Frequency-modulated (FM) radio transmits the information by varying the frequency of the radio wave; amplitude-modulated (AM) radio transmits by varying the amplitude of the wave (Fig. 13.6).

Radio astronomy is based on the radio waves produced by natural processes in the universe. Free electrons (electrons unattached to atoms) moving in magnetic fields produce most radio waves from space. Recall (Chapter Twelve) that any charged particle moving through a magnetic field feels a sideways force. A force that always acts sideways is just what is needed to bend the paths of the moving electrons into circles, just as the sideways (i.e., downward) force by gravity on a horizontally moving staellite is just what is needed to bend the satellite's path into a circle. So individual electrons, moving in the magnetic fields that often exist in space, go in circles. This circular motion produces electromagnetic waves in much the same way that vibrating charged particles produce electromagnetic waves.

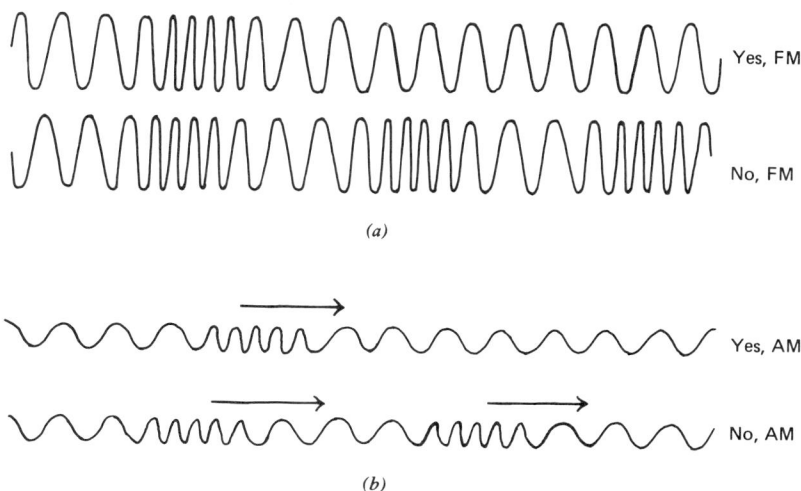

FIGURE 13.6 A possible way to transmit yes or no via (a) frequency modulation, (b) amplitude modulation.

Radio waves from space can be picked up on Earth by large radio receivers called **radio telescopes.** This radio information provides an important supplement to the visual information received by visual telescopes.

Today's astronomers use every portion of the electromagnetic spectrum. Recent years have seen the birth of infrared astronomy, X-ray astronomy, and gamma-ray astronomy. This revolution in the technology of astronomy has occurred only since the development of much of the necessary radio and radar technology during World War II. These advances have produced an explosion in human knowledge about the universe.

At the short-wavelength end of the spectrum, most **X rays** are emitted by individual atoms, as are most ultraviolet and visible rays. Because they are only weakly absorbed by biological matter, these radiations are able to penetrate a human body and expose a specially prepared film on the other side. This is the principle of the X-ray machines used in medicine and in industry.

Most **gamma rays** are emitted by charged particles inside a nucleus. The nuclear force that holds the nucleus together is much stronger than the electromagnetic force that holds the electrons in their orbits, so nuclear energies are much larger than the energies associated with the orbiting electrons, and, therefore, gamma rays are very energetic.

Checklist

electromagnetic radiation	chlorofluorocarbon
electromagnetic spectrum	catalyze
six regions of the spectrum	operation of a refrigerator
thermal vibrations	freon
infrared photography	AM and FM radio
trace molecule	radio astronomy
ozone layer	

Further Thought Questions

7. Rattlesnakes can detect infrared radiation. Would you say that rattlesnakes can see in the dark? (Objects emit infrared radiation even at night, since thermal vibrations never cease.)

8. Intelligent life from other places in the universe might be biologically equipped to use what are to us nonvisible regions of the spectrum. Suppose they use 1-to-2-m radio waves. Would you then say that they *see* our FM radio broadcasts?

Further Exercises

6. List several differences between sound waves and radio waves.

7. Are radio frequencies higher or lower than visible frequencies?

8. About how many times greater is the frequency of a 1000-kilohertz (kHz) AM radio wave than that of a medium-range sound wave?

9. If you charge a comb by rubbing it through your hair, and then shake it up and down, are you producing electromagnetic waves? With what frequency would you have to shake the comb to produce AM radio waves? Would you have to shake faster or slower to produce visible light?

10. Referring to Thought Question 8, would these creatures be able to detect (using their radio receivers) small objects like a speck of dust? Could they detect a penny? Assuming that they are capable of receiving only radio waves and no other type of radiation, would you guess that these creatures are much smaller, much larger, or about the same size as we are? (Recall that waves cannot be used to detect objects that are much smaller than one wavelength.)

11. Electromagnetic waves do not generally penetrate metals. Recalling our discussion of the atomic structure of metals in Chapter Twelve, explain why this is so. *Hint:* Recall that the electrons in a metal are easily moved; what happens to the energy of an electro-magnetic wave that is trying to penetrate a metal?

12. Suppose your eyes could see infrared radiation. What would the room you are in look like? What would be the brightest objects?

13. Why do the heating elements of an electric stove get red when they get hot? Does the light have only one wavelength?

Recommended Reading

You can keep up to date on such science-related issues as atmospheric heating and ozone depletion by reading any of the excellent science news magazines: *Bulletin of the Atomic Scientists* (one of my favorites); *Coevolution Quarterly; Discovery; Environment; Futurist; Mother Earth News; Nature; Not Man Apart; Omni; Science* (another of my favorites); *Science 80; Science News; Technology Review.* For example, you'll find a report on the 1979 study of atmospheric heating in *Science,* November 23, 1979. You will find reports on the 1979 study of the ozone problem in *Science,* December 7, 1979, and January 25, 1980.

A Life No scientist has made such a mark on our century as Einstein. He was certainly the greatest physicist of his generation. Perhaps more important, he was—and is—universally admired as a symbol of wisdom and humanity. The following pages recount Einstein's life in pictures.

(Left) Einstein's mother Pauline Einstein, née Koch, and father Hermann Einstein. Their son Albert was born on March 14, 1878.

(Right) Earliest known photograph of Einstein.

Albert with his sister Maja, ages about three and five *(left)* and at age fourteen *(above)*.

(Below) Einstein during his student days in Zurich.

(Above, left) The graduating class of the Swiss high school at Aarau, 1896. Einstein, age nineteen, is seated at the left.

(Left) A famiy portrait with Einstein's wife Mileva and son Hans Albert, Bern, Switzerland, 1904.

(Right) The young patent examiner in his office in Bern, Switzerland, in 1905. His job classification was "technical expert, third class." In his spare time, he was working on the papers that would revolutionize physics. This photo was taken at about the time of the publication of his papers setting forth the special theory of relativity.

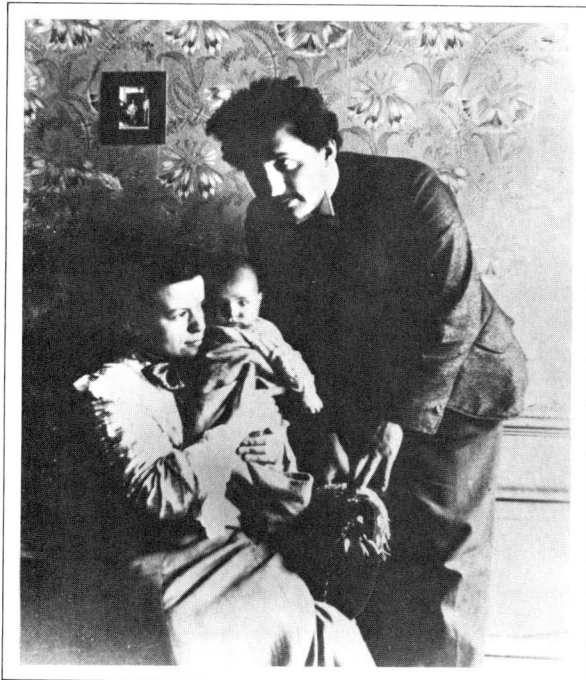

(Left) Einstein in 1916, two years after he moved to Berlin and just after he had published his paper on the general theory of relativity.

(Below) News clipping, 1934, shortly after Einstein immigrated to the United States. His doubts about atomic energy were to be proved wrong in 1942 when the world's first nuclear reactor began producing energy.

The quantitative aspects of Einstein's work are expressed as mathematical equations. The photos show a lecture in 1931 (above), and the last page of calculations made by Einstein (right).

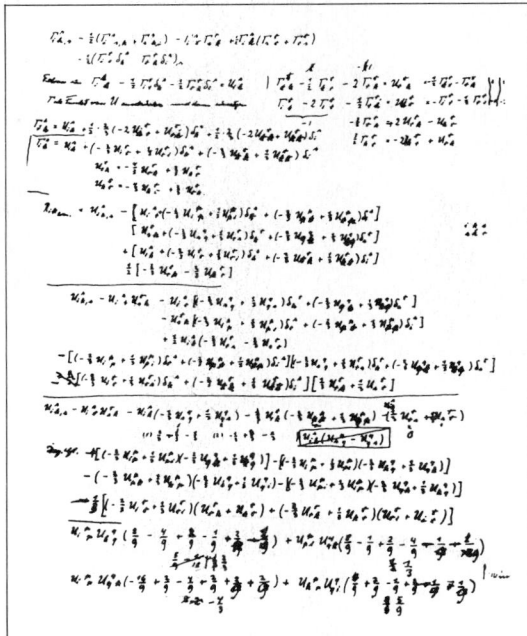

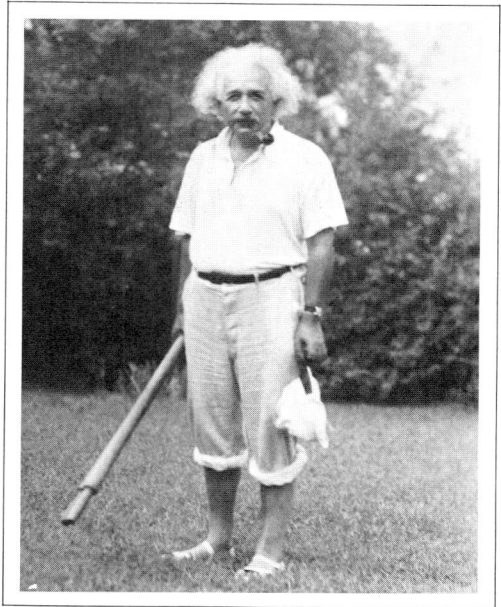

Einstein was a man of varied interests. Here, he is shown playing the violin in 1927, bicycling in 1933, sailing on Saranac Lake in New York in 1936, and at his summer home at Huntington, New Jersey, in 1936.

Albert Einstein
Old Grove Rd.
Nassau Point
Peconic, Long Island

August 2nd, 1939

F.D. Roosevelt,
President of the United States,
White House
Washington, D.C.

Sir:

Some recent work by E.Fermi and L. Szilard, which has been communicated to me in manuscript, leads me to expect that the element uranium may be turned into a new and important source of energy in the immediate future. Certain aspects of the situation which has arisen seem to call for watchfulness and, if necessary, quick action on the part of the Administration. I believe therefore that it is my duty to bring to your attention the following facts and recommendations:

In the course of the last four months it has been made probable - through the work of Joliot in France as well as Fermi and Szilard in America - that it may become possible to set up a nuclear chain reaction in a large mass of uranium,by which vast amounts of power and large quantities of new radium-like elements would be generated. Now it appears almost certain that this could be achieved in the immediate future.

This new phenomenon would also lead to the construction of bombs, and it is conceivable - though much less certain - that extremely powerful bombs of a new type may thus be constructed. A single bomb of this type, carried by boat and exploded in a port, might very well destroy the whole port together with some of the surrounding territory. However, such bombs might very well prove to be too heavy for transportation by air.

-2-

The United States has only very poor ores of uranium in moderate quantities. There is some good ore in Canada and the former Czechoslovakia, while the most important source of uranium is Belgian Congo.

In view of this situation you may think it desirable to have some permanent contact maintained between the Administration and the group of physicists working on chain reactions in America. One possible way of achieving this might be for you to entrust with this task a person who has your confidence and who could perhaps serve in an inofficial capacity. His task might comprise the following:

a) to approach Government Departments, keep them informed of the further development, and put forward recommendations for Government action, giving particular attention to the problem of securing a supply of uranium ore for the United States;

b) to speed up the experimental work,which is at present being carried on within the limits of the budgets of University laboratories, by providing funds, if such funds be required, through his contacts with private persons who are willing to make contributions for this cause, and perhaps also by obtaining the co-operation of industrial laboratories which have the necessary equipment.

I understand that Germany has actually stopped the sale of uranium from the Czechoslovakian mines which she has taken over. That she should have taken such early action might perhaps be understood on the ground that the son of the German Under-Secretary of State, von Weizsäcker, is attached to the Kaiser-Wilhelm-Institut in Berlin where some of the American work on uranium is now being repeated.

Yours very truly,

A. Einstein
(Albert Einstein)

(Above) The possibility that Hitler's Germany might acquire atomic bombs finally persuaded Einstein, an outspoken pacifist, to send this letter to President Roosevelt.

Photograph taken during the 1940's.

Never one to take himself too seriously, Einstein stuck out his tongue when asked to "smile" on his seventy-second birthday, in 1951.

(Right) Einstein in his study.

Albert Einstein died on April 18, 1955.

14

THE RELATIVISTIC CENTURY

The electromagnetic field and the electromagnetic wave theory of light took us beyond the Newtonian view that the universe consists of material particles whose machinelike motions are responsible for all natural phenomena. The ***theory of relativity,*** one of the two great theories of twentieth-century physics, takes us further from the Newtonian framework. The new theory modifies the Newtonian view of such basic concepts as time, space, velocity, mass, force, and energy. Even more fundamentally, a certain subjectivism creeps into the new theory, in the sense that the theory emphasizes the importance of the observer. This emphasis on the observer will become a central feature, perhaps *the* central feature, of the ***quantum theory,*** the other great theory of twentieth-century physics (Part 4).

The Education of a Genius

Albert Einstein got fed up with high school and dropped out in his mid-teens. Nobody was especially surprised by this move, least of all his teachers, for the boy had been a daydreamer and a mediocre student since he began elementary school. Before that he had been a slow child, learning to speak at a much later age than average.

His teachers were glad to see him go. One of them informed Einstein that he would "never amount to anything," and another suggested that he leave school because his presence in class destroyed the discipline of the students. Einstein was most happy to accept the suggestion. He left Munich and spent the next few months hiking and loafing around the Alpine regions of northern Italy. He was a model dropout.

His father's business was not going well and it became clear that young Einstein would soon have to support himself, so he decided to study engineering at the Swiss Federal Polytechnic University in Zurich. Unfortunately, he failed his entrance exams. It seems he had problems with biology and French.

To prepare for another try, he spent a year at a Swiss high school in Aarau. He flourished at this congenial, democratically run school, and he recalled later that it was here that he had his first ideas leading to the relativity theory.

The university admitted Einstein on the basis of his diploma from Aarau. At the university, Einstein was known as a charming but indifferent student who attended cafes regularly and lectures sporadically. He managed to pass the necessary exams and eventually to graduate with the help of his close friend Marcel Grossmann, who shared his systematic notes with his nonconforming comrade.

College was an unpleasant experience for Einstein. As he put it:

It had such a deterring effect upon me that, after I had passed the final examination, I found the consideration of any scientific problems distasteful to me for an entire year. . . . It is little short of a miracle that modern methods of instruction have not already completely strangled the holy curiosity of inquiry, because what this delicate little plant needs most, apart from initial stimulation, is freedom; without that it is surely destroyed. . . . I believe that one could even deprive a healthy beast of prey of its voraciousness, if one could force it with a whip to eat continuously whether it were hungry or not. (Quoted in "Einstein and some Civilized Discontents," by Martin J. Klein, in *Physics Today*, January 1965. This section and the next are based on Professor Klein's article.)

After his graduation in 1900 Einstein applied for an assistantship to do graduate study. It went to someone else. He looked for a teaching position in Switzerland, but couldn't find anything. Finally, in 1902, Einstein's friend Grossman came to the rescue by helping him land a job as a patent examiner with the Swiss Patent Office at Bern. Einstein held this job for 7 years and often referred to it as "a kind of salvation." Its primary virtues were that it paid the rent and it occupied only 8 hours of the day. His job left him time to ponder nature's riddles.

And ponder he did. During the twentieth century three significant physical theories have been introduced: *the special theory of relativity,* the **general theory of relativity,** and the **quantum theory.** Albert Einstein invented the first two almost single-handedly and played an important role in the development of the third.

It is appropriate to close this sketch of Einstein's education with a remark he made near the end of his life. "If I were a young man again and had to decide how to make a living, I would not try to become a scientist or scholar or teacher. I would rather choose to be a plumber or a peddler, in the hope of finding that modest degree of independence still available under present circumstances." (Quoted in the *Physics Today* article cited above.)

Thought Question 1
What is your response to Einstein's criticism of education? Are these criticisms valid today at the grade school, high school, or college level?

Thought Question 2
Will genius such as Einstein's always eventually come to the surface and be recognized in American society, or does it sometimes remain undeveloped or unnoticed?

The Scientist as Citizen

Before turning to the theory of relativity, let's follow one other important theme in Einstein's life, the theme of the scientist as social agent.

During most of his life, the irrepressible Einstein was active in one public cause or another. He was especially influential in three undertakings. The first was the international Zionist movement that led eventually to the establishment of the state of Israel. Einstein spent the years 1913–1933 as research director at the Kaiser Wilhelm Institute in Berlin. Being Jewish, he was quite aware of the anti-Semitism around him, an awareness that encouraged his support of the Zionist movement. In 1933, Hitler came to power and the racial purges began in Germany. Einstein decided that it was impossible for him to continue his work in Berlin (indeed, the Nazis had already confiscated his property), so he accepted an offer to join the Institute for

Advanced Study at Princeton, New Jersey, where he remained until his death in 1955.

Einstein's second cause was peace. In 1914, shortly after the outbreak of World War I, a number of Germany's prominent intellectuals published a militaristic and chauvinistic manifesto, proclaiming its signers' full support of Germany's war effort. Einstein responded to this document in characteristic style. He was among the few German citizens willing to sign a sharply worded rebuttal calling for an end to war and the establishment of a world organization for international cooperation. During the years after World War II he labored to establish a world government that would abolish war once and for all. He was one of the founders of the International Pugwash movement of scientists, on behalf of international arms control. He tried to impress upon the world the fact that the next war would have no winners, that the next war could destroy not only civilization, but humanity as well.

Einstein's third major influence on public policy was the letter that he wrote to President Roosevelt in 1939 in which he advocated a U.S. nuclear bomb project. By 1939 it was clear that Germany had embarked on a war to conquer the world. Add to this the facts that Germany had stopped the sale of uranium from the Czechoslovakian mines in its control and that many of the world's best scientists were German. The danger was obvious: Hitler's Germany might be building a nuclear bomb. With such a bomb Hitler truly could conquer the world. Prompted by this fear, Einstein urged that the United States develop the bomb without delay, in order to pre-empt the German bomb. As it turned out, Germany did have an atomic bomb project under way, but one that never got close to its goal of a weapon of mass destruction. The U.S. project, on the other hand, was successful, only not in time for use against the enemy whom Einstein had feared.

Einstein's entries into the public stage were controversial. He was criticized for his pacifist role on the grounds that he was a scientist, rather than a specialist in international affairs, and that he was using his prestige to promote causes that he didn't really understand. He was criticized for his role in the atomic bomb project on the grounds that a self-proclaimed man of peace should not be promoting weapons development.

Thought Question 3
What are your views on these two criticisms? Should scientists be involved in international affairs? Was it inconsistent for Einstein to promote the atomic bomb project?

Einstein's role in public affairs is typical of a trend among twentieth-century scientists. Throughout the century, science and technology have assumed an increasing social importance. Many individual scientists and technologists today feel that they are obliged to accept some responsibility for the inevitable social impact of science. So scientists have become involved in social, economic, political, and moral questions relating to such issues as weapons development, energy resources, and preservation of the environment. Albert Einstein was one of the early examples of such "public scientists."

Thought Question 4
Should these public scientists be involved in nonscientific matters or should they stick

to pure science? Should the public treat a scientist's statements in nonscientific matters with any special regard?

217
The Relativistic
Century

Relativity Before Einstein

Turning now to the theory of relativity itself, I would like to introduce Velma and Mortimer, the relativistic duo. They will help clarify some aspects of the theory of relativity. Generally, they will be in relative motion, that is, one will move **relative to** the other, or, in other words, the distance between them will be changing.

Exercise 1
Velma and Mortimer are both driving northward at 55 mph. Are they in relative motion? (Answer below.)

The answer is no. They are in motion relative to the ground, but they are not in motion relative to each other.

A **theory of relativity** is any theory that correlates the observations of two observers who are in relative motion. For example, if Velma reports that she sees a hawk flying east at 5 m/sec (as measured by Velma), a theory of relativity should be able to tell us what Mort sees.

Einstein was certainly not the first person to think about relative motion. Scientists have dealt with relative motion at least since the time of Galileo. Pre-Einstein ideas about relative motion are called the **Galilean theory of relativity.** Galilean relativity seems reasonable and straightforward and was generally accepted until it was revised by Einstein in 1905. To understand Einstein's revision, called the special theory of relativity, we should first understand Galilean relativity.

Suppose Mort is standing beside a railroad track and Velma is in a train headed north past Mort at 20 m/sec* (45 mph). From Mort's point of view, or *relative to Mort,* Velma moves north at 20 m/sec. But there are other points of view besides Mort's. Velma's, for instance. After all, Copernicus long ago destroyed the idea that all motion must be measured relative to Earth, as though our planet were the one object in the universe that was truly at rest. Motion can be measured relative to any reference: the sun, the moon, Velma, and so on. Relative to Velma, it is Mort who is moving, and he moves south at 20 m/sec. In other words, measurements made by Velma, who is in the train, would show Mort to be 20 m farther to the south of Velma every second (Fig. 14.1).

Now suppose that Velma throws a rock toward the front of the train. Her throwing speed is 5 m/sec. That is, when she throws a rock it moves away from her at 5 m/sec, whether she is standing on the ground, inside a train, or anywhere else. Relative to Velma, the rock must move north at 5 m/sec. Thus measurements made by Velma would show the rock to be 5 m farther to the north of Velma every second. Mort looks into the passing train through a window. What does Mort see?

Have you thought about it? The answer is that Mort sees the rock move north at 25 m/sec, because the train is already moving north at 20 m/sec and the rock is moving 5 m/sec faster than the train (Fig. 14.2).

*We will often use m/sec, rather than the more familiar km/hr or mph, for speeds.

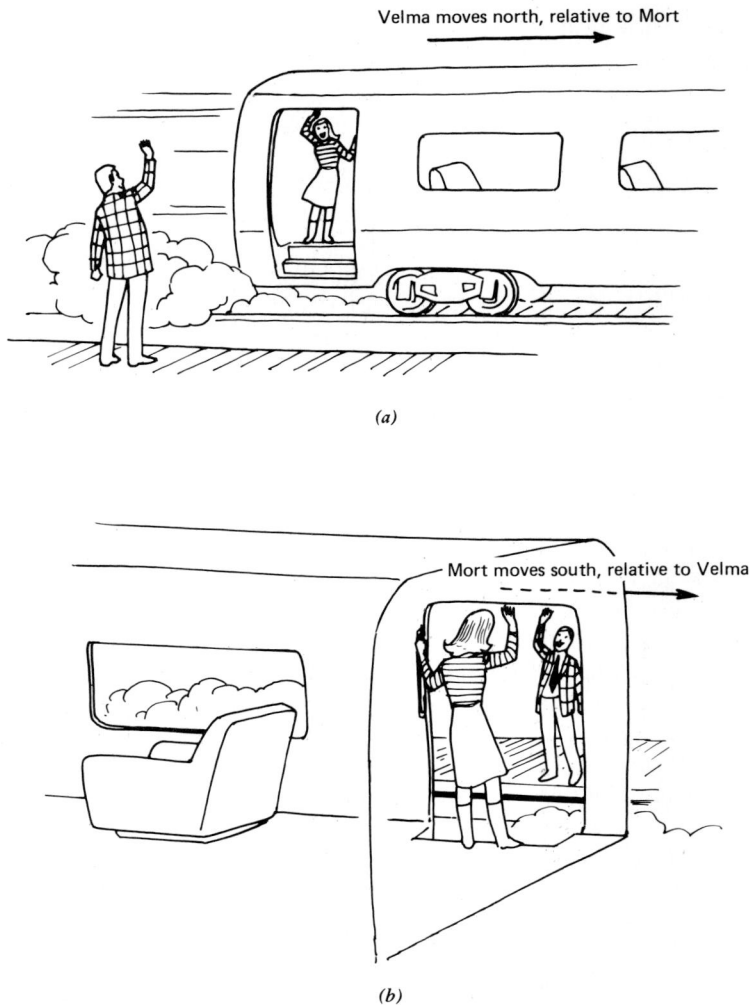

FIGURE 14.1 (a–b) Velma moving north in a train, from two points
of view.

Let's try some more questions along this line.

Exercise 2
Velma throws a rock south (toward the rear of the train) at 5 m/sec relative to Velma.
What does Mort see?

Exercise 3
Mort throws a rock north at 3 m/sec. (a) What does Velma see? (b) What will Velma
see if Mort throws a rock south at 3 m/sec?

Let's break Exercise 2 down to its essentials. Nothing in the examples depends
on Velma being inside the train. She could just as well be riding on an open flatcar,

provided that we neglect wind and air resistance. In fact, nothing in the examples depends on Mort and Velma being on Earth. They could just as well be on the moon. Or they could just as well be in outer space, in two spaceships. The essentials are: Velma moves at 20 m/sec *relative to Mort*, and Velma throws a rock backward at 5 m/sec *relative to Velma*. Mort will see the rock move in the same direction as Velma at 15 m/sec. Measurements made by Mort are said to be made in Mort's **reference frame,** and measurements made by Velma are made in Velma's reference frame. So, in Velma's reference frame the rock thrown by Velma moves backward at 5 m/sec, whereas in Mort's reference frame the rock thrown by Velma moves forward at 15 m/sec.

Let's look at a different sort of example, one that involves the motion of a wave, such as a sound wave or a water wave. Both types of wave are disturbances that travel through a certain medium, air or water (Chapter Twelve). Without air, there would be no sound, and it would be hard to imagine water waves without water. These waves have a definite, fixed speed relative to the medium through which they are moving. For instance, the speed of sound is 340 m/sec (750 mph) relative to the air through which the sound wave is passing.

Try these:

Exercise 4
Velma is in an *enclosed* passenger car in a train headed north at 20 m/sec. She stands in the rear of the car and shouts toward the front of the car. (a) How fast does the sound wave move relative to Velma? (*Note:* The air in the car is moving along with the car). (b) How fast does the sound move relative to Mort (who is still standing beside the tracks)?

Exercise 5
Mort stands at the south end of the railroad platform and shouts toward the north end. Assume that no wind is blowing. (a) How fast does the sound wave move relative to Mort? (b) How fast does it move relative to Velma?

FIGURE 14.2 Velma throwing a rock, from Mort's point of view.

Our analysis of the problems in this section would have seemed reasonable to Gaileo and seems reasonable to most people. Moreover this analysis is borne out by experiment: measurements would agree with the predictions of Exercises 2 through 5.

Historical Background for Einstein's Relativity

Physicists have always had difficulty in understanding the nature of light. One problem involved an experiment conducted in 1887 by Albert Michelson and Edward Morley. In order to understand Einstein's theory of relativity, it is helpful to begin by reviewing the nineteenth-century description of light and by studying the Michelson-Morley experiment. It is worth noting, however, that while he was developing his theory, Einstein himself was not especially concerned with these debates about the nature of light or with the Michelson-Morley experiment.

It has been known since the seventeenth century that light moves through vacuum* at about 300,000 km/sec. That's 186,000 miles per second. In every second, a light beam moves a distance equal to eight times around Earth!

Historically, there have been two points of view about the nature of light. According to one point of view, light is a stream of tiny particles shot out from the light source; according to the other, light is a wave produced by the light source. By the late nineteenth century both experiment and theory indicated that light is a wave rather than a stream of particles, and all scientists accepted this point of view. Chapter Twelve presented this **electromagnetic wave theory of light.**

But if light is a wave, then what is it a wave *in*? It is difficult to imagine a wave that isn't a wave in some medium. Can you imagine a water wave without the water? It boggles the mind to think of a wave without a medium . . . like the smile on Lewis Carroll's Cheshire cat without the cat. Scientists thought that there had to be a substance to "support" light waves, just as water supports water waves and air supports sound waves. Because light travels to Earth from the sun and from distant stars, this medium had to fill all space, even those regions far from Earth that have no ordinary matter (no air, no atoms). Thus this substance which supports light waves could be no ordinary material made of atoms. Furthermore, this substance would have to be able to vibrate very fast, for the vibrations associated with light waves were known to be extremely rapid. And it would have to be very dilute (i.e., low-density, or *tenuous*) because the planets and the moons and so forth had to continually move through this material, and these bodies would surely be slowed down appreciably and eventually stopped if this substance were not very tenuous.

Thus was born the idea of the light-bearing **ether****: a tenuous medium that occupied all space and whose vibrations were light waves.

Light is one of the more fundamental phenomena in nature. By the late nineteenth century it was known that light is a direct consequence of the electromagnetic force, one of nature's fundamental forces. Because the ether filled up all space and supported light waves, it was thought of as the natural reference frame for physical phenomena. According to this view, objects moving through the ether were truly in motion; an object was truly at rest only if it was not moving relative to the ether.

*Light moves a little slower through air and considerably slower through solids, such as glass.
**No relation to the gas used as an anasthetic in medicine.

The ether was difficult to detect directly. By the mid-1880s nobody had succeeded in measuring any of its properties or even in directly observing it. In 1887, Michelson and Morley thought of a way to observe the ether and a way to measure the speed of Earth as it moved through the ether. In other words, they proposed to measure the absolute motion of Earth, the speed of Earth relative to the ether.

Michelson and Morley were good at experimenting with light. In 1880, Michelson made the best measurement of the speed of light that had ever been made, a feat that earned him America's first Nobel Prize in physics. The 1887 experiment of Michelson and Morley was based on the idea that since light was a wave in the ether, light would necessarily have a definite, fixed speed relative to the ether. Every observer moving through the ether would find a slightly different speed for light depending on the observer's motion through the ether. Any wave in any medium behaves this way. A sound wave, for instance, has the fixed speed of 340 m/sec relative to the air through which it is moving; an observer moving through the air will find this sound wave to have a different speed (see Exercises 4 and 5 above).

Now, let's return to Velma.

Exercise 6

Velma is riding on a train in the open, on top of a flatcar headed north. She would like to find out how fast she is moving. She has with her an accurate device for measuring the speed of sound waves, and she is aware that sound moves at 340 m/sec relative to the air. She measures the speed of a sound wave as it moves past in her direction of motion (north). Will the outcome of the measurement be greater than, less than, or equal to 340 m/sec (Fig. 14.3)?

Exercise 7

Continuing the preceding exercise, suppose Velma finds that the sound wave passes her at 310 m/sec, that is, the speed of the sound wave relative to Velma is 310 m/sec. How fast is the train moving relative to the air?

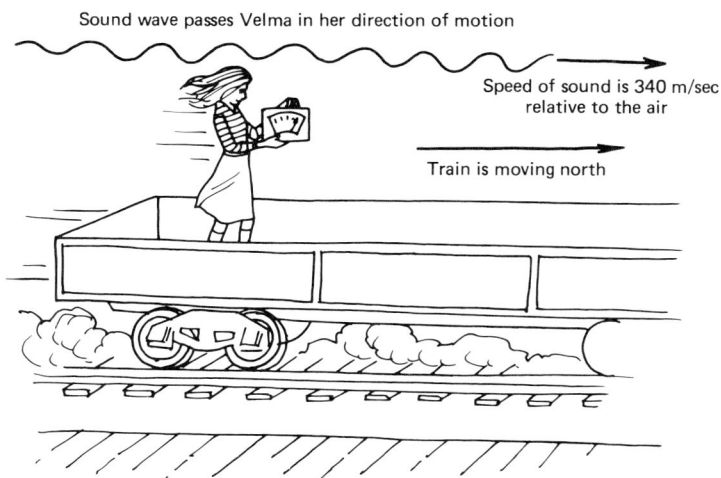

Sound wave passes Velma in her direction of motion

Speed of sound is 340 m/sec
relative to the air

Train is moving north

FIGURE 14.3 How fast is Velma moving relative to the air?

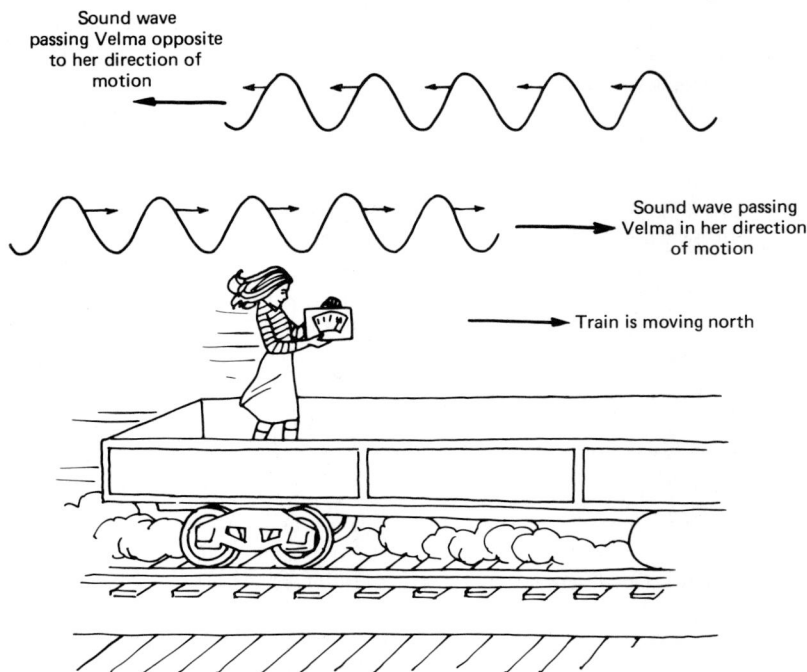

FIGURE 14.4 One way to find your velocity relative to the air.

Exercise 8

Now suppose that Velma doesn't know the precise speed of sound relative to the air. (The speed of sound happens to be somewhat different on different days, depending on the air temperature). Can you suggest a way in which she can determine her own speed relative to the air by measuring the speeds of sound waves? Keep in mind that Velma is out in the open, on a flatcar moving through the air, so it is impossible for her to directly measure the speed of sound in still air. (Answer below, in Exercise 9.)

Exercise 9

Continuing the preceding exercise, Velma hits on the idea of measuring the speed of sound waves passing her in two directions (Fig. 14.4). She finds that sound waves moving north pass her at 310 m/sec, and sound waves moving south pass her at 360 m/sec. Find both the speed of sound relative to the air and the speed of Velma relative to the air.

Michelson and Morley attempted to determine Earth's speed through the ether by measuring the speeds of light waves, much as Velma (Exercise 9) determines the train's speed through the air by measuring the speeds of sound waves. Light waves were supposed to move at some fixed speed relative to the ether. Because Earth was thought to move through the ether, a light beam moving in one direction past Earth would have to move at a different speed (relative to Earth) than a light beam moving in a different direction past Earth, just as sound waves pass Velma at different speeds in Exercise 9. Michelson and Morley had perfected a device, called an interferometer, that could accurately record differences in the speeds of light beams moving in different directions. If light beams passing Earth in different directions moved at different speeds relative to Earth, the interferometer would surely detect the difference.

Michelson and Morley performed their experiment in 1887. The result: they could detect no difference in the speeds of light beams in different directions! They repeated their experiment at various times of the day and at various times of the year, when Earth moved in different directions with respect to the stars. In every case, no effect of Earth's motion on the speed of light beams could be detected, despite the fact that their interferometer should have been able to detect any such difference. More recent measurements, with increasingly precise measuring instruments, have yielded the same result.

How can light have the same speed relative to all observers, regardless of the motion of the observer? This question was perplexing in 1887, and in an important sense it is still perplexing. Naturally, scientists made valiant attempts to explain the Michelson-Morley result in terms that would conform to the ether concept and Galilean relativity. All these attempts were eventually discarded, usually because they were disproved by experiment.

The attempt to observe the ether by measuring Earth's speed relative to the ether ended in failure. It seemed that nature was conspiring to prevent us from detecting the ether. What was the answer?

Einstein's Solution

Enter an unpromising graduate of the Polytechnic University in Zurich, now an obscure clerk at the Patent Office in Bern. As I mentioned earlier, Einstein wasn't initially concerned about the Michelson-Morley experiment. He was interested in more purely theoretical matters, particularly the logical consistency of the theory of electricity and magnetism. From this interest he developed, in 1905, a new theory that not only laid to rest the debates over the ether, but revolutionized our understanding of two of the most basic concepts in physics: space and time.

Einstein's idea was simple but radical. He dispensed with the ether idea. This might sound like a simple idea, but it is strange. Einstein's proposal is that light waves, unlike every other known wave phenomenon, are not waves *in* anything. They are waves without a medium, like the smile on the Cheshire cat without the cat.

Furthermore, Einstein proposed that we simply accept the seemingly paradoxical result of the Michelson-Morley experiment. That is, he proposed that every observer, no matter how he or she is moving, will see every light wave (in vacuum) pass by at precisely the same speed relative to the observer. At first glance, it is a ridiculous idea. Ridiculous, but apparently true, because it agrees with experiment.

Einstein used these ideas as the starting point for his ***special theory of relativity***. He summarized these ideas in the form of two fundamental principles, which I'll state here and discuss later:

The Principle of Relativity. The laws of physics are the same in all nonaccelerating reference frames.

The Constancy of Lightspeed.* The speed of light in empty space is the same for all nonaccelerated observers, regardless of the motion of the light source or the motion of the observer.

---

*We'll call this particular speed, 300,000 km/sec (more precisely, 298,000 km/sec), "lightspeed."

The first principle says, as we will see below, that there is no preferred reference frame, no ether.

You'll notice that these principles apply only to nonaccelerated reference frames and observers, that is, those which are not speeding up, slowing down, or turning a corner relative to such "fixed" objects as the stars. The special theory of relativity applies only to such nonaccelerated reference frames and observers. That is, in fact, why it is called the special (or restricted) theory of relativity. Earth's motion is slightly accelerated, because a point fixed on Earth is spinning around Earth's axis and rotating around the sun. But the speeds involved in these motions are extremely slow compared with lightspeed, so Earth's motion is nearly nonaccelerated. As a reasonable approximation, we'll neglect Earth's acceleration so that we can apply Einstein's theory to Earth-based observers.

Some 10 years after inventing the special theory of relativity, Einstein developed a more general theory that applied even to accelerated reference frames and observers. This theory is known, appropriately enough, as the **general theory of relativity**. For now we'll look only at the special theory.

The Principle of Relativity

The idea of Einstein's first principle is that any physical experiment carried out entirely within a self-contained laboratory will come out the same, regardless of the motion of the laboratory (as long as it is not accelerating). For instance, if two magnets attract each other with a certain force in a laboratory fixed in Des Moines, Iowa, then the same two magnets should attract each other with precisely the same force in a laboratory in a rocket ship moving past Earth at half the speed of light, provided only that the experiment doesn't depend on anything outside the lab.

If you've ever done any simple physics experiments in an airplane, train, or car moving at constant velocity, you have experienced this principle. While riding in an automobile, if you hold one hand directly above the other and drop a penny from the upper hand, the penny falls straight down (as seen by you in your "moving" reference frame) into your lower hand, just as it would if you were standing still on Earth.

Exercise 10
Where would the penny have landed if the car was speeding up? If it was slowing down? If it was turning a corner to the right? Note that your reference frame is accelerated in each of these three cases, so Einstein's principle isn't expected to apply. Try these experiments the next time you are a passenger in a car, train, or airplane.

The principle of relativity simply says that "everything is normal" in a nonaccelerated reference frame. For example, a flight attendant has no difficulty serving coffee on a smoothly moving (i.e., nonaccelerated) jet airplane. The coffee behaves just as it would in a restaurant on the ground. Everything is normal inside the airplane. The only way to tell that you are moving is by looking out a window or asking the pilot.

The principle of relativity leads us to dispense with the ether concept. To see why, let's return to our friend Mort, who is now in an isolated laboratory. *Isolated* means that Mort cannot make contact with anything outside the lab. Suppose Mort wants to know how fast he is moving with respect to Earth. He can't simply open a window and look out to see how fast he is moving past Earth, for he would not then be isolated. He can't telephone or communicate in other ways with anything outside the laboratory. As long as he remains isolated, Mort can only carry out experiments within his own lab. But the principle of relativity tells us that every experiment he performs inside the lab must have an outcome that is independent of his speed past Earth. So, there is no way for Mort to determine his own speed.

In addition, Mort cannot detect the ether, for if Mort could detect the ether he could measure his own speed with respect to the ether. Using this information, he could determine his speed relative to any other object (such as Earth) whose speed through the ether was known. So the principle of relativity implies that any universal reference frame, such as the ether, is undetectable. In science, there is no point in talking about something that is undetectable. Thus so far as science is concerned, the ether is nonexistent.

Before Einstein, it was believed that objects at rest relative to the ether were really at rest and that objects moving relative to the ether were really moving. Einstein dispenses with this universal reference frame, so that there is no longer any sense in which one observer is really moving and another is really at rest. All we can say is that one observer is moving *relative to* the other. The question of which reference frame is at rest is entirely arbitrary. It is conventional to say that objects are at rest when they are fixed relative to the ground, but this is mere convention. According to Einstein, no reference frame is any more at rest than any other.

The Constancy of Lightspeed

Your experience riding in moving vehicles may make it easy to reconcile the first principle with your intuition. The second principle is harder to swallow. According to the second principle, it doesn't matter how fast you are moving when you see a light beam or how fast the source of light is moving, the beam's speed relative to you will always be 300,000 km/sec. To see how shocking this idea is, let's consider a few examples.

Let's present Mort with a laser. For our purposes, a laser is simply a source of a narrow and bright beam of light. If Velma and Mort are both standing on Earth and Mort shines a laser beam past Velma, she will naturally find the tip of the beam moving past her at 300,000 km/sec (Fig. 14.5a). Nothing unusual about that.

Now let's put Velma in a spaceship moving at 200,000 km/sec (two-thirds of lightspeed!) relative to Earth. Velma is moving toward Mort, who is standing on Earth. He shines his laser beam toward Velma's approaching spaceship. How fast does the beam pass Velma, as measured relative to Velma? (Fig. 14.5b).

It seems obvious that Velma sees the beam pass her at 500,000 km/sec. After all, the beam is coming at her at 300,000 km/sec, and she is moving toward the approaching beam at 200,000 km/sec. This answer is given by Galilean relativity and seems reasonable to most people's intuitions. According to Einstein's theory, however, this answer is wrong. Einstein's principle of the constancy of lightspeed says that every nonaccelerated observer sees every light beam move at 300,000 km/sec.

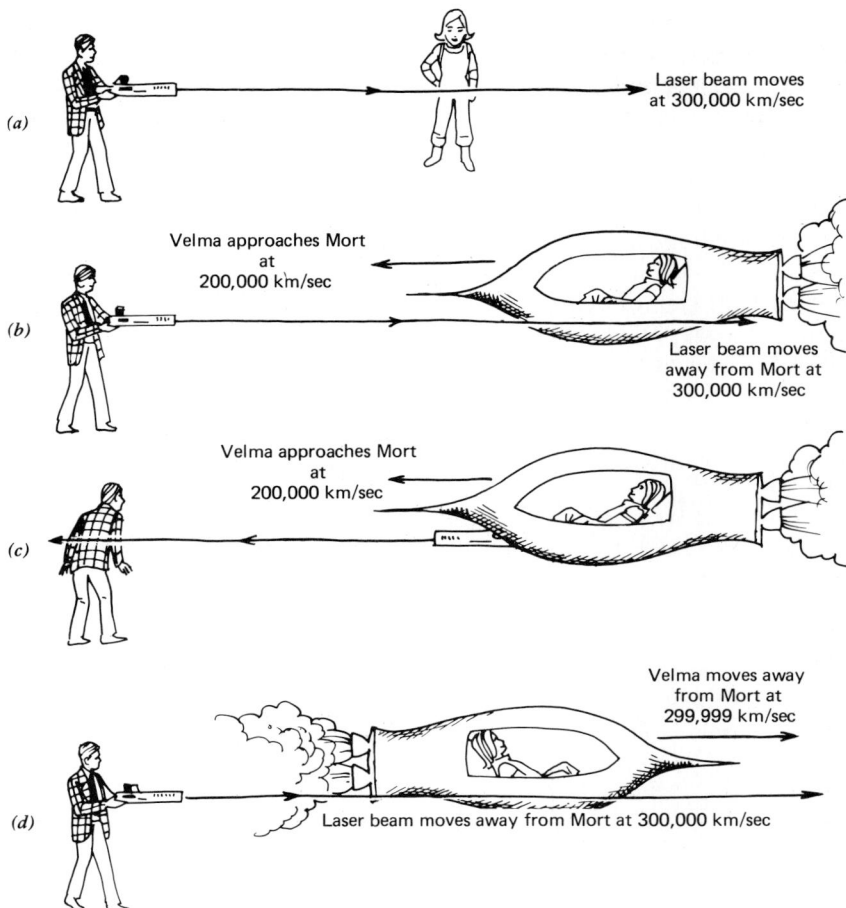

(a)

(b)

(c)

(d)

FIGURE 14.5 a–d Velma, Mort, and a laser beam.

According to this principle, Mort should see the beam move away from him at 300,000 km/sec, and Velma should see the beam move past her at 300,000 km/sec.

That seems shocking to me.

There are many variations on this example. We can give the laser to Velma in the spaceship and let her shine the beam toward Mort (Fig. 14.5c). Despite what Galilean relativity and our intuitions might tell us, Velma sees the beam leave her spaceship at 300,000 km/sec and Mort sees the beam pass him at 300,000 km/sec.

We can give the laser back to Mort and let Velma move away from him. (Fig. 14.5d) at nearly lightspeed, at perhaps 299,999 km/sec. According to Mort, Velma is moving only 1 km/sec slower than a light beam. If Mort shines his laser beam toward Velma's rapidly receding spaceship, he sees the beam leave his laser with a speed of 300,000 km/sec. Galilean relativity and our intuitions tell us that Velma sees the beam creep past her at only 1 km/sec because she is already moving at nearly the speed of light, but Einstein's relativity says that Velma sees the beam pass her at 300,000 km/sec! From her point of view, she is not even coming close to keeping pace with the light beam.

Strange as these ideas may seem, they are verified experimentally. The nonintuitive flavor of Einstein's theory is typical of twentieth-century physics. Although such theories as relativity and quantum mechanics correctly predict the results of experiments, we may feel that these theories in some sense fail us: We have difficulty understanding them in the intuitive way in which we hope to understand nature; we cannot picture them. The easy-to-visualize mechanical universe of Newtonian physics has vanished in our century.

Checklist

Einstein's three public causes	the ether
relative motion	principle of relativity
a theory of relativity	principle of the constancy of lightspeed
Galilean relativity	special theory of relativity
reference frame	nonaccelerated observer
lightspeed	general theory of relativity
Michelson-Morley experiment	

Further Thought Questions

5. Does it seem to you that a thing (e.g., the ether) is nonexistent simply because it is scientifically undetectable?

6. As explained in Chapter Three, we cannot see atoms even with the aid of light-based microscopes. Does this mean that, so far as science is concerned, atoms do not exist? Are there any means by which we can detect atoms? Do atoms have any observable effects?

7. What is the point of a scientific theory if it doesn't explain things in an intuitively meaningful, understandable way? What does it mean to explain a thing scientifically?

Further Exercises

11. A World War II fighter plane moving north at 400 mph fires a machine gun having a muzzle velocity of 1200 mph (in other words, the bullets move at 1200 mph relative to the plane). (a) Assuming that the gun is fired forward, find the speed and direction of the bullets relative to the ground. (b) Repeat part (a) assuming that the gun is fired backward. (*Note:* Use Galilean relativity. Einstein's theory gives the same answers at these speeds which are much lower than lightspeed.)

12. Suppose the airplane in the preceding exercise was a jet airplane moving at 1200 mph, mounted with the same machine gun (muzzle velocity of 1200 mph). Rework Exercise 11. Describe the subsequent motion of the bullets after firing in case (b).

13. Although everything is normal in a smoothly moving jet airplane, things are not normal when the ride is bumpy—you spill your coffee and so on. Does this situation conflict with Einstein's principle of relativity? Explain.

•

Recommended Reading

1. B. Hoffman, *Albert Einstein, Creator and Rebel,* The Viking Press, New York, 1972. A good biography that emphasizes Einstein's scientific work.

2. P. A. Schilpp, editor, *Albert Einstein, Philosopher-Scientist,* Library of Living Philosophers, New York, 1949. See especially the "Autobiographical Notes" by Einstein.

3. Louise B. Young, editor, *The Mystery of Matter,* Oxford University Press, New York, 1965. Part 4, "What is the Secret of Atomic Energy?" contains Einstein's letter to President Roosevelt and Einstein's reflections on the bomb several years after Hiroshima.

4. *Einstein.* A 42-minute film, marketed by Time-Life Multimedia. The story of Einstein as told by his friends and colleagues. A beautiful and moving film.

15

TIME AND SPACE ACCORDING TO EINSTEIN

Einstein's two principles are the starting point for the special theory of relativity, just as Newton's three laws are the starting point for Newtonian physics. As in all scientific theories, it is impossible to be certain that these basic principles are true. Nevertheless, these two principles lead to many specific predictions about nature, which can be experimentally checked.

In this chapter we'll study the more important predictions of the theory. They may be hard to believe, but they seem to be true. They have been experimentally tested under many different circumstances. All the tests have been positive. Today all physicists accept the validity of the special theory of relativity.

Time Is Relative

The most fundamental and the most fascinating of the predictions of special relativity concerns time. Time is such an all-pervasive concept that it is difficult to think about. What do we mean by *time?* Let's be more specific: Suppose Velma (remember Velma from Chapter Fourteen?) has an ice cream cone that is guaranteed to melt in precisely 5 minutes. What does 5 minutes mean?

Thought Question 1
Think about that.

The answer is deceptively simple: *Five minutes* just means that a device called a clock has moved ahead by five 1-minute marks. Not a very surprising idea. But, when coupled with the two principles of relativity, an idea that has profound consequences.

Let's consider a simple experiment using the idea that *time* means "that which is measured by clocks" and using the two principles of relativity to predict the outcome of the experiment.

Suppose that Mort and Velma are in relative motion. To be specific, imagine Mort standing on Earth. Keep in mind, however, that there are no "chosen" reference frames; Mort could just as well be moving past Earth at half of lightspeed without changing any of the essentials of our experiment.

We want to put Velma in a reference frame that is moving rapidly relative to Mort. So, imagine Velma in a spaceship moving past Mort at three-fourths of lightspeed.

Mort and Velma each carry one of the special 5-minute ice cream cones. They are going to measure the time elapsed while their cones melt.

To do this they'll each need a clock. They could each use an ordinary spring-wound wristwatch, but wristwatches are fairly complicated objects, depending as they

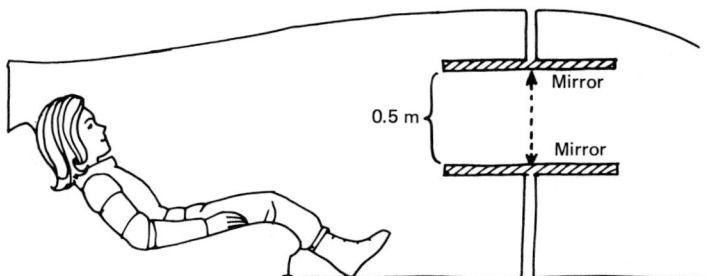

FIGURE 15.1 Velma's light clock as seen by Velma.

do on the elastic properties of springs and so forth. So Mort and Velma construct an especially simple clock, one whose behavior is easy to predict with only the help of Einstein's two principles.

Velma (and Mort, too) constructs her clock with two mirrors and a light beam. The two mirrors are placed opposite and facing each other, ½ m apart, as shown in Figure 15.1. The light beam reflects back and forth between the two mirrors. According to the principle of the constancy of lightspeed, Velma sees the light beam travel 300,000 km or 3×10^8 m in 1 sec. Since Velma sees the light beam cover 1 m in each round trip from one mirror to the other and back again, the beam must complete 3×10^8 round trips during 1 sec. Thus Velma can keep track of the time by counting the number of round trips completed by the beam—3×10^8 round trips is one second. Suppose that the clock ticks after 3×10^8 round trips (1 sec). Mort and Velma each carry one of these **light clocks.**

The principle of relativity tells us that one reference frame is just as good as another for carrying out physics experiments. The experiment of timing an ice cream cone while it melts is a physics experiment—after all, an ice cream cone is made of molecules, subject to the principles of physics—so Velma and Mort must get the same result, namely 5 minutes, when they measure the melting time of their cones. When Velma performs this experiment with her cone, in her reference frame (the spaceship), using her clock, she must find that the cone melts in precisely 5 minutes, or 300 ticks of her light clock; and when Mort performs this experiment with his cone, in his reference frame, using his clock, he must also find that his cone melts in 300 ticks of his light clock. Nothing very remarkable here—after all, both cones are guaranteed 5-minute cones.

But strange things happen when they observe each other's clocks and cones.

First, consider Mort observing Velma's clock. What does he see as he looks at her bouncing light beam? He doesn't see Velma's light beam simply bouncing up and down because Velma's spaceship is moving past Mort while her light beam is bouncing. Figure 15.2 shows three "snapshots" of Velma's spaceship at three different

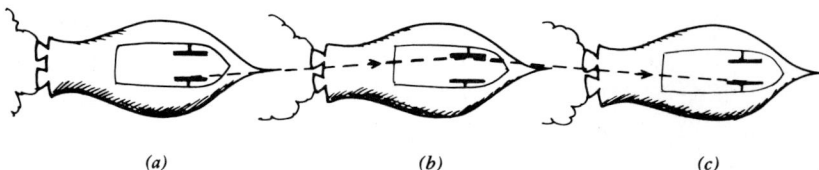

(a)   (b)   (c)

FIGURE 15.2 (a–c) Velma's light clock as seen by Mort.

instants as seen by Mort. In snapshot (a) Velma's light beam is just leaving the lower mirror; in (b) it is bouncing from the upper mirror; in (c) it has gotten back down to the lower mirror. The three snapshots show one complete round trip of Velma's light beam as observed by Mort.

It is precisely ½ m from the bottom mirror of Velma's clock to the top mirror. Thus, as you can see from the diagonal path in the diagram, the distance covered by the beam between snapshots (a) and (b) must be more than ½ m as observed by Mort. In other words, the beam moves farther as seen by Mort than it does as seen by Velma, simply because Mort sees the beam moving forward (along with the spaceship) as it moves upward, while Velma only sees it move upward.

Thus during one complete round trip of Velma's light beam, Mort observes Velma's beam to move farther than 1 m. But according to the principle of the constancy of lightspeed, Mort sees all light beams move at 3×10^8 m/sec: he sees his own beam move that fast and he sees Velma's beam move that fast. Thus he must see Velma's beam arrive back at Velma's bottom mirror *after* his own beam has completed one round trip, because (according to Mort) Velma's beam travels farther than his own.

But Mort is using his own clock to keep track of the time. So, according to Mort, Velma's clock is slow!

Thus, starting from Einstein's two postulates, we can predict that Mort observes that Velma's clock is slow. This observation is not the result of any defect in Mort's or Velma's clocks. It happens because time does not behave the way we thought it did. Time flows differently for different observers if those observers are in relative motion. ***Time is relative.***

Time is perhaps the most fundamental quantity in the natural world. If time doesn't behave the way we thought it did, then probably a lot of other things don't behave the way we thought they did.

It's possible to go through all of this in a more mathematical fashion. The mathematical analysis yields a formula, a prescription, that tells you how slowly Mort observes Velma's clock going for any given speed of Velma's spaceship relative to Mort. If Velma is moving past Mort at three-fourths of lightspeed, as we are assuming, this formula predicts that Mort sees Velma's clock running about two-thirds as fast as his own. For instance, when Mort sees his own clock registering 1.5 sec (4.5×10^8 complete bounces), he sees Velma's clock registering only 1.0 sec (he has seen Velma's light beam bounce 3×10^8 times).

Now let's turn the tables and look at all this from Velma's point of view. Velma is standing in her spaceship watching the beam in her light clock bounce up and down. Her clock keeps correct time, for her. Figure 15.3 shows three snapshots of Mort and his clock as observed by Velma at three different instants. Since Velma is moving to the right relative to Mort, she must observe Mort moving to the left relative to her. Thus these snapshots proceed from right to left. In snapshot (a) Mort's beam is leaving his lower mirror; in (b) the beam is bouncing from the upper mirror; in (c) it's back down to the bottom mirror. During the complete round trip, Velma sees Mort's light beam go farther than 1 m.

Exercise 1
According to Velma, is Mort's clock slow, fast, or just right? (Answer below.)

To Velma, Mort's clock is slow, just as to Mort, Velma's clock was slow, because

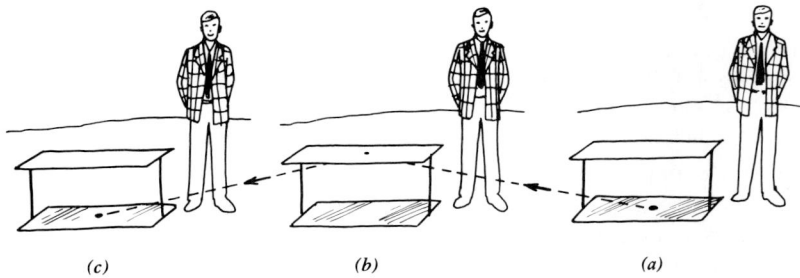

(c)   *(b)*   *(a)*

FIGURE 15.3 *(a–c)* Mort's light clock as seen by Velma.

Velma sees Mort's beam move farther than 1 m during the round trip, just as Mort saw Velma's beam move farther than 1 m during one round trip.

This is perplexing. Ordinarily, if I say that my clock is running slower than yours, you will say that your clock is running faster than mine. But Velma and Mort are not in an ordinary situation. They are moving past each other at three-fourths of lightspeed. We don't ordinarily see clocks moving that fast. Furthermore, both clocks are keeping correct time—time itself is responsible for these strange effects, not defective clocks.

Mort moves to the left, relative to Velma, at three-fourths of lightspeed. The formula referred to above tells us once again that Velma sees Mort's clock running two-thirds as fast as her own. For instance when Velma sees her clock registering 1.5 sec, she sees Mort's clock registering 1.0 sec. Figure 15.4 shows the essential results on the observations of the two clocks, from each of the two points of view.

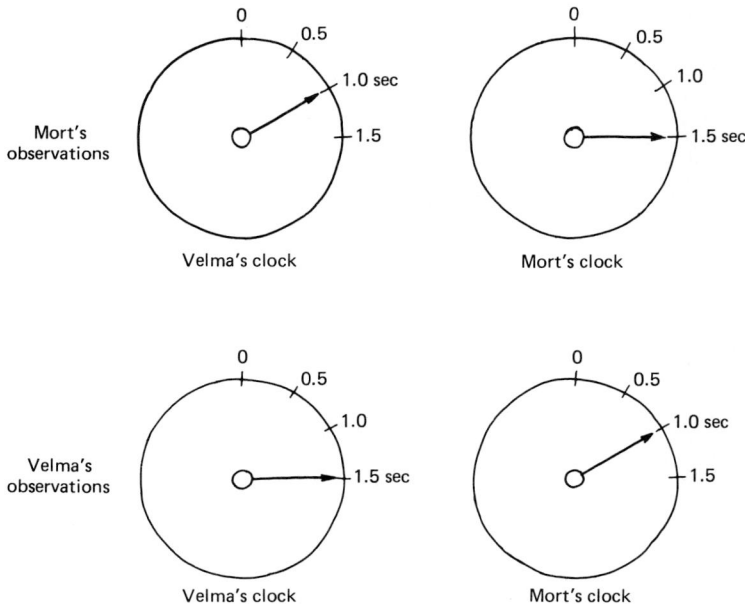

Mort's observations — Velma's clock / Mort's clock

Velma's observations — Velma's clock / Mort's clock

FIGURE 15.4 The two clock dials, as observed by Mort and Velma.

Exercise 2
Consider the ice cream cones. How long does Mort say it takes for his cone and for Velma's cone to melt? How long does Velma say it takes for her cone and for Mort's cone to melt? Remember that Mort says Velma's clock runs two-thirds as fast as his own and vice versa. (Answers below.)

Mort naturally says that his own cone melts in 5 minutes. But when Mort observes that his clock registers 5 minutes (i.e., when his cone melts), he'll observe Velma's clock to register only ⅔ of 5 minutes, or 3.33 minutes (Fig. 15.4). Furthermore, Velma's cone can't melt until Velma's clock reads 5.0 minutes—it's a 5-minute cone just like Mort's. Thus Mort must wait until his clock reads 7.5 minutes before he will observe Velma's cone to melt (⅔ of 7.5 = 5.0). Mort says that Velma's cone is running slow, just like her clock.

Let's turn the tables again and look at this from Velma's point of view. From Velma's point of view, her own cone melts in 5.0 minutes and Mort's cone melts in 7.5 minutes. From her point of view, it is Mort's cone that runs slow.

Thus Mort and Velma disagree about whose cone melts first. Mort says that his cone melts 2.5 minutes before Velma's, and Velma says that her cone melts 2.5 minutes before Mort's. It's all rather strange. But true. There isn't space here to describe the experiments conducted to check these predictions. Suffice it to say that there have been many experiments, and it seems that Einstein was right.

Relativistic Biological Aging

Suppose you could raise a special short-lived breed of frogs, guaranteed to have a life span of only 5 minutes. Despite the many differences between 5-minute ice cream cones and 5-minute frogs, both frogs and cones are made of atoms and molecules, subject to the principles of physics. So 5-minute frogs should behave the same way as 5-minute ice cream cones when it comes to such purely physical (i.e., nonbiological) principles as the relativity of time.

Exercise 3
Velma and Mort each have one of the 5-minute frogs. According to Velma, how long does each frog survive? What does Mort say about the life span of each frog? (Answers below.)

The answers must be the same as the answers to Exercise 2. Velma says her frog is an ordinary 5-minute frog but that Mort's frog manages to stretch out its life to 7.5 minutes. Mort says that his frog lives 5 minutes but that Velma's frog lives 7.5 minutes.

Let's move along the biological chain to human beings. Velma and Mort, for instance. Suppose that both Velma and Mort have a life span of precisely 70 years. What was true of 5-minute frogs must also be true of 70-year humans, since the relativity of time is a purely physical principle, valid for all collections of atoms and molecules. Velma and Mort, for instance.

Thought Question 2

The above paragraph may appear to take an *extremely materialistic* stance with regard to human beings. What is your response to this? For starters, you might want to consider these three positions: (a) Human beings really are nothing but collections of molecules, entirely explainable in terms of physics; (b) Human beings are collections of molecules, subject to all the principles of physics, but humans are something more than only this; (c) Human beings are basically different from ice cream cones, and even from frogs, so the principles of physics don't apply to human beings.

Exercise 4

According to Velma, how long does she live and how long does Mort live? According to Mort, how long do each of them live? (Answers below.)

The answers to Exercise 4 are just like the answers to Exercises 2 and 3, only scaled up from 5 minutes to 70 years. Velma says that she lives her normal 70 years but that Mort "melts slowly," that is, ages slowly. According to Velma's clock, Mort lives for 105 years. Unfortunately, Velma will be dead by then, because she can only live 70 years. Mort says that he lives for 70 years but that Velma lives for 105 years.

Exercise 5

Velma observes that she lives for precisely 70 years, traveling the entire time in her spaceship in a straight line at three-fourths of light-speed relative to Mort. How old does Velma say Mort is when she is dying? (Answer below.)

Exercise 6

Turn the preceding Exercise around. How old does Mort say Velma is when he is dying? (Answer below.)

Both Velma and Mort say that the other person ages slowly. Velma says that when she is 70 Mort is only 47 (two-thirds of 70); he will really be 47 according to Velma. If he is a physically normal person, he will have all the biological features associated with 47-year-olds. Mort says that when he is 70 Velma is 47. They disagree about who dies first. Time is different for the two observers. There is no single, universal time. Instead, there is "Mort's time" and "Velma's time."

Now let's bring Velma and Mort back together: What would happen if Velma took off from Earth in a spaceship, traveled for a long time at a high speed, and returned to Earth? If Velma took off just as she and Mort were being born, how old would they both be when Velma returned? Surely, now that they are both standing next to each other on Earth they will agree about each other's ages! Happily, this actually is the case. They do agree. However, it turns out that they are not both the same age, even though they were born at the same time. In fact, working through this problem mathematically, the prediction turns out to be that both observers agree that Mort has aged more than Velma. If Velma were moving fast enough, she might return at 20 years of age and find Mort, at 70, a senior citizen.

What is different about this situation, as opposed to the previous situation in which each observer thought that the other observer was younger? The difference is that Velma is now an *accelerated observer;* she accelerates at several crucial points during the trip. Mort is nonaccelerated throughout the problem because he stays behind on Earth. So we can use the special theory of relativity to predict his observations.

Exercise 7

List the places along the trip at which Velma must accelerate if she is to leave from Earth and return to Earth.

This sort of thing has actually been checked experimentally, not with humans, but with atomic clocks flown around Earth on jet airplanes. As predicted, the clock that took the trip returned "younger" (it registered less time elapsed) than the one that remained behind on Earth. It really seems to be true.

Space Is Relative

Rather than try to show how the other predictions follow from Einstein's two principles, I'll simply state and discuss the predictions.

Let's install a square window, 1 meter wide by 1 meter high, in Velma's spaceship. When we say "1 meter by 1 meter" we mean that when Velma measures the window using her own measuring instruments she finds it to be 1 m on each side.

Interesting things happen when Mort looks at the window in Velma's spaceship. Mort uses his own measuring instruments to measure the dimensions of the window as Velma passes him. He finds that the window is contracted along the direction of motion of the spaceship. For example, if Velma passes him at three-fourths of lightspeed, the quantitative prediction is that Mort finds Velma's window to be 67 cm wide ($\frac{2}{3}$ m) by 1 m high (Fig. 15.5).

Once again the same result works the other way. Suppose that Mort's laboratory on Earth has a square window 1 m on each side. Velma, viewing this window from her passing spaceship, will find it to be contracted along the direction of motion. At a speed of three-fourths the speed of light, she'll find the dimensions of Mort's window to be 67 cm wide by 1 m high.

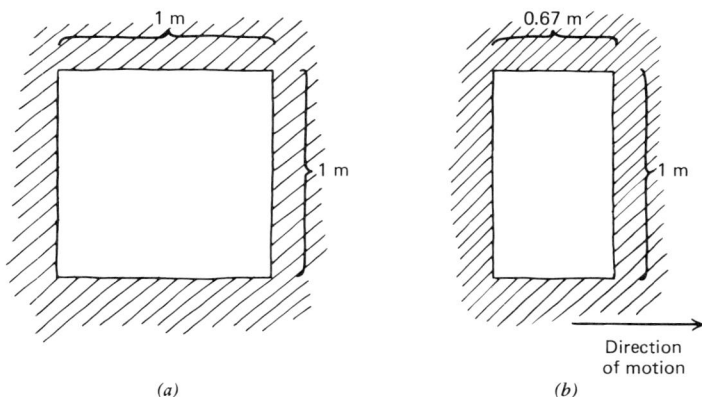

FIGURE 15.5 The window in Velma's spaceship, as observed by (a) Velma, and (b) Mort. Velma is moving toward the right past Mort.

So, not only time, but also space, is relative to the observer. Different observers, such as Velma and Mort, see different widths for the same window, just as they measure different time spans for the melting of the same ice cream cone. There is no single, universal space; there is only "Mort's space" and "Velma's space."

Non-Galilean Velocity Addition

We have seen (Chapter Fourteen) that when light beams are involved, speeds don't add "normally." What about situations that don't involve light beams?

Recall the example (Chapter Twelve) of Velma throwing a ball inside a moving railroad car. If the car moves at 20 m/sec, and if Velma throws the ball at 5 m/sec relative to Velma, then the Galilean theory of relativity predicts that Mort, standing beside the tracks, will see the ball going 25 m/sec. This was the standard answer to this sort of question prior to Einstein, and the answer that seems intuitively sensible.

However, Einstein's theory predicts that Mort sees the ball moving just a little slower than 25 m/sec. For speeds as slow as these (20 m/sec and 5 m/sec), the difference between Galileo's prediction and Einstein's prediction is so small that it cannot be measured with present measuring instruments. At higher speeds the difference between Galilean relativity and the special theory of relativity becomes sizable. And it is the special theory of relativity that gives the experimentally correct answers.

Let's put Velma back in her spaceship moving past Mort at three-fourths of lightspeed. Suppose she has a ball launcher that launches baseballs forward at one-half of lightspeed, relative to the spaceship.

Exercise 8

Find the Galilean prediction for the speed of one of these baseballs, relative to Mort. (Answer below.)

The Galilean theory predicts that Mort sees these baseballs move at $3/4 + 1/2 = 5/4$ times lightspeed. That's 25 percent faster than a light beam! The special theory predicts a speed that is slower than lightspeed. Quantitatively, the prediction turns out to be 91 percent of lightspeed.

If Velma, still moving at three-fourths of lightspeed, launches her baseball at three-fourths of lightspeed relative to the spaceship, Mort will see the ball pass him at 96 percent of lightspeed. If Velma launches the ball at 90 percent of lightspeed, Mort will see it pass at 99 percent of lightspeed. If Velma launches the ball at nearly lightspeed, Mort will see it pass at nearly lightspeed! Very much like the laser beam.

Mass Is Relative

Newton's second law relates an object's acceleration to the object's mass and to the force on the object. But acceleration is related to space and time, concepts that are radically revised by Einstein's theory. Thus it isn't surprising that the theory of relativity leads to a revision of Newton's second law.

For example, suppose that Velma, who is once again in her spaceship moving past Mort at three-fourths of lightspeed, is carrying a 1-kg rock. Recall (Chapter Six) that "1 kilogram" means that when Velma puts this rock on a balance scale in her spaceship, the rock just balances a standard 1 kg mass.

Suppose Velma applies a force of 5 N to her 1-kg rock. The quantitative form of Newton's second law predicts that this force will cause the rock to accelerate at 5 m/sec^2. Einstein's theory gives the same prediction.

Now let's look at this rock from Mort's point of view. The rock is in Velma's spaceship, so it is moving at three-fourths of lightspeed relative to Mort. Suppose that Mort now applies a 5-N force to Velma's rock. Newton's second law once again predicts that the rock has an acceleration of 5 m/sec^2. But Einstein's theory now gives a different prediction. It predicts that the rock's acceleration, as measured by Mort, is only 1.5 m/sec^2. The answer that agrees with experiment is, as you may have guessed, Einstein's.

As you can see from this example, Einstein's theory predicts that it is harder to accelerate a fast-moving object than it is to accelerate a slow-moving object. For a 1-kg rock starting from rest, the theory predicts that a 5-N force produces an acceleration of 5 m/sec^2; for a rock starting from three-fourths of lightspeed, the prediction is that a 5-N force produces an acceleration of only 1.5 m/sec^2. At speeds much slower than lightspeed, Einstein's and Newton's theories agree, and both agree with experiment; at high speeds, Einstein's theory agrees with experiment, and Newton's doesn't.

The general lesson here is that the principles of Newtonian physics break down at speeds that are a significant fraction of lightspeed. Newton's second law, Galilean relativity, and our intuitive concepts about space and time, are good approximations at lower speeds, but they do not work at higher speeds. Newton's theory is limited to low speeds. Einstein's theory works at all speeds, low and high. We do not, however, simply discard Newtonian physics on this account. Scientists use Newtonian physics for most low-speed phenomena, because Newtonian physics agrees with experiment in this range and because Newtonian physics is much simpler than Einstein's physics. The low-speed world is "Newtonian." Thus airplanes, space satellites, stationary structures such as buildings and bridges, falling apples, and speeding bullets are Newtonian. The high-speed world, on the other hand, is distinctly "Einsteinian."

Thought Question 3

We saw in Chapter Two that Copernicus's circular-orbit cosmology has its limitations, that this theory is valid only up to a certain accuracy. Kepler's elliptical-orbit cosmology is an improvement, but it, too, is limited to situations in which we don't need to take into account the gravitational forces between planets. Newton's physics is limited to low speeds. Einstein's physics is an improvement, but we will find in Part 4 of this book that even Einstein's physics has its limitations, that at the atomic and subatomic level new phenomena occur that Einstein's theory could not have predicted. Does it seem as though the human race is progressing toward an ultimate "scientific truth" in all of this? Do you think that we will ever find a scientific theory that has *no* limitations?

The concept of inertia (Chapter Six) is helpful in clarifying Einstein's revision of Newton's second law. Because Mort finds it difficult to accelerate the rock, he says that the rock has a large inertia (resistance to acceleration). The inertia of the rock is much greater for Mort than it is for Velma. This resistance to acceleration, or inertia, gets larger (as measured by Mort) as the rock goes faster relative to Mort. To Velma, the rock's mass is still 1 kg. Because the rock is at rest relative to Velma, we say that it has a ***rest mass*** (mass when measured by an observer for whom the rock is at rest) of 1 kg. But the rock seems to Mort to be more massive than 1 kg, because it is so hard to accelerate. We say that the ***inertial mass*** (resistance to acceleration) of the rock gets larger as the rock goes faster.

Along with time and distance, mass (inertial mass) is also relative.

It is worth emphasizing the fact that, even though Velma is moving at three-fourths of lightspeed past Mort (or even if she were moving at 99.9 percent of lightspeed past Mort), nothing is unusual to Velma inside her own spaceship. Her clock runs normally, her ice cream cone melts normally, her frog ages normally, her window looks normal, and objects in her spaceship are no more massive than usual. She lives out her normal 70-or-so years and gets the normal amount of "living" in each day. In fact, according to the principle of relativity (Chapter Fourteen), everything has to be normal to Velma inside her own spaceship, because the principles of physics are the same in her reference frame as they are in any other reference frame. It is Mort, observing Velma, who notices unusual effects in Velma's reference frame. He sees her clock running slow, he sees her age slowly, and so forth.

Lightspeed as the Limiting Velocity

You've seen that the faster an object goes, the harder it is to accelerate that object. Einstein's relativity predicts that it becomes infinitely* hard to accelerate an object as its speed gets closer and closer to lightspeed. In other words, an object's inertia gets larger and larger, without limit, as the speed of the object gets closer and closer to lightspeed.

Therefore, it is impossible to accelerate an object right up to precisely lightspeed. An infinitely large force would be needed to do this, and there aren't any infinite forces.

Now ask yourself: Is there anything in this universe that goes as fast as light?

• *Well, yes, there is. Namely, light itself.*

Now try this one: Does the fact that light travels at lightspeed contradict Einstein's prediction about the impossibility of accelerating an object right up to precisely lightspeed?

• *No. This does not contradict Einstein's prediction.*

Einstein's theory says that an object cannot accelerate up to lightspeed. In other words, we can't take an object moving slower than lightspeed and accelerate it until it reaches lightspeed. But we don't need to accelerate a light beam up to lightspeed, because it's already moving that fast! In a vacuum, light beams always move at lightspeed. They never speed up from some slower velocity, they never slow down, and they certainly never stop. It is lightspeed or nothing.

There may even be objects that always go faster than lightspeed, objects that never slow down to lightspeed. Such objects would not contradict the predictions of

*The word ***infinite*** means "without limit," or "larger than any number."

Einstein's relativity. Scientists have looked in a variety of places for such faster-than-light objects, but to no avail. Thus as of this writing, such particles are purely hypothetical: they don't seem to conflict with the known laws of physics (although there is considerable debate on this score), yet they have never been observed. Despite their hypothetical nature, they have been given a name: *tachyons*, meaning "fast ones."

Mass-Energy Equivalence: $E = mc^2$

When you throw a ball you give it kinetic energy. According to the theory of relativity, you also slightly increase the ball's inertial mass, because of relativistic mass increase. The energy that you give to the ball shows up not only as an increase in the ball's speed, but also as an increase in its inertial mass. Einstein was able to show that the quantitative relation between the (inertial) mass* increase in kilograms and the energy increase in joules is

$$\text{Mass increase} = \frac{\text{energy increase}}{\text{lightspeed squared}}$$

Exercise 9
Recall that lightspeed is 300,000 km/sec, or 3×10^8 m/sec. In metric units (using meters and seconds), how big is the square of lightspeed?

Exercise 10
Suppose that we throw a 1-kg rock, imparting 90 J of kinetic energy to it in the process. Find the increase in the rock's mass.

Note the tiny mass increase in Exercise 10: 10^{-15} kg! The rock, which had a mass of 1 kg at rest, has a mass of 1.000000000000001 kg while moving. This increase in mass is far too small to be measured. For ordinary velocities, even up to several thousands of miles per hour, the mass increase is an extremely small fraction of the total mass—so small in fact that it is usually unmeasurable. Significant mass increases occur only at speeds that are a significant fraction of lightspeed.

There are many other ways to increase a system's energy. For example, the elastic energy of a spring increases when it is stretched; the thermal energy of a cup of coffee increases when it is heated. In every case, Einstein asserted that the system's mass increases right along with its energy and that the mass increase and the energy increase are related in the way stated above: Every energy increase shows up as a corresponding mass increase.

Einstein carried this idea further, and concluded that the above relationship applies not only to an object's increase in mass and energy, but also to its total mass and energy. Thus, for example, the 1-kg rock of Exercise 10 has a total energy of $1 \times$ the square of lightspeed, or 9×10^{16} J, even when it is at rest. Any system having inertial mass has energy, and any system having energy has inertial mass. We might as well say that energy *is* mass, and mass *is* energy. In symbols,

$$m = E / c^2$$

*Henceforth the word **mass** means "inertial mass."

where *c* stands for lightspeed. Another way of saying the same thing is:

$$E = mc^2$$

A few examples:

Suppose you heat a cup of coffee by adding 1 kcal of thermal energy to it. Einstein's theory predicts an increase in the coffee's mass even though no coffee has been added. Let's calculate the size of this increase. One kcal is about 4000 J (Chapter Nine), so the predicted mass increase is 4000 J / 9 × 10^{16}, which is 4.6 × 10^{-14} kg. That's .000000000000046 kilograms! This is entirely unmeasurable because it is so small compared with the coffee's mass.

Suppose you stretch a spring. Einstein's theory predicts an increase in the spring's mass because you added energy to the spring. Suppose that you did 500 J of work during the stretching process. This added 500 J of energy to the spring. The additional mass is thus 500/9 × 10^{16} or 5.5 × 10^{-15} kg, an entirely unmeasurable fraction of the original mass of the spring.

Now we'll look at an example in which the change in mass is not negligible. Consider a helium nucleus, composed of two protons and two neutrons. Suppose that we separate these four particles. Because the nucleus is held together by the strong nuclear force, we must do work on this system in order to separate the four particles. This work must increase the system's energy. But if the energy is increased, so is the mass. In other words, the helium nucleus must have a smaller mass than the total mass of the four separate particles that constitute this nucleus. Quantitatively, the measured mass of a helium nucleus is 6.64 × 10^{-27} kg, whereas the total mass of two separate protons and two separate neutrons is 6.69 × 10^{-27} kg. That's a mass difference of nearly 1 percent: a significant difference, and one that is experimentally detectable. The experimental result agrees with Einstein's prediction.

This example is typical of nuclear energy. Nuclear forces are so strong that any alteration of a nucleus involves large amounts of energy and therefore also involves large increases or decreases in mass, so large, in fact, as to represent a measurable change in the mass of the nucleus. Thus the formula $E = mc^2$ has become associated especially with nuclear physics. The formula applies to all energy transformations, including heated coffee and stretched springs, although the resulting mass changes are not directly measurable for heated coffee and stretched springs.

How much energy is there in a penny? A penny has a mass of about 3.5 g or .0035 kg. Thus it has a total energy of .0035 × 9 × 10^{16} J, or 32 × 10^{13} J. To grasp the significance of this number, let's calculate how many people could be lifted through a height of 1 km, if all this energy could be put into lifting. Recall (Chapter Nine) that

Gravitational energy = weight × height

In the present case we want to lift people through 1000 m, so the weight, in newtons, that can be lifted with 32 × 10^{13} J of energy is 32 × 10^{13} / 1000, or 32 × 10^{10} N. Assuming that a person weighs about 600 N, that's over 500 million people! Obviously even small amounts of mass are equivalent to large amounts of energy. Although there is no convenient way to use the enormous energy residing in a penny, we have found ways of using a certain fraction (about 1 percent) of the energy residing in, for example, uranium, plutonium, and hydrogen (see Chapter Twenty-one).

Checklist

light clocks

time

moving clocks run slow

relativistic aging

moving objects are contracted

relativistic velocity addition

rest mass

inertial mass

relativistic mass increase

limitations of Newtonian physics

tachyons

$E = mc^2$

nuclear energy

Further Thought Questions

4. Mort says that Velma's clock runs slow, and Velma says that Mort's clock runs slow. Who is right?

5. Copernicus said that the earth goes around the sun; Ptolemy said that the sun goes around the earth. Who was right?

6. I say that jazz music is better than country music; you say that country music is better than jazz. Who is right?

7. Is this the relativistic century? Is there a tendency today to regard truth in such nonscientific areas as taste, life-styles, morals, and religion, as relative to the observer? If so, is this tendency related in any direct or indirect way to Einstein's theory of relativity?

Further Exercises

11. State whether there would be a change in mass and, if so, whether the change would be an increase or decrease, in each of the following situations: (a) an automobile battery is charged; (b) a red-hot steel bar is allowed to cool down; (c) a rubber band is stretched; (d) hydrogen and oxygen combine (or "burn") to form water inside a sealed, insulated container; (e) hydrogen and oxygen combine to form water inside a container that allows the heat to escape; (f) the two atoms in an O_2 molecule are pulled apart; (g) two oxygen atoms join together to form an O_2 molecule.

12. As a meter stick that has a rest mass of 1 kg moves past you, your measurements show it to have a mass of 2 kg and a length of 1 m. What is the orientation of the stick?

13. Is it possible, within the principles of physics, for a child to be biologically older than his or her parents?

14. If you were in a rocket ship moving away from Earth at nearly lightspeed, would you notice any change in your pulse? Would you notice any change in your size? In your mass?

15. How fast are you moving right now? What meaning does this question have?

Recommended Reading

1. Poul Anderson, *Tau Zero*, Berkley Publishing Corporation, New York, 1970. Science fiction about a spaceship that is doomed to accelerate forever. Inside the spaceship life proceeds more-or-less normally as the occupants observe the strange effects predicted by Einstein that are going on outside the spaceship.

2. Nigel Calder, *Einstein's Universe*, Penguin Books, Ltd., Middlesex, England, 1980. Special and general relativity made plain.

3. George Gamov, *Mr. Tompkins in Paperback*, Cambridge University Press, London, 1965. A series of fantasies by a noted nuclear physicist and popularizer of modern physics. Gamov exaggerates actually existing relativistic and quantum phenomena to such an extent that they can easily be observed by the hero of the stories, C.G.H. Tompkins, a bank clerk interested in modern science.

PART 4

The Post-Newtonian Universe

16

A NEW RADIATION THEORY AND A NEW MECHANICS

Nineteenth-century physics was self-satisfied. Newtonian physics had stood the test of two centuries and had explained an enormous variety of observed phenomena. It had the unquestioning respect of nearly every scientist. The other major theory known at the time was electromagnetism, a theory that matured during the nineteenth century and that seemed to explain nearly everything that had been left unexplained by Newtonian mechanics and gravitational theory. Physicists of the late nineteenth century felt that these theories could in principle explain all observed phenomena, that all the truly fundamental principles were already known.

Albert Michelson (Chapter Fourteen) made a famous statement of this position in 1894:

> While it is never safe to affirm that the future of Physical Science has no marvels in store even more astonishing than those of the past, it seems probable that most of the grand underlying principles have been firmly established and that further advances are to be sought chiefly in the rigorous application of these principles to all the phenomena which come under our notice. It is here that the science of measurement shows its importance, where quantitative results are more to be desired than qualitative work. An eminent physicist has remarked that the future truths of Physical Science are to be looked for in the sixth place of decimals.

This attitude would be destroyed in a few years by quantum mechanics and relativity. Today the fundamental principles of physics are still in flux. The contemporary physicist feels that the fundamental principles are still not known and that our present laws of nature are subject to change.

Nevertheless, the arrogance* of nineteenth-century science might persist today among scientists who believe that all the basic principles of physics will be known someday and that they will in principle be capable of explaining all observable phenomena. This nineteenth-century attitude might also persist among people who believe that our social problems, and perhaps our personal problems as well, can be solved if we will only resort more fully to the rational and objective methods of science.

Thought Question 1
What is your opinion? Are the methods of science rational and objective? Is the scientific approach capable of solving all our problems? Example: You will (excluding sudden, accidental death) someday be faced with the problem of accepting and experiencing your own death. Can science solve this problem for you? Can it help?

---

*I use the word *arrogance* carefully here. Although the individual nineteenth-century scientist was not necessarily arrogant, it is true that science as a whole was arrogant in the sense that all the basic principles of physics were supposed to be known.

Thought Question 2

Is contemporary science arrogant in any respect? How?

By 1900, Lord Kelvin (Chapter Eight) felt bold enough to say that there remained only two minor clouds dimming the brilliant sky of science. These two clouds were:

1. The troubling outcome of the Michelson-Morley experiment and the related controversy over the nature of the ether.
2. Certain difficulties involving the electromagnetic radiation emitted by heated objects (discussed below).

The minor clouds turned out to be major. The first, as we know, was cleared up only by the advent in 1905 of Einstein's theory of relativity. This theory revolutionized the most fundamental of all physical concepts: time and space.

The second cloud led to the quantum idea. This concept, suggested by Max Planck (Figure 16.1) in 1900, the year of Lord Kelvin's statement, led to the development, during the years 1900 to 1930, of the **quantum theory.** This theory replaced (or radically revised, depending on how you want to look at it) both of the older fundamental theories of physics: Newtonian mechanics gave way to quantum mechanics; and the electromagnetic wave theory of radiation gave way to the

FIGURE 16.1 Three giants of modern physics. (a) Max Planck.

FIGURE 16.1 (*b*) Einstein receiving the Planck Medal, 1929.

FIGURE 16.1 (*c*) Erwin Schroedinger (Ullstein Bidendienst, West Berlin).

quantum theory of radiation. Even more fundamental than this, the new theory leads to questions about the meaning of science, about the meaning of scientific observation, about the reality of such objects as electrons and protons, and about the extent to which we can know the world scientifically.

The new theories, Einstein's relativity and the quantum theory, define the twentieth-century physicist's way of looking at nature. We have seen that Einstein's relativity involves a radical departure from the tidy and intuitively reasonable nineteenth-century view of nature. Be prepared—quantum theory is even more radical. Einstein's theory upset our concepts of time and space. The quantum theory makes us ask fundamental questions about the idea of an objectively existing natural world!

Radiation from Heated Objects

The agitation, or thermal motion, of atoms and molecules causes every object to emit infrared radiation (Chapter Thirteen). This agitation also extends to the electrons within atoms, which emit visible, ultraviolet, and X-ray radiation. Thus thermal motion causes every object to emit radiation over most of the electromagnetic spectrum—infrared, visible, ultraviolet, and so on. At ordinary room temperatures this *thermal radiation* is quite small in every wavelength range except the infrared; thus we can't see room-temperature objects at night without lights. At higher temperatures, however, the atoms and electrons are sufficiently agitated to emit larger amounts of radiation in the visible range, so that, at sufficiently high temperatures, objects begin to emit visible light (i.e., they glow).

It is possible to measure the amount of energy radiated by any object at any particular temperature and to measure the amount of energy emitted in any particular wavelength range. For example, we could design an experiment to measure the amount of energy radiated by the electric grill on a stove, at a temperature of $150°$ C, in the wavelength range 7.3×10^{-7} m to 7.4×10^{-7} m. When many such measurements are made over many different wavelength ranges for a given object at a given temperature, the results can be summarized in the form of a graph, or *radiation curve,* for that object at that temperature. The experimental curve of Figure 16.2 is a typical radiation curve. It shows the amount of radiation emitted at various wavelengths by the measured object. The infrared range, visible range, and so on are indicated along the wavelength axis.

Exercise 1
Is the object for which this curve is drawn emitting much visible light, that is, is it glowing?

Exercise 2
Judging from the preceding discussion, how would the radiation curve for this object be altered if the object were cooler?

Because radiation is an electromagnetic phenomenon, one would think that the electromagnetic wave theory of radiation could correctly predict the shapes of the radiation curves for objects heated to various temperatures. During the nineteenth century several attempts were made to use the electromagnetic theory to predict

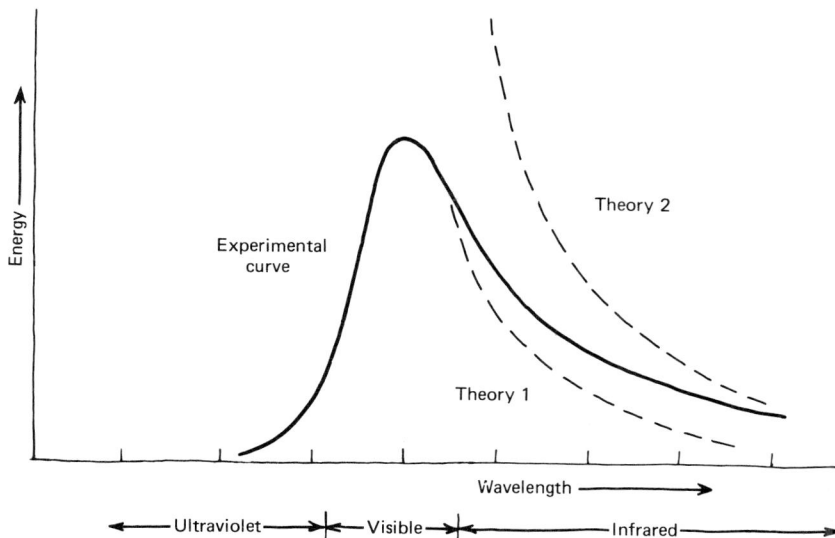

FIGURE 16.2 Comparison of a typical experimental radiation curve with predictions from two pre-twentieth-century theories.

these curves. The predictions arising from two of these attempts, based on somewhat different assumptions, are shown as the dashed curves in Figure 16.2. Neither prediction fits the experimental facts. All such attempts failed. Just as nineteenth-century physics proved unable to explain the Michelson-Morley experiment, it was unable to explain the radiation curves.

This difficulty was one of Lord Kelvin's clouds on the scientific horizon.

Scientists, always reluctant to let go of a theory that has proved useful over the years, struggled to reconcile electromagnetic theory with the radiation curves. Figure 16.2 illustrates the failure of these attempts. This discrepancy between theory and experiment finally disproved the electromagnetic theory of radiation, because if the electromagnetic theory is true then it should provide a correct prediction of the radiation curves. Experiments can never prove a theory, but they can disprove a theory. This failure of the electromagnetic theory of radiation paved the way for a new theory of radiation.

Planck initiated the theory that did correctly predict the radiation curves. The new theory is called the **quantum theory of radiation.** It was the first appearance of the "new physics," the counterintuitive, nonmechanistic physics of the twentieth century (Einstein's theory didn't appear until 5 years later). Appropriately enough, Planck announced his new theory at a scientific meeting in 1900, the year of our century's birth.

Before studying Planck's theory, let's look at a second phenomenon the details of which cannot be explained by the electromagnetic wave theory of radiation.

The Photoelectric Effect

Under certain conditions, electromagnetic radiation hitting a piece of metal causes the metal to eject electrons from its surface. This phenomenon is called the **photoelectric effect** (Fig. 16.3).

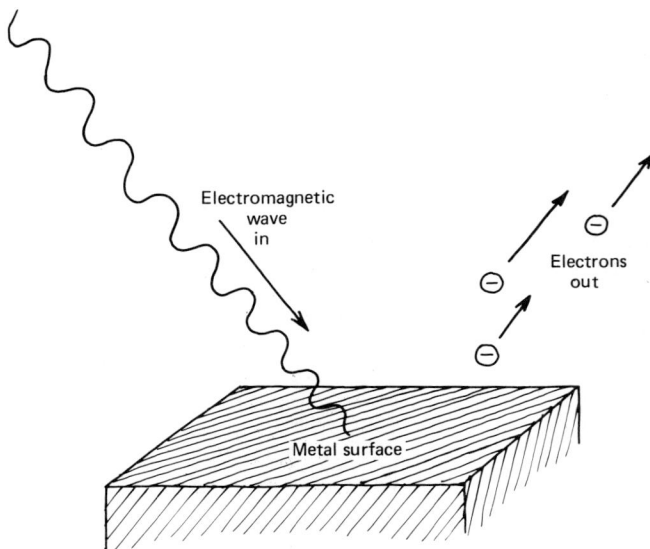

FIGURE 16.3 The photoelectric effect, as described by the electromagnetic wave theory of radiation.

Exercise 3
When the photoelectric effect occurs, what type of energy is present initially and what type is present finally, that is, what type of energy conversion occurs?

Exercise 4
Try to think of one or two technological devices that are based on the photoelectric effect. *Hint:* What devices convert light (radiant energy) into electric current (electric energy)? (Answer below.)

The closely related ***photovoltaic effect*** converts electromagnetic radiation into a flow of electrons parallel to the surface of certain materials such as thin silicon crystals. This effect is the basis for the solar cells that might someday play a major role in the use of solar energy (Chapter Eleven). The photoelectric effect is also the basis for the photographer's light meter and for other light-sensing devices.

Although the photoelectric effect was discovered in the 1880s, it was not properly explained until 1905. The problem was that many of the details of this phenomenon were inconsistent with the standard electromagnetic wave theory of radiation. Einstein finally provided a consistent explanation that was based on Planck's new quantum theory of radiation rather than on the standard electromagnetic wave theory of radiation.

Einstein's wide-ranging ability in physics is typical of the greatest scientists. The father of the theory of relativity was also a participant in the birth of the quantum theory!

We'll look at just one of the several perplexing aspects of the photoelectric effect in order to understand why this effect is inconsistent with the electromagnetic wave theory of radiation.

Any electron in a metal is "bound" to the metal in the sense that the electron must receive some energy from somewhere in order to escape from the metal. If this were not true, metal surfaces would be spontaneously ejecting electrons all the time. Even the most loosely bound electrons in the metal, those nearest the surface and possessing the most energy to begin with, need at least a small "kick" from outside to escape the metal. For any given piece of metal, it is possible to discover experimentally just how energetic this kick must be if it is to dislodge any electrons.

According to the electromagnetic wave picture of radiation, radiant energy hitting the metal surface arrives in the form of a continuous wave. Energy is spread out continuously along this wave, and the surface gradually absorbs this energy as more and more of the wave hits the surface. This radiant energy is the energy that provides the kick necessary to knock electrons out of the surface. According to the electromagnetic wave picture, the wave delivers this kick by shaking the electrons until, eventually, one of them shakes loose.

Although this explanation of the photoelectric effect certainly sounds reasonable, it turns out to conflict with the observed details of the actual experiment. The electromagnetic wave theory implies that there must be a delay time between the first arrival of the wave at the surface and the ejection of the first electron. This delay time must be at least long enough for the absorbed energy to build up until it is equal to the amount of energy needed to dislodge one electron. The required delay time can be calculated, and it turns out in some circumstances to be quite long. If the incoming radiation is weak, the necessary delay time is several seconds.

On the other hand, when the experiment is actually carried out, there is no observable delay between the arrival of radiation at the surface and the emission of the first electron. This spells trouble for the electromagnetic wave theory of radiation.

The Quantum Theory of Radiation

We've seen that the electromagnetic wave theory of radiation gives incorrect predictions about the radiation from heated objects. In 1900, Planck announced a new theory of radiation, a theory in direct contradiction to the electromagnetic wave theory of radiation but that correctly predicted the radiation from heated objects.

Planck's idea was that a vibrating charged particle emits radiation only when its state of vibration changes from one amplitude (or width) of vibration to a different amplitude (Fig. 16.4). This idea contradicts the electromagnetic wave theory, which says that every vibrating charged particle emits a continuous electromagnetic wave during the entire time the charge is vibrating. Planck further asserted that a vibrating charged particle cannot gradually change its width of vibration from some initial state, such as state (a) of Figure 16.4, to some final state, such as state (b) of Figure 16.4, but instead must instantaneously jump from the initial to the final state. Such sudden transitions are called **quantum jumps.**

Exercise 5
In Figure 16.4, which of the two states appears to have the most energy? According to the principle of energy conservation, how are the energies of the particle in states (a) and (b) related to the energy of the emitted radiation?

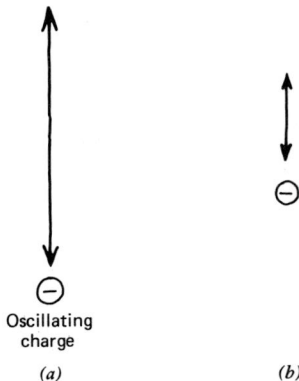

Oscillating
charge

(a)   (b)

FIGURE 16.4 Two states of vibration of a charged particle.
The two states have different amplitudes.

Planck found that in order for this theory to lead to a correct prediction of the radiation curves for heated objects, he needed to make one more assumption, that the energy of the radiation emitted during a quantum jump was proportional to the frequency of the vibrating charged particle.* To obtain agreement with experiment, Planck found that the "proportionality constant" relating frequency to energy had to be 6.6×10^{-34}, if energy and frequency were expressed in metric units. That is, Planck's relation was:

$$\text{Energy of emitted radiation} = (6.6 \times 10^{-34})$$
$$\times \text{ (frequency of oscillator)}$$

Einstein, who took an interest in Planck's theory, was the first to realize that if radiation is emitted only during instantaneous quantum jumps, then the radiation itself must come in the form of tiny "bundles" rather than as a long, continuous wave. Thus was born the **photon** concept, the idea that electromagnetic radiation consists of tiny particles, or photons, of radiant energy.

Even though according to the new theory the emitted radiation is not a wave but a tiny particle or photon, Einstein associated a frequency with this photon, namely the frequency of the vibrations that emitted the photon. According to Planck's relation above,

$$\text{Energy of a photon} = (6.6 \times 10^{-34})$$
$$\times \text{ (frequency of that photon)}$$

The proportionality constant, 6.6×10^{-34}, is known, not surprisingly, as **Planck's constant.**

To summarize:

The Quantum Theory of Radiation. A vibrating charged particle emits electromagnetic radiation only when it makes a quantum jump from one state of vibration to a different state of vibration. This transition occurs instantaneously. The emitted radiation appears as a tiny particle, or photon. The energy of a photon is proportional to its frequency.

---

*The frequency of the vibrations was presumed to be the same in the initial and final states regardless of the width of vibration of these states, just as the frequency of a clock's pendulum is the same regardless of the width of the pendulum's swing.

Photons are not like familiar, everyday particles, such as grains of dust, speeding bullets, or even speeding electrons. We can speed up, slow down, or stop bullets or electrons, but we know from experiments that electromagnetic radiation always moves at 3×10^8 m/sec and thus photons must always move at 3×10^8 m/sec. We can't speed them up, slow them down, or stop them. A vibrating charge creates a photon at the instant the charge makes a quantum jump, and from the instant of its creation the photon moves at lightspeed.

Another funny thing about photons—their rest mass (Chapter Fifteen) is zero. According to the special theory of relativity, the inertial mass of a fast-moving object is much larger than its rest mass and increases without limit as the object's speed approaches lightspeed. For any object having a nonzero rest mass, the inertial mass would be infinite any time the object was moving at lightspeed. To prevent this disaster, we must assume that a photon's rest mass is zero.

As mentioned, the new theory correctly predicts the radiation curves from heated objects. That's why Planck invented it. The details of Planck's argument are mostly long and mostly mathematical, so we won't delve into them here.

The photoelectric effect provides fairly convincing support for the new theory. You'll recall that as soon as radiation hits a metal surface, the surface emits electrons. The electromagnetic wave theory of radiation cannot adequately explain this experiment because the wave theory predicts a measurable delay time between the radiation's arrival and the electron's emission. The new particle theory of radiation does provide an adequate explanation. If radiation comes in the form of separate particles and if each of these particles happens to have sufficient energy to knock an electron out of the metal surface, then as soon as the first photon is absorbed by the surface an electron will be knocked out (Fig. 16.5). Simple.

Planck's theory has other nice features. For example, Planck's assumed proportionality between a photon's frequency and its energy tells us that higher-frequency radiation contains higher-energy photons. The following exercises demonstrate a few of the many implications of this idea.

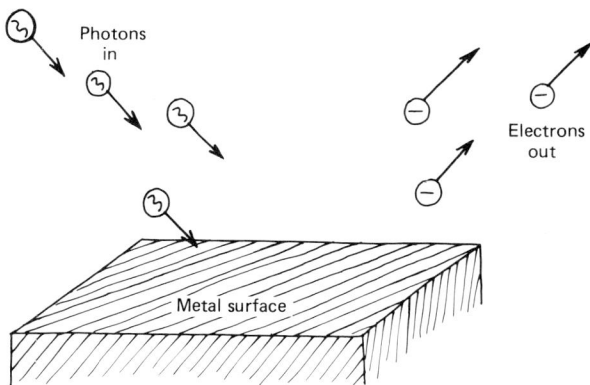

FIGURE 16.5 The photoelectric effect, as described by the quantum theory of radiation.

Exercise 6
Which has the higher frequency, shorter-wavelength light or longer-wavelength light? Which color light beam contains the highest-energy photons: red, yellow, or violet? Which color light beam has the lowest-energy photons?

Exercise 7
Which type of radiation would you guess has the greatest ability to do biological damage to the cells of your body: visible, ultraviolet, X ray, or gamma ray? Why?

Exercise 8
Explain why chlorofluorocarbon molecules (Chapter Thirteen) must rise to great altitudes before they are broken apart by the sun's radiation. Why doesn't the radiation near Earth's surface break down these molecules?

A Wave Is a Particle

It may seem that the success of the quantum theory of radiation requires the wholesale replacement of the older theory by the new theory. Unfortunately, things are not this simple.

Although Planck's theory gives a satisfactory explanation of the radiation curves and the photoelectric effect, it is the electromagnetic wave theory that gives a satisfactory explanation of the many interference experiments that can be done with radiation. We have seen (Chapter Twelve) that the interference effects produced by light beams are direct evidence that these beams are continuous waves rather than a stream of tiny particles. A purely particle theory of radiation, such as Planck's theory, cannot explain these effects

So the wave theory of radiation explains interference experiments but is inconsistent with the radiation curves and the photoelectric effect. The particle theory of radiation explains the radiation curves and the photoelectric effect but is inconsistent with the interference experiments. What is radiation? Is it a wave or is it a particle? Some experiments support, and some experiments contradict, either point of view. It's a dilemma.

We'll call it the **wave-particle dilemma** about electromagnetic radiation.

The Electron Interference Effect

Warning: Our dilemma is about to deepen.

Consider the electron. Or consider any other material particle of the atomic and subatomic world: proton, neutron, atom, molecule, and the like. By **material object** we will mean any object having rest mass (Chapter Fifteen), as opposed to radiation (the photon), which has no rest mass. This distinction between **matter** (which has rest mass) and **radiation** (which doesn't) is useful.

The ideas that we will study in this section apply not only to atomic and subatomic objects, such as electrons, but also to familiar objects, such as baseballs. But the quantum effects discussed here are more easily observed for smaller, less massive

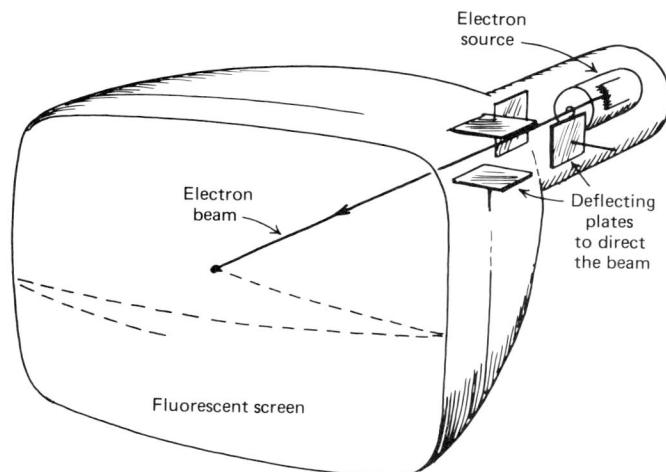

FIGURE 16.6 The television tube.

objects. Because the electron seems to be the least massive piece of matter around,* it will be our featured example.

Many different experiments support the view that electrons are tiny particles. For example, if a stream, or beam, of electrons is directed at a fluorescent screen (a screen designed to emit a small flash of light whenever a particle strikes it), the screen emits tiny flashes at different points. The flashes are apparently caused by particles hitting the screen. If the electron beam is weaker, the flashes are fewer and farther between. If the beam is sufficiently weakened, a point comes at which, for example, only one flash occurs every minute. Surely this is evidence that the beam is a stream of small particles—one particle every minute in the case of the very weak beam.

You have, in fact, seen the above experiment in action. A television tube (Fig. 16.6) is a large, evacuated tube with a fluorescent coating on the inside of the face. A narrow beam of electrons goes from the back to the front of the tube, where each electron makes a flash of light as it strikes the coating. The beam makes a narrow line of flashes across the face, then moves down and traces out another line of flashes just below the first, and so on, all the way down the face, forming a complete picture. The beam covers the entire face in about four-hundredths of a second. The pattern of light spots and dark spots is controlled by the electromagnetic television wave sent from the transmitting station (Chapter Thirteen).

By the end of the nineteenth century, experiments had convinced scientists that electrons are small, electrically charged particles. Then, in the 1920s a new phenomenon turned up, one that was at odds with the particle theory of electrons. Many different experiments demonstrate the new phenomenon. The most instructive of these is the double slit experiment with electrons.

The double slit experiment with electrons is similar to the double slit experiment with light (Chapter Twelve). A source of electrons emits a beam of electrons toward a pair of narrow, parallel slits cut in a screen. Part of the beam passes through the slits. A second screen, with a fluorescent coating, is placed at some distance behind the first screen. Figure 16.7 shows the setup. In the actual experiment the width of

*Except possibly for a particle known as the neutrino. See the footnote on p. 271 (Chapter 17).

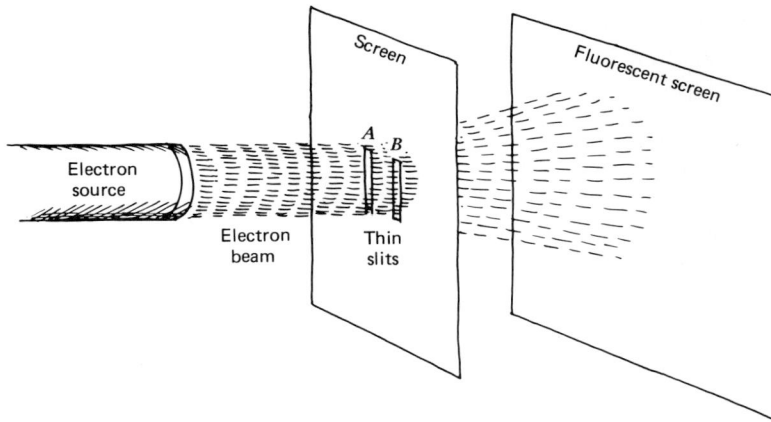

FIGURE 16.7 The electron double slit experiment. What will we observe on the screen?

each slit is about 0.00003 cm, and the distance between the two slits is about 0.00010 cm. The main problem in performing this experiment is making slits this small, and they need to be this small in order to observe the new phenomenon. That's why this phenomenon wasn't discovered until the 1920s.*

Nothing very surprising happens when only one slit, say slit A (the one on the left in the figure), is open. The electrons that pass through this slit create a pattern of flashes on the screen that has roughly the same shape as the slit. The pattern is somewhat broader than the slit, but this is not surprising because the beam is expected to spread somewhat after passing through the slit since all the electrons are not likely to be moving in precisely parallel directions. With an intense electron beam, there are so many flashes on the screen that a complete image of the slit appears (Fig. 16.8a). As the beam is weakened, the screen begins to show isolated and brief flashes at various points within the previous pattern, rather than a continuous solid image of the slit. These individual flashes are further evidence that the pattern is caused by tiny particles.

If slit A is closed and slit B is opened, the same pattern occurs except that it is shifted to lie behind slit B (Fig. 16.8b).

There are no surprises in any of this. It's what we expect, if electrons are tiny particles.

What happens with both slits open?

Figure 16.9 shows the experimental result. It is an interference pattern! Figure 16.10 is a photograph of this pattern. The photograph was made by placing a chemically prepared plate at the position of the screen in Figure 16.9. Each grain in the photographic negative is produced by a single electron. Compare Figure 16.9 and Figure 16.10 with the similar interference pattern for light, Figures 12.11 and 12.12.

Now, that is surprising. We thought that an electron beam was a stream of particles, not a wave. How can a stream of particles form an interference pattern?

According to the standard particle theory of electrons, any electron that strikes

*This particular experiment wasn't performed until 1961; however, other experiments demonstrating the new phenomenon were carried out in the 1920s.

the fluorescent screen must have come through either one slit or the other, not through both. After all, one electron is much smaller than either slit. Suppose an electron gets through slit A. Then surely it will do just what electrons did in the single slit experiment when they got through slit A. Namely, this electron should contribute to the pattern shown in Figure 16.8a; and any electron getting through slit B should contribute to the pattern shown in Figure 16.8b. The overall pattern, with both slits open, should simply be a superposition of the two patterns of Figure 16.8. Figure 16.11 shows this expected pattern.

But we don't get the expected pattern. We get Figure 16.10 instead. It looks like

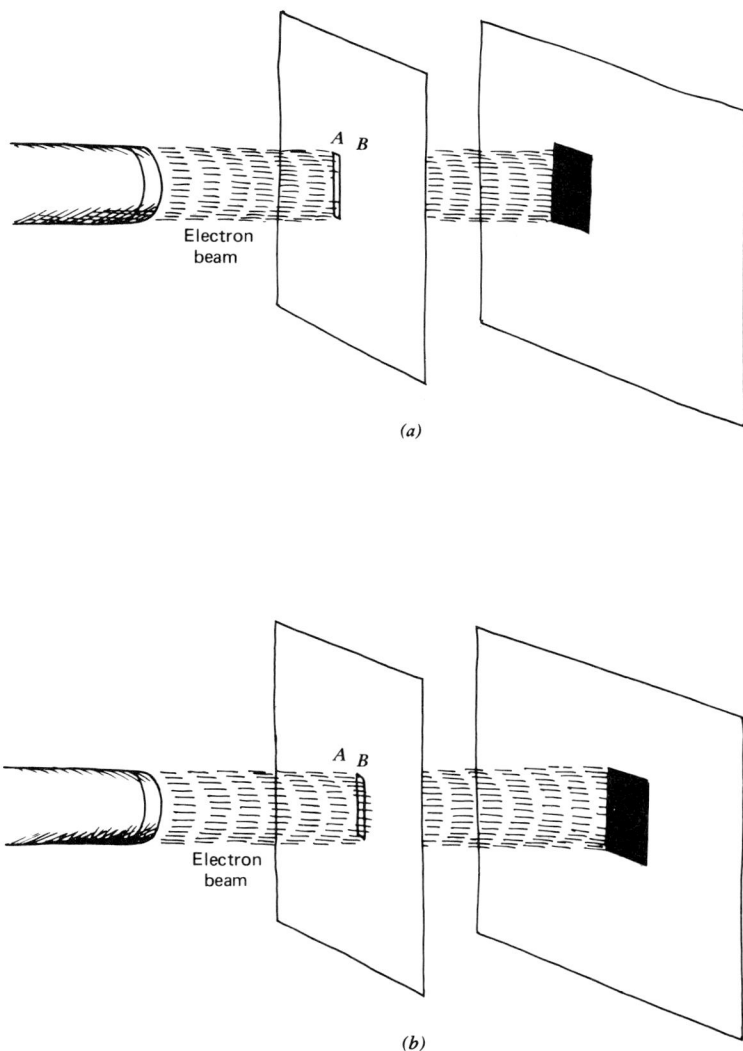

(a)

(b)

FIGURE 16.8 An electron beam experiment with only one slit open. (a) Slit A only. (b) Slit B only.

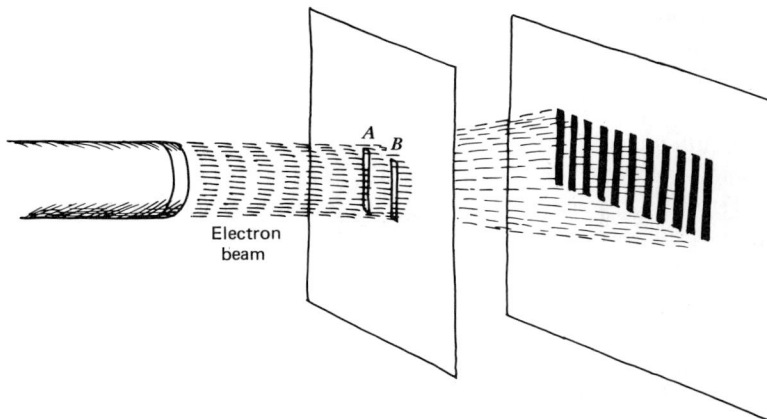

FIGURE 16.9 The outcome of the double slit experiment with electrons.

an interference pattern. It is the kind of thing we would expect if a *wave* were coming through the two slits.

It is hard to explain this pattern. We might suppose that the pattern is caused by forces acting between the electrons coming through slit *A* and the electrons coming through slit *B*. In order to investigate this possibility we could repeat the experiment using a very weak beam, say one electron per minute. In this case only one electron comes through at a time and the electrons cannot exert forces on each other. Surely, under these conditions we will get the pattern shown in Figure 16.11. But this expectation is wrong. If we leave the chemically treated plate in place for several hundred minutes in order to allow several hundred electrons to hit the screen so that we can see a recognizable pattern, the result once again turns out as shown in Figure 16.10.

This pattern contradicts the particle theory of electrons. For example, consider the dark regions in the interference pattern, the unexposed dark lines in Figure 16.10. When both slits are open electrons never strike within these regions. In a wave theory they would be called regions of destructive interference (Chapter Twelve). With only slit *A* open some electrons will strike within these regions, and with only slit *B* open

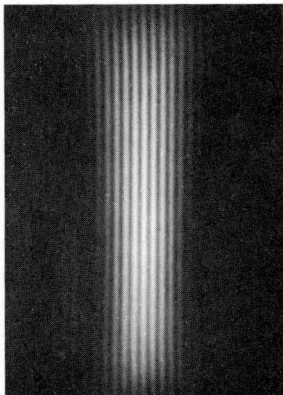

FIGURE 16.10 The double slit interference pattern made by electrons. This photograph was provided by Professor Claus Jönsson, who performed the original experiment.

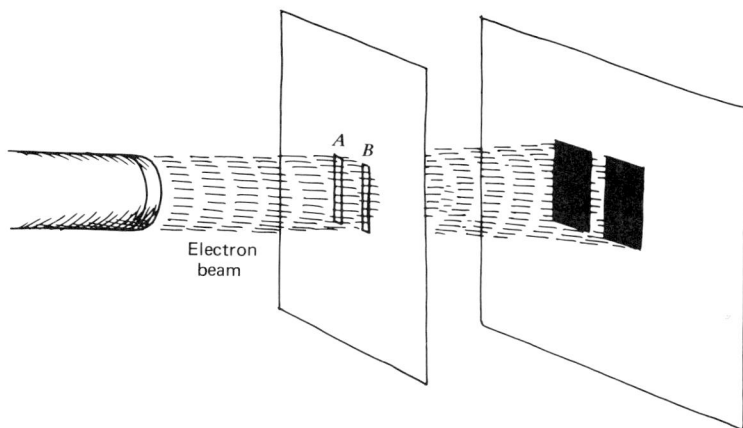

FIGURE 16.11 The two patterns of Figure 16.8 superimposed on each other. This is what we expect to observe in the double slit experiment, but it's not what we get.

some electrons will strike within these regions. How does an individual electron, coming through one or the other slit, "know" that it is supposed to avoid these dark regions when both slits are open? How does the electron know that the other slit is open? Or does each individual electron somehow manage to go through both slits? It is impossible to reconcile this result with a strict particle picture of electrons.

Prior to the discovery of this **electron interference effect,** scientists believed that all matter was composed of tiny particles. And there is certainly some truth to this view: protons, neutrons, and electrons do seem to exist, and the small flashes they produce on fluorescent screens do seem to show that these objects are small particles. But the electron interference effect complicates this simple picture. The interference pattern shows that individual electrons (or protons, or neutrons) are not *just* tiny particles. These objects sometimes act like waves.

The electron interference effect poses fundamental problems for the nineteenth-century's particle theory of matter, just as the radiation curves and the photoelectric effect pose fundamental problems for the nineteenth century's wave theory of radiation.

The Quantum Theory of Matter

In the 1920s physicists invented a theory that describes and correctly predicts the wave pattern. The theory doesn't really explain this pattern, it just accepts it, much as Planck's quantum theory of radiation accepted the existence of discrete bundles of radiation without trying to explain their existence or reconcile them with the wave theory of radiation, and much as Einstein accepted the constancy of lightspeed.

In view of the electron double slit experiment, it isn't surprising that the new theory is a wave theory of matter. The new **quantum theory of matter,** or **quantum mechanics,** asserts that matter is always associated with a wave. The wave is usually called a **matter wave.** The centerpiece of the new theory is a formula, a prescription, for the matter wave in any particular situation. Erwin Schroedinger, one of the architects of the new theory, invented this formula in 1926. You can view his famous *Schroedinger equation* in Figure 16.12.

The Schroedinger equation is one of the many so-called differential equations that physicists use to describe nature. This particular equation describes the behavior of matter waves. The Greek letter Ψ (psi) on Debra's T-shirt is the symbol for this wave. We won't delve into any of the details of the equation. The main things to understand about it are: (1) This equation enables scientists to predict the motion of the matter wave in any particular experimental situation; and (2) This equation implies that Ψ exhibits typical wave effects such as interference.

Scientists have tested the quantum theory of matter, and in particular the Schroedinger equation, under a variety of experimental conditions. The predictions of the theory have turned out to be accurate.

FIGURE 16.12 Debra is wearing the Schroedinger equation.

Like most scientific principles, the Schroedinger equation has limitations. It applies only to phenomena in which all speeds are much slower than the speed of light. The equation does not apply to situations involving high speeds, in which the special theory of relativity plays an important role. It was not until the 1930s that scientists began to develop a theory incorporating insights from both of the major twentieth-century theories—quantum theory and relativity.

The widely used **electron microscope** is not based on light waves or on electromagnetic waves (such as X-rays) of any sort. It is based on the matter waves associated with an electron beam. The wavelength of these waves is much smaller than the wavelength of visible light. Recall that light's relatively long wavelength prevents experimenters from viewing atoms or molecules with visible light microscopes. The matter waves associated with electrons are, on the other hand, comparable in size to atomic dimensions. So these waves can "see" (i.e., respond to) objects as small as atoms. "Photographs" of atoms, such as Figure 3.5 of Chapter Three, are actually patterns made by electron waves.

A Particle Is a Wave

The mystery deepens. Earlier in this chapter we saw evidence that radiation is made of tiny particles, contradicting other evidence that radiation is a wave. Now we have the electron interference experiment as evidence that matter is a wave, contradicting other evidence that matter is made of tiny particles.

What is matter anyway? What is an electron? Or a proton? Or an atom? An electron behaves sometimes as a particle, sometimes as a wave. Could it be both? But a particle is a totally different sort of thing from a wave. Could it be sometimes one and sometimes another? How can we picture, or even think consistently about, something with such contradictory properties? We have another dilemma: a wave-particle dilemma about matter.

In the next chapter we will try to get some insight into these difficult but important questions about the underlying structure of the universe. In the opinion of many scientists, these questions have great human significance. They may in fact lead our culture toward a new understanding, a less materialistic and less object-oriented consciousness, of the natural world around us.

Checklist

the quantum theory	wave-particle dilemma
thermal motion	distinction between matter and radiation
thermal radiation	operation of a television tube
radiation curve	double slit experiment with electrons
photoelectric effect	electron interference effect
photovoltaic effect	quantum theory of matter
Max Planck	Erwin Schroedinger
quantum theory of radiation	Schroedinger's equation
photon	matter waves
Planck's constant	
electron microscope	

Further Thought Questions

3. Does it seem strange to you that a photon, which is a tiny particle, can have a frequency attached to it? Waves and vibrating objects have frequencies, but how can a photon, a particle moving in a straight line at the speed of light, have a frequency? (*Note:* If this perplexes you, don't be alarmed—it *is* perplexing.)

4. Discuss the similarities between matter and radiation. Discuss the differences.

Further Exercises

9. Describe the shape of the radiation curve for an object that is heated until it begins to glow red.

10. Explain, in terms of radiation curves, why high-temperature objects glow nearly white. *Hint:* White light is composed of all wavelengths in the visible part of the spectrum.

11. Energies of atomic and subatomic phenomena are sometimes measured in small units called electron volts. Suppose that 3 electron volts of work must be done in order to remove just one of the most weakly bound electrons from the surface of a metal plate. Suppose that a certain weak beam of radiation is calculated, on the basis of the electromagnetic wave theory, to carry 1.5 electron volts of energy into the metal surface each second. According to the electromagnetic wave theory, how long must it be before the first electron is emitted?

12. Continuing the preceding exercise, suppose that we recalculate using the quantum theory of radiation. Suppose that this theory tells us that each photon in the beam has an energy of 3.5 electron volts. Will this beam knock any electrons out of the surface? How long will it be before the first electron is emitted? How much energy might this electron have, once it gets off of the metal surface?

13. Rework the preceding exercise, assuming that each photon in the beam has an energy of only 2.5 electron volts. Does this correspond to longer-wavelength (redder) or shorter-wavelength (more violet) radiation than the 3.5 electron volt photons?

14. Why does a television tube need to be evacuated?

Recommended Reading

1. Barbara Cline, *Men Who Made a New Physics*, Signet, New York, 1966. Biographies of the leading figures in the development of the quantum theory.
2. Robert H. March, *Physics for Poets*, McGraw-Hill, New York, 1978. A beautifully written physics text for liberal arts students that emphasizes the philosophical aspects of science. Cahpters 13 through 19 deal with atoms and quantum physics.

17

QUANTUM THEORY: THE GHOST IN THE MACHINE

*The universe is not only queerer than we
suppose; it is queerer than we can suppose.*

J. B. S. Haldane, biologist and author

What is the universe really made of? Nature seems determined to thwart all our researches into this question. Matter and radiation present different, contradictory aspects to us at different times, acting now like tiny particles and now like waves. Physicists have never really resolved this dilemma. They have simply learned to accept it. Scientists have invented a logically consistent theory, the quantum theory, which correctly describes the experimentally observed properties of matter and radiation. When the experiment exhibits particle phenomena, the theory predicts particle phenomena; when the experiment exhibits wave phenomena, the theory predicts wave phenomena. But the quantum theory doesn't make the dilemma vanish because the theory is just as difficult for our intuitions to grasp as the experiments were in the first place.

This chapter is a discussion of the meaning and significance of the quantum theory. The problem of understanding the quantum theory of radiation is in many ways similar to the problem of understanding the quantum theory of matter. For instance the wave-particle dilemma makes its appearance in each theory. We will concentrate only on the quantum theory of matter.

The Unpredictability of Nature

Suppose you fire a rifle bullet at a distant wall. The bullet strikes the wall at a particular place. If you repeat the experiment precisely, the bullet will again hit the same point. A precise repetition would be difficult—you would need to clamp the rifle in place, control all air currents, and so on. But if you can control all of the variables, you will find that the bullet hits the same target point. The path of the bullet is predictable: identical causes (identical rifle firings) produce identical effects (identical target points).

A belief in the predictability of nature lies at the heart of Newtonian physics and permeates pre–twentieth-century science. Newton's physics is a scheme for precisely predicting the future behavior of objects from knowledge of their present behavior. The predictability of nature underlies the idea that the universe is a grand machine running like clockwork according to pre-determined laws (Chapter Eight). In the minds of many, the validity of the entire scientific enterprise is a consequence of the predictability of the natural world. It seems that in a capricious universe, one in which identical causes produce different effects, science would be impossible. For example, Einstein, who never entirely believed the quantum theory even though he made

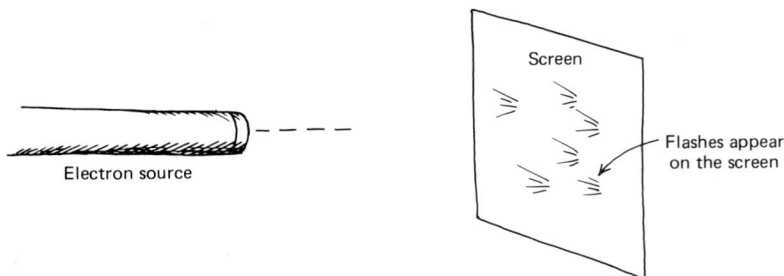

FIGURE 17.1 No matter how carefully we arrange our experiment, the flashes always occur at different points.

important contributions to it, remarked that "God does not play dice with the universe."

Nevertheless, experiments seem to show that, at least at the atomic and subatomic level, God does play dice.

Let's return to the electron beam experiment. To keep matters simple, suppose we just direct the beam at a fluorescent screen that has no slits in front of it. Let's weaken the beam to the point that only one electron is emitted every minute. (Aren't these imaginary experiments nice? We can set them up any way we please, so long as the setup doesn't violate any laws of nature.) Naturally, we'll observe one small flash on the screen every minute. Suppose we try to control the experiment in such a way as to make every electron strike the same point on the screen. The first flash occurs at some particular point. We find experimentally that the second flash nearly always occurs at a different point, even though the experimental setup is unchanged. Our first thought is that we haven't controlled all the experimental variables well enough. So we evacuate all the air from the region of the electron beam, clamp the source of electrons, the "electron gun," rigidly in place, and so on. All such efforts are fruitless, as the second flash still occurs at some distance from the first flash (Fig. 17.1). The outcome is not predictable: identical causes produce different effects.

There have been many experiments like this one, involving many sorts of atomic and subatomic particles in many sorts of situations. In every case, identical repetitions of the experiment produce results that differ in some respect. Attempts to control the variables are fruitless in reducing the uncertainty of the outcome.

Experimental outcomes such as those depicted in Figures 16.9 and 16.10 are direct evidence for this randomness, or uncertainty, in nature at the atomic level. These patterns were formed by a large number of electrons striking many different points on the screen. Yet every one of these electrons came from the same electron gun and passed through the same double slit aparatus. No matter how we may try to control the experiment, we always get a scattered pattern of hits on the screen.

And yet a certain amount of predictability remains. Although individual flashes on the screen are not predictable, the *overall pattern* is predictable. If we allow 1 billion electrons to hit the screen and if we then repeat the experiment by allowing another billion electrons to hit the screen, we'll find that the same overall pattern is formed both times.

Because the pattern resulting from a large number of individual events is predictable but the individual events are not predictable, the quantum theory renounces the possibility of precisely predicting individual atomic events and focuses instead on predicting the overall patterns arising from a large number of individual events. In fact, the Schroedinger equation was designed for precisely this purpose. The wave pattern, the psi of Schroedinger's equation, is the pattern arising from a large number of individual atomic events.

As an example, consider again the double slit experiment with electrons, shown in Figures 16.9 and 16.10. Figure 17.2 is a graph of the pattern produced in a typical such experiment. The graph shows the intensity of the pattern at various points on the screen as viewed from above in Figure 16.9. You might want to turn your book 90 degrees clockwise to read this graph, because the intensity of the pattern is graphed toward the left versus the position on the screen, which runs vertically. In terms of the particle theory of electrons, the pattern was formed by a large number of individual electrons striking different points on the screen. The graph is thus a record of the number of hits at any point on the screen. Although the quantum theory is unable to predict the point at which any particular electron will strike the screen, the theory is able to predict this graph. Scientists use the Schroedinger equation to make this prediction. For example, it is possible to predict that no flashes will occur at the points marked x on the graph, that the largest number of flashes will occur near the point marked a, that a smaller number of flashes will occur near b and so on. But nobody can predict which flashes will occur at which points. We cannot predict where the first flash will occur, for example. We can only predict that after a large number of flashes have occurred, the overall pattern will agree with the graph in Figure 17.2.

The Cosmic Dice

An element of chance seems to be operating at the atomic level. For instance, in the double slit experiment with electrons the odds, or probabilities, seem to favor flashes occurring at points a, b, and c rather than at the points marked x. The situation

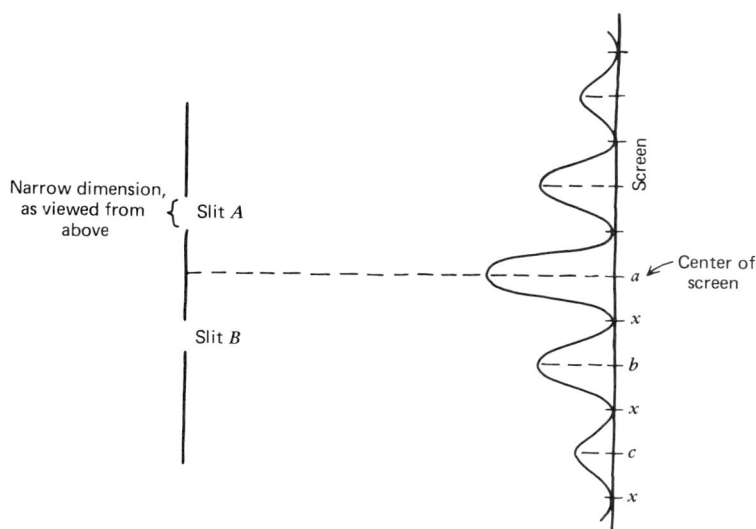

FIGURE 17.2 Graph of the pattern of flashes produced in a double slit experiment with electrons.

is analogous to an ordinary game of chance. For example, in a "fair" toss of a coin we cannot predict the outcome of individual tosses, but we can predict the overall pattern of a large number of tosses. We usually refer to this pattern as the odds, or probabilities, for heads or tails. For a fair coin toss we predict that heads will occur with a probability of 50 percent and that tails will occur with a probability of 50 percent. That is, in a long series of tosses, roughly 50 percent of the outcomes will be heads and 50 percent will be tails. We cannot predict which tosses will yield which outcome, we cannot predict the outcome of the first toss, the second toss, or any subsequent toss. Only the pattern is predictable.

Scientists interpret the wave patterns of the atomic world in precisely this probabilistic way. The wave pattern that appears in experiments like the electron double slit experiment and that appears in the quantum theory as the psi of Schroedinger's equation is regarded as a probability pattern. It gives the odds of a flash occurring in the vicinity of, say, point *a* in Figure 17.2.

Although the coin toss and the electron double slit experiment both involve probabilities, there is a crucial difference between the two. The motion of the coin is, in principle, completely predictable. If we knew the precise initial speed, direction of motion, position, air conditions, and the like, and if we had a computer large enough to perform the numerical calculations, we could use Newton's laws to predict the precise motion of the coin and thus to predict the outcome. And this prediction would be correct: Newtonian physics works just fine for human-scale objects like coins. We generally don't have access to all this information and computer power. Nevertheless, "nature knows" how these objects will move, even if human observers, in their ignorance, do not.

Atomic and subatomic objects are different. They obey quantum mechanics, not Newtonian mechanics. It seems that even nature doesn't know the outcome of experiments with such tiny objects. No amount of detailed information, no computer, no laws of physics, allow the outcome to be predicted. The quantum theory can predict only the overall probability pattern.

Thus the wave pattern, the psi in Schroedinger's equation, is a probability wave. We'll call it the **psi-wave.** It is a fairly abstract beast, a little like the electromagnetic waves of Chapter Thirteen. A psi-wave is not a wave *in* anything. It isn't a wave of any physical thing in the first place. It is simply a way of describing the odds for such events as the flashes at various points on a screen. It seems that nature has arranged for these odds to be predictable, and for psi, which represents these odds, to obey the Schroedinger equation.

To be more quantitative, here is the meaning of the psi-wave as shown by experiment for the case of an electron striking a screen. The psi-wave has some numerical value at each point on the screen (different values at different points). The square of the value of the psi-wave at any particular point turns out to be proportional to the probability that the electron will hit the screen at that point. For example, suppose that the psi-wave is twice as large at *a* as at *b* in Figure 17.2. Then a flash is four times as likely to occur near *a* as near *b*.

Exercise 1

What value of the psi-wave should the Schroedinger equation predict at the points marked *x*? If nine times as many flashes occur near *a* as occur near *c*, what should the Schroedinger equation predict about the relation between the psi-wave at *a* and that at *c*?

Exercise 2

(a) Give the probability distribution for fair tosses of a single die. (b) Under ordinary circumstances, can human observers predict the outcome of a single toss of a pair of dice? (c) Does the uncertainty in this experiment have anything to do with quantum theory?

The Human Connection

Quantum theory gives a correct description of the wave-particle dilemma, but the Schroedinger equation and our probabilistic interpretation of the psi-wave only describe the mystery without dispelling it. We would still like to know: What is the universe made of? What is an electron beam made of? What is really coming through the two slits? Particles? A wave?

Werner Heisenberg (Fig. 17.3), one of the founders of the quantum theory, gave considerable thought to these questions. His idea was that the answers might lie in a careful consideration of the measurement process.

Suppose you want to measure some properties of an atomic or subatomic particle, such as an electron. If you want to make accurate predictions about the future behavior of an electron, two things you will need to measure are its position and its velocity. After all, you need to know where it is and how fast (and in what direction) it is moving if you want to know where it will be in the future. So you set out to measure an electron's position and velocity.

One way to measure an object's position is to look at it. In other words, to reflect visible light off the object and into your eyes (or into some more precise detector) so that you can see where the object is (Fig. 17.4). This method won't quite work on an electron because light waves can't "see" objects, such as electrons, which are much smaller than the wavelength of the light. Shorter-wavelength radiation has to be used to accurately determine the electron's position.

Exercise 3

(a) Of the following types of radiation, which will provide the most accurate and which will provide the least accurate position measurement for an electron? Ultraviolet, visible, gamma ray, infrared, X ray. (b) Which type of radiation has the highest energy per photon? Which has the lowest energy per photon?

The incoming radiation acts like a stream of photons as well as like a wave. When one of these photons hits the electron, it gives the electron a jolt, just as one billiard ball gives another billiard ball a jolt when the two collide. Naturally, the electron's speed and direction of motion are altered by this jolt. In the process of trying to measure the electron's position you have disturbed its velocity.

As you might suppose, this disturbance makes it difficult to carry out a simultaneous measurement of the exact position *and* exact velocity of the electron. Impossible, in fact. The problem is that at least one photon must bounce off the electron if we are to have any idea of the electron's whereabouts. That photon must have a short wavelength if this position measurement is to be accurate. But the short wavelength photons are precisely the ones that disturb the velocity the most, because they are the ones with the most energy.

(a)

FIGURE 17.3 Two founders of the quantum theory. (a) Werner Heisenberg in 1926, at about the time he developed the uncertainty principle.

It appears that there are limits on the accuracy with which we can measure the position and velocity of small particles like electrons. The observer trying to measure these quantities is caught in a trade-off: Attempts to reduce the inaccuracy in position necessarily increase the inaccuracy in velocity.

This trade-off between inaccuracies may seem to be a minor point, but it is the sort of minor point that is likely to shake the foundations of science. For science is founded on measurement. The measurement process is the most basic of all "connections" in science. It connects the human observer to the natural world. Any theory that puts fundamental limitations on the measurement process is likely to have consequences for the entire scientific enterprise.

The Uncertainty Principle

Heisenberg used the quantum theory to analyze the measurement process quantitatively. He found that *every* process of measurement disturbs the measured object, no matter what the process and no matter what the object. This disturbance causes inaccuracies, or uncertainties, in the position and velocity of the measured

FIGURE 17.3 (b) Neils Bohr's engagement photograph in 1911. Two years later he developed the first accurate theory of atomic structure.

FIGURE 17.3 (c) Heisenberg (left) and Bohr (right) in 1934. (AIP Neils Bohr Library.)

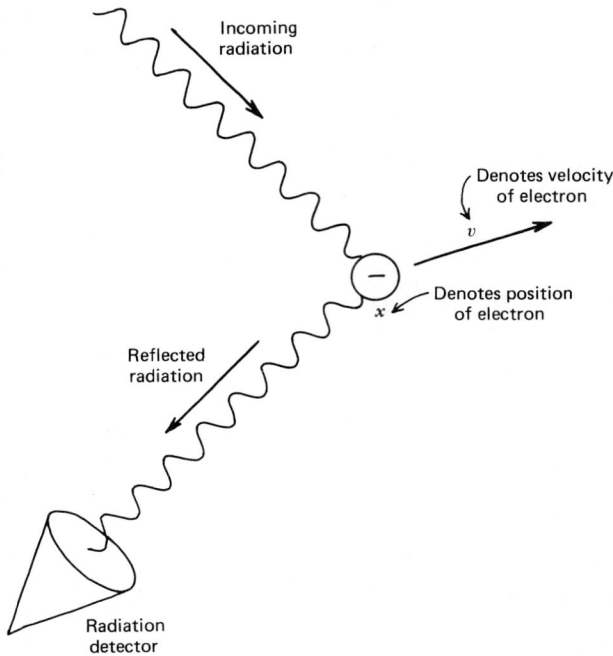

FIGURE 17.4 How to "see" an electron.

object. He found a relation between these uncertainties that states quantitatively the idea discussed in the preceding section. According to Heisenberg's relation, there is a trade-off between the uncertainties in position and velocity of a particle. Any attempt to reduce the position uncertainty will, beyond a certain point, disturb the system so much that a large velocity uncertainty will be produced. And vice versa. Attempts to reduce the velocity uncertainty will produce large position uncertainties.

Before going further, we need to see just what Heisenberg meant by the "uncertainty" in position and in velocity.

Exercise 4

Suppose that a friend informs you that it is "250 miles to Kansas City, give or take 10 miles." How large is the uncertainty in the distance to Kansas City? Does this uncertainty have anything to do with the quantum theory? (Answers below.)

Your friend informed you that it is between 240 and 260 miles to Kansas City, so the statement contains an uncertainty of 20 miles. This uncertainty has nothing to do with quantum mechanics. You could get much more accurate information than this without coming into conflict with any principles of quantum theory. Only measurements made on objects of molecular, atomic, or smaller size are affected by the quantum theory.

Exercise 5

You measure the speed of a hydrogen atom as it moves down an evacuated (emptied of air) tube in a physics experiment. You find its velocity down the tube to be between 12,200 m/sec and 12,400 m/sec. How large is the uncertainty in the atom's velocity?

Exercise 6

Suppose that at some particular instant the atom in the preceding exercise is found to be between 0.04527 m and 0.04531 m from the end of the tube. What is the uncertainty in the atom's position?

Heisenberg found that there is a restriction on the product of the two uncertainties; that is, the position uncertainty multiplied by the velocity uncertainty can be just so small and no smaller. His result is known as

The **Heisenberg Uncertainty Principle.** We cannot accurately predict, or even measure, both the position and velocity of a material particle. Quantitatively, the uncertainties must be large enough so that the product of the position uncertainty times the velocity uncertainty is greater than Planck's constant divided by the particle's mass.

Briefly,

$$\text{(Uncertainty in position)} \times \text{(uncertainty in velocity)} > \frac{6 \times 10^{-34}}{\text{particle's mass}}$$

where the symbol $>$ means "is greater than."

Thus a highly accurate position measurement (small uncertainty in position) must be accompanied by a large uncertainty in velocity, and vice versa. We can't make both uncertainties small at the same time.

Notice the way in which the particle's mass enters into this relation. For larger masses the right-hand side of the relation is smaller, which means that the uncertainties can be smaller for more massive particles. For an extremely low-mass particle we must have a large uncertainty in position, velocity, or both, but for a high-mass particle we can have small uncertainties in both position and velocity. In other words, less massive objects are more affected by the uncertainty principle. This is exactly what we expect, because less massive objects have less inertia and so they are disturbed more by the measurement process.

The lightest particle of matter known is the electron.* Being the lightest and hence most easily disturbed particle, it should be the one most affected by the uncertainty principle. If we use the electron's mass, about 10^{-30} kg, the right-hand side of Heisenberg's relation boils down to 6×10^{-4}, or 0.0006. In other words, for an electron, the product of the two uncertainties, expressed in metric units (m and m/sec), must be at least as big as 0.0006.

As an example, suppose that we measure an electron's position with an uncertainty equal to the 6×10^{-7} m wavelength of yellow light. The smallest allowable uncertainty in velocity for this electron then turns out to be 1000 m/sec. Any velocity uncertainty smaller than 1000 m/sec would conflict with the uncertainty principle. Both these uncertainties are fairly sizable: 6×10^{-7} m is large compared to the size of an electron, and 1000 m/sec is about 2200 mph.

---

*Recent flash! The particles known as neutrinos (Chapter Nineteen) had been thought to have zero mass and were thus classified as radiation rather than as matter. Recent experiments indicate that the neutrino's mass might be nonzero, but very small. If so, then the neutrino is the lightest material particle known. Isn't science fun?

Thus the quantum theory puts important restrictions on the accuracy with which we can know the position and velocity of small, low-mass objects. For more massive objects, the inaccuracies are not required to be as large. For example, if the object has a mass of 1 kg, the right-hand side of Heisenberg's relation becomes 6×10^{-34} —a very small number. So the position and velocity uncertainties are both allowed to be extremely small.

Exercise 7

Arrange these objects in order, beginning with the object having the largest "quantum mechanical uncertainties" (i.e., uncertainties resulting from Heisenberg's principle) to the object having the smallest uncertainties: proton, helium atom, electron, water molecule, grain of dust, baseball, glucose molecule, automobile.

The lesson here is that Heisenberg's principle is important only for objects of molecular or smaller size. The quantum theory puts no significant restriction on experimental accuracies for larger objects, such as baseballs or even grains of dust, because quantum mechanical uncertainties for such objects are much smaller than can be detected by the precision of any measuring instrument.

The feature of quantum physics that distinguishes it most sharply from Newtonian physics is its description of nature in terms of probabilities rather than certainties. Heisenberg's principle shows that these probabilities arise because the observer always disturbs the measured object. This disturbance produces an irreducible uncertainty in the observer's knowledge of the object. The best the observer can do is describe what he or she *does* know in probabilistic terms.

Since Heisenberg's principle has only a negligible effect on everyday objects, it isn't surprising that the entire quantum theory has only negligible consequences for such objects. When dealing with human-scale objects, like baseballs and cars, we can forget about quantum physics and use old-fashioned Newtonian physics instead. Strictly speaking, quantum physics still correctly predicts the behavior of these objects, but quantum physics becomes identical with Newtonian physics as we pass from the molecular world to our larger-scale world. So we use Newtonian physics, a simpler theory than quantum physics, to describe the world of everyday experience. This is just like the relation of Einstein's relativity to the older Galilean relativity. Einstein's relativity becomes identical with Galileo's relativity as we pass from high speeds to low speeds.

Summing up: The new physics of quantum theory and relativity theory is important only in the realms of the very small or the very fast, realms that are far from everyday human experience.

That's why the new physics is so surprising, so nonintuitive. We live in a world far removed from the world of the small and the fast. People are a couple of meters tall, whereas atoms are 10^{-10} meters across. People walk at a few meters per second, but light moves at 3×10^8 meters per second. Our minds are not prepared for these distant worlds.

Whatever Became of the Natural World?

The uncertainty principle presents science with an interesting dilemma. Science is supposed to study the natural world. Now, what can "natural world" mean? If it means anything at all, it surely means the world outside subjective human experience, outside the human imagination, outside human feelings. It surely means the objective world, those phenomena which can be observed by all humans (provided they possess the proper microscopes, telescopes, etc.), those events which occur "out there" in the universe rather than only "in here" in our minds.

In the effort to objectively study the natural world, scientists have always moved the subjective human observer as far out of the way as possible. Scientists try to employ only objective, quantifiable, universally agreed-on concepts, such as frequency and temperature, rather than subjective, qualitative concepts, such as color and warmth. Science separates the subject (the observer) from the object (the observed) and studies only those items lying on the object side. This subject-object split is central to science and perhaps to all of Western culture.

Thought Question 1

Does it seem to you that our culture does in fact promote a separation between human feelings on the one hand and the "real world" of, for instance, automobiles and business dealings, on the other hand? If so, in what ways does this separation affect your life?

Scientists seek the "objective realities" behind such "appearances" as color and warmth. This search for objective reality has penetrated to the world of molecules, beyond that to the world of atoms, and beyond that to the world of protons and electrons. Thus science discovers that color is "really nothing but" an electromagnetic wave of a certain frequency interacting with certain charges in the eye's retina. Thus science discovers that warmth is "really nothing but" molecules moving silently in empty space.

And finally, just as we begin to think that here, at the level of the atoms, science has at last found that objective natural world beyond mere appearances and beyond mere human imagination, we find the human observer staring back at us out of our microscopes, out of the atoms. We have not escaped the human element at all. Instead we have come full circle, and returned to the human world. There is no way to divorce object from subject. For the world observed is the world changed. The natural world, separate from and uninfluenced by humankind, is a concept that itself lies outside of science. Science deals only with what can be observed, and the uncertainty principle tells us that any such natural world is not observable.

Thought Question 2

We have seen before that things that cannot be observed should not be said to exist scientifically. How seriously should we take this? Should we conclude that the natural world does not exist? Or that it only partially exists?

As the great theoretical physicist John A. Wheeler put it in 1973:

No theory of physics that deals only with physics will ever explain physics. I believe that as we go on trying to understand the universe, we are at the same time trying to understand man. . . . The physical world is in some deep sense tied to the human being.

The scientific enterprise seems to have contradicted itself. The entire undertaking was based on the idea that there is a natural world, a world uninfluenced by the observer, and now quantum physics, the most fundamental description we have of this natural world, informs us that in fact it is a basic principle of science that no such natural world can be observed.

Most physicists who give any serious thought to these philosophical matters would agree with Neils Bohr's thinking. Bohr (Fig. 17.3) was one of the founders of the quantum theory and the architect of an early and still widely used model of atomic structure. He felt that electrons and atoms are so far removed from our familiar world of baseballs and automobiles that our minds are totally unable to conceptualize them. So any attempt to picture what is happening in for example the double slit experiment is doomed to failure: a particle picture is doomed; a wave picture is doomed; any intermediate wave-particle picture is doomed. All we are allowed to say is: an electron gun fired, flashes appeared, and the flashes formed an interference pattern. The quantum theory correctly predicts the pattern, but the theory does not tell us what is really going on.

Bohr's point is that we're not allowed to form any picture of what is going on at the atomic level. In the double slit experiment, we can say that an electron source is operating that and flashes are appearing, but any picture of what is going on behind the scenes, between the firing and the flashes, is bound to be wrong. If we assume that a stream of tiny particles is moving from the source to the screen, we cannot explain the interference pattern that is seen when both slits are open. If we assume that a wave is moving from the source to the screen, we cannot explain the tiny flashes appearing on the screen. Electrons and atoms are only inventions of the human mind, mental constructs that are useful in helping us think about the atomic world. But if we begin to take these concepts too seriously, if we begin to think that nature is really made of these objects, we soon run into contradictions with experiment. Any single picture we form will soon run into contradictions with experiment.

In view of the tantalizing ambiguity of the quantum theory, it is not surprising that when Bohr was knighted as an acknowledgement of his achievements in science and his contributions to Danish culture, he chose as a suitable motif for his coat-of-arms the Chinese symbol of *t'ai-chi*, representing the complementary relationship of the archetypal opposites yin and yang (Fig. 17.5). The inscription above the symbol reads, *Contraria sunt complementa*, "Opposites are complementary." Thus did the quantum theory's most profound philosopher acknowledge the harmony between such seeming opposites as waves and particles, a harmony contemplated in the ancient Eastern teaching that the essence of all natural and human phenomena lies in the dynamic interplay of complementary opposites.

Werner Heisenberg conveyed the tone of these ideas in his autobiography, when he emphasized that

> atoms are not *things*. The electrons which form an atom's shells are no longer things in the sense of classical physics, things which could be unambiguously described by concepts like location, velocity, energy, size. When we get down to the atomic level, the objective world in space and time no longer exists, and the mathematical symbols of theoretical physics refer merely to possibilities, not to facts. (W. Heisenberg, *Der Teil und das Ganze*, Munich, 1969).

The quantum theory is today's deepest description of the natural world. Yet the theory remains essentially mysterious. What implications does this mystery have for the validity and significance of physics, and indeed of the entire scientific enterprise? A few possible answers: (1) These philosophical considerations are irrelevant to science and to human affairs. (2) The materialistic position, that the real world consists

FIGURE 17.5 Neils Bohr's coat of arms. (AIP Neils Bohr Library.)

only of tiny particles moving in empty space, is correct. However, the quantum theory shows that humans cannot completely know this microscopic world. (3) The dilemmas of quantum physics show that the quantum theory cannot be correct. A better theory will someday be found. (4) Science is a correct description of at least a part of the real world. However, the quantum theory shows that scientific descriptions are incomplete in the sense that there are other valid ways of viewing reality. (5) These considerations show that science is not a valid way of viewing the world.

Thought Question 3

With which of these positions do you agree, and with which do you disagree? What is your view?

In my opinion, quantum physics marks the end of the Newtonian interpretation of the world as clockwork mechanism and marks the rejection by science of the vision of a universe based on atoms moving in empty space. Allen Wheelis labeled it well in the title of his remarkable book, *The End of the Modern Age* (Ref. 6). The modern age, by which Wheelis means the age initiated by the Copernican revolution and characterized by Newtonian physics, has been increasingly modeled in the image of science as the way to truth and progress. Today, in our twentieth-century post-modern age, quantum physics seems to be telling us that science is at best only one part of "the way."

The Newtonian vision treats the universe as an object that can at least in principle be precisely measured and precisely predicted. This vision of the universe as object has led us to expect that the external, natural world is completely knowable and that with sufficient time and effort we can answer the question "What is the universe really made of?"

Quantum theory regards the world observed as the world changed. Since all scientific knowledge comes through observation, the quantum view relinquishes the hope of completely knowing the natural world. In fact, quantum physics makes us question the existence of any such totally independent objective world, because any picture we try to form of such a world turns out to conflict at some point with our observations. The difficulty is that nature will not hold still for our observations, that the objects of the external world insist on interacting with us.

The message of quantum physics is that there is a ghost in the Newtonian world machine, and the ghost is ourselves.

The quantum theory gives correct answers, correct predictions, as far as it goes. But it seems that the theory does not go, cannot go, to the end of all our questions. To such questions as "What is the universe really made of?" and "What is an electron?", the quantum theory replies that the essential connection between the human observer and the rest of the universe prevents us from ever knowing, and in fact may even render such questions meaningless.

Checklist

predictability versus unpredictability	uncertainty in position
probability	uncertainty in velocity
probability wave	Heisenberg's uncertainty principle
psi-wave	Neils Bohr
Werner Heisenberg	

Further Thought Questions

4. Criticize the following statement: The true nature of an electron is intermediate between that of a particle and of a wave.

5. Which of the two possibilities for the universe, Newtonian determinism or quantum randomness, do you find most appealing? Or is there some other alternative which you prefer?

6. List as many contrasts and as many similarities as possible between Newtonian physics and quantum physics.

7. Mao Tse-Tung, despite his unfamiliarity with quantum mechanics, once said, "If

you want to know the taste of a pear, you must change the pear by eating it."
Discuss.

8. Referring to the section on the world as machine in Chapter Eight, discuss the implications of Newtonian physics and of quantum physics for the question of individual free will. Do you feel that physics has anything to do with such questions as free will?

9. The mathematician, physicist, and writer Sir James Jeans, commenting on twentieth-century physics, once said, "The universe begins to look more like a great thought than like a great machine." Is modern science leading us away from the materialistic philosophy?

Further Exercises

8. Do you disturb the temperature of a cup of coffee when you use a cool thermometer to measure its temperature? Would you disturb a lake if you used a sonar beam (high-frequency sound wave) to measure its depth? Do we disturb the moon when we measure the distance to the moon using radar? Can you think of a measurement that doesn't disturb the measured object?

9. We found that, for an electron, the product of the position and velocity uncertainties had to be at least as big as 0.0006 (using metric units) or 6×10^{-4}. A hydrogen atom's mass is, roughly, 1000 times larger than an electron's mass. Does the uncertainty principle require the uncertainties for a hydrogen atom to be bigger than, or smaller than, the uncertainties for an electron? How much bigger or smaller?

10. Exercises 5 and 6 described a hydrogen atom having a velocity uncertainty of 200 m/sec and a position uncertainty of 0.00004 m, or 4×10^{-5} m. Find the product of these uncertainties. Is this a physically possible situation, or does this situation violate the principles of the quantum theory (see Exercise 9)?

11. In one example in the text, a measurement of an electron's position using yellow light produced a 1000 m/sec uncertainty in the electron's velocity. Would red light have produced a larger or smaller velocity uncertainty? What about violet light? Ultraviolet light? Which of these would give the smallest uncertainty in position?

12. Do we alter that which we attempt to measure in a public-opinion survey? Does this effect have anything to do with the quantum theory?

Recommended Reading

1. A. Baker, *Modern Physics and Antiphysics*, Addison-Wesley Publishing Company, Reading, Mass., 1970. Readable material on the quantum theory and relativity theory.

2. Fritjof Capra, *The Tao of Physics*, Shambhala Publications, Berkeley, Calif., 1975. Capra explores the parallels between modern physics and Eastern mysticism, including excellent accounts, suitable for nonscientists, of the foundations of our scientific understanding of the atomic world. The author is a high-energy physicist.

3. George Gamov, *Thirty Years that Shook Physics: The Story of Quantum Theory*, Doubleday, Garden City, N.Y., 1966. An account of the development of quantum mechanics, including many anecdotes about the persons involved, by one who was there when it happened.

4. Werner Heisenberg, *Encounters and Conversations*, Harper & Row, New York,

1970. Gives a feeling for the thought processes involved in the development of quantum mechanics.

5. Arthur Koestler, *The Roots of Coincidence*, Random House, New York, 1972. The author is a well-known novelist and popularizer of science. This book is a speculation on the possible implications of contemporary physics, particularly quantum physics, for parapsychology.

6. Allen Wheelis, *The End of the Modern Age*, Harper & Row, New York, 1973. A scientifically and poetically literate account of the rise and fall of Newtonian physics and of its mechanistic underpinnings.

18

A CLOSER LOOK AT ATOMS

Despite its mysteries, the quantum theory is quite definite from the strict scientific and mathematical point of view. That is, the theory yields verifiable predictions about the numerical outcomes of a wide variety of quantitative experiments.

And the theory works. When the predictions are checked experimentally, they turn out to be correct to a high accuracy. In fact, the quantum theory of the electromagnetic force, a theory known as quantum electrodynamics, is the most quantitatively accurate physical theory ever devised.

Most of our knowledge about atoms has been obtained by studying the radiation they emit. When the light from a particular element is spread out into its different frequencies, the pattern observed is called the **spectrum** of that element. Figure 18.1 shows the visible portions of the spectra of hydrogen and helium. These patterns are made when the light emitted by each element is spread out by sending it through a prism (see Figure 18.8, below) or through some other device that can separate a light beam into its component frequencies. Such spectra present a wealth of precisely measurable detail about atoms. In this chapter we'll see how quantum theory meets the challenge of explaining these spectra.

Neils Bohr invented the first quantitative theory of atomic structure, a theory that gave a reasonable account of atomic spectra. He proposed his theory in 1913, some 13 years after Planck's introduction of the quantum theory of radiation but more than a decade prior to the discovery of the electron interference effect and the invention

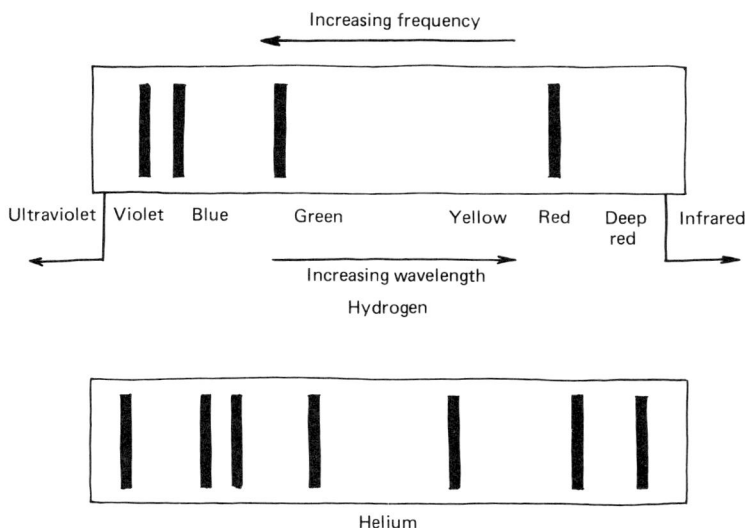

FIGURE 18.1 The spectra of hydrogen and helium.

of the quantum theory of matter. Bohr's theory was semi-Newtonian, born out of the older mechanistic planetary model of the atom (Chapter Three) and incorporating some of the new quantum ideas. We won't dwell on Bohr's theory here but will instead go directly to the modern quantum description of the atom. I mention Bohr's theory because of its historical importance.

The Quantum Description of the Hydrogen Atom

Hydrogen, consisting of just one proton and one electron and constituting 70 percent of all the atoms in the universe, is the simplest and most prevalent element. Thus most of this chapter will be devoted to hydrogen. The ideas developed for hydrogen are also valid for other atoms and molecules.

According to the planetary model of the atom, a hydrogen atom consists of one small particle, called a proton, located at the center, and one even smaller particle, called an electron, moving in circular or elliptical orbit around the proton.

Exercise 1
According to the preceding two chapters, this picture is wrong in several respects. What is wrong with it? (Answer below.)

Although the simple planetary model is useful in explaining a few properties of the atom and in providing a framework for our thoughts, it doesn't begin to explain the full range of phenomena that we observe experimentally. This clear-cut, mechanical description of the atom is the kind of thing we might expect if we lived in a Newtonian universe, but it seems that we live in a quantum mechanical universe. Large things act essentially as though they obey Newtonian physics, but small things, like atoms, don't even come close to the Newtonian picture. Protons and electrons cannot be said to be located at any particular place; they cannot be said to have any specific speed or direction of motion, so they can't be said to move in a circular or elliptical orbit; and they cannot even be said to be tiny particles, for they often act like waves.

Let's look at the contemporary quantum description of the atom.

The electron, rather than the nucleus, is the main participant in most atomic phenomena. For instance, chemical reactions depend on the electron structure, and not the nuclear structure, of atoms. So to keep things simple, our quantum description of the hydrogen atom will treat the proton as a tiny particle, at a definite position, at rest at the center of the atom. We will use the quantum theory only to describe the electron.

According to the quantum theory, only probabilities are predictable. The theory doesn't describe definite, specific orbits for our electron; in a very real sense, such orbits do not exist. What is describable by quantum theory is the psi-wave, the probability wave (Chapters Sixteen and Seventeen) associated with the electron. We've seen that the psi-wave does not tell us where the electron is or where it's going; the wave gives us only probabilities as to where the electron might be and where it might be going.

Physicists determine, or predict, the psi-wave for each atom by using the Schroedinger equation as it applies to each type of atom. The Schroedinger equation is the basic tool for studying atomic structure.

The psi-wave is not a picture of the electron or of the atom, any more than the statement that a certain nickel "has a 50-50 chance of coming up heads" is a picture of the nickel. Nevertheless, we can best picture an atom by visualizing the psi-wave. Recall that this wave can be calculated mathematically from the Schroedinger equation for the hydrogen atom. The psi-wave has a specific numerical value at each point near the atom; its value is nearly zero at points very far from the atom. The meaning of the psi-wave is that, in any experiment designed to locate the electron, the electron is more likely to be found where the value of the psi-wave is larger.

A graph showing the values of the psi-wave at each different point near the atom would be a good way to visualize the psi-wave, but a simpler way is with a picture, such as Figure 18.2, showing the psi-wave as a cloud. The psi-wave is large wherever the cloud is especially dense, and the psi-wave is small wherever the cloud is diffuse (nondense). Thus the electron is more likely to be found at high-density points and less likely to be found at low-density points.

Exercise 2

Suppose that the psi-wave is three times as large at point a as at point c in Figure 18.2, and twice as large at b as at c. Suppose that there is a 0.1 percent probability (1 chance in 1000) of finding the electron near point c. Recalling that probabilities are proportional to the square of psi, what is the probability of finding the electron near point b? Near point a?

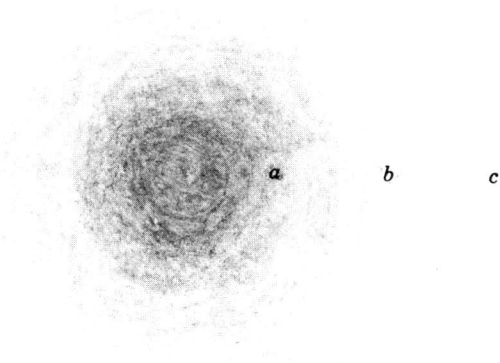

FIGURE 18.2 One way to visualize the nonvisualizable. The cloud is a representation of one of the possible states of the hydrogen atom.

It turns out that not one but several different psi-waves satisfy the Schroedinger equation. In other words, the equation predicts that the electron can be represented by any one of a number of different psi-waves. Figure 18.3 pictures several of these permitted psi-waves. At any one time the electron is represented by only one of these psi-waves. These various possible psi-waves are called the **allowed states** of the hydrogen atom.

Let's compare some of the allowed states. Some states are concentrated closer to the center, closer to the nucleus, than other states. In the more concentrated states, the electron is more likely to be found close to the nucleus. The most concentrated state in Figure 18.3 is state 1, so this is the state in which the electron is most likely to be found near the proton. In other states, such as state 2, the electron is more likely to be far from the proton.

Recall that the force that holds the electron in the atom is the electric force between the positively charged proton and the negatively charged electron. This force is stronger when the electron and proton are closer together and weaker when they are farther apart. So this binding force is stronger in the more concentrated states, such as state 1. Because state 1 is the most concentrated of all the allowed states, it is the state in which the electron is most strongly bound to the proton, the state in which it is most difficult for some external agent to pull the electron out of the atom. In other words, the atom is most stable when it is in this state. This stablest state is called the **ground state**.

Quantum Jumps

The Schroedinger equation allows the quantum state of the atom to change from time to time. For instance, an atom in state 1 of Figure 18.3 could switch to state 2. According to the quantum theory, such a change of state occurs instantaneously. To picture this process, imagine that psi-cloud 1 suddenly vanishes and is replaced by psi-cloud 2.

Let's compare the energy of the two states 1 and 2. In state 1 the electron is likely to be found close to the nucleus, and in state 2 it is more likely to be found farther from the nucleus; so when the atom switches from state 1 to state 2, the electron moves away from the nucleus. Energy must be supplied from the outside in order for this to happen. Some outside agent has to do work on the atom in order to pull the electron some distance away from the proton in the face of the electric attraction between proton and electron. According to the work-energy principle (Chapter Nine) this work done on the atom must increase the atom's energy. You can think of the electric attraction between the proton and the electron as, in effect, a spring holding the electron in toward the proton. When the electron is farther from the proton, this "spring" is more stretched and the atom has more energy.

Thus the atom must have more energy in state 2 than it has in state 1. We can extend this reasoning to all the allowed states of the atom. Those states in which the electron is more likely to be found farther from the nucleus are states of higher energy.

So the ground state, the most stable state, is also the state of lowest energy. Any other state is called an **excited state,** because the atom must gain some energy, or become excited, to go into one of these states starting from the ground state.

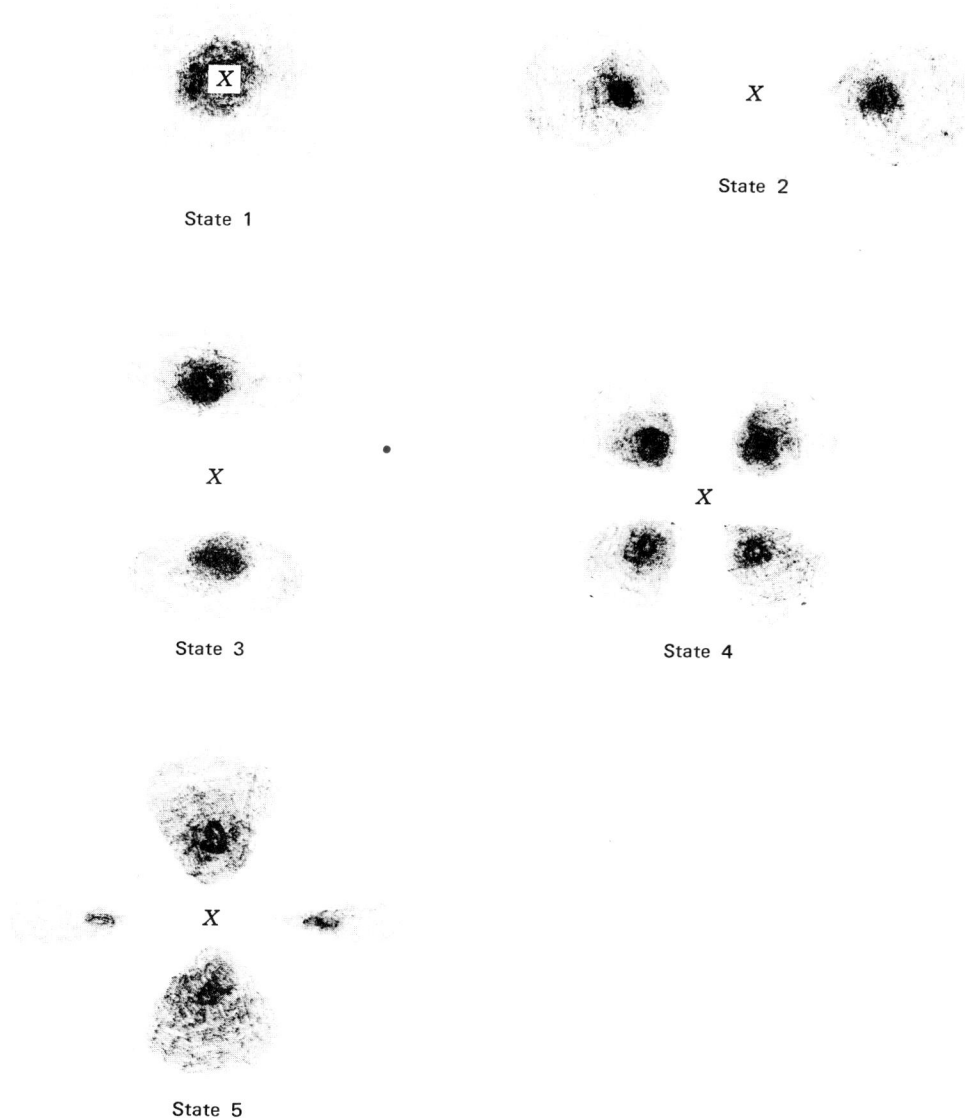

State 1

State 2

State 3

State 4

State 5

FIGURE 18.3 Drawings of the psi-waves for several different possible states of the hydrogen atom. Each psi-wave shows a different probability description for the electron. The position of the nucleus is indicated by an x.

An atom in an excited state can spontaneously (without outside help) switch to a less excited state, simply because the proton is pulling inward on the electron. For instance, a hydrogen atom might spontaneously jump from state 4 to state 2 and then to state 1. Once it gets to the ground state no further spontaneous changes are possible because no lower energy states are available.

Most atoms spend most of their time in the ground state because any time they happen to be in any other state they have a natural tendency to jump spontaneously to the ground state, but they can't jump "up" out of the ground state without outside help to provide energy. In a collection of hydrogen atoms, the majority of them are normally in the ground state.

We have made much out of the fact that the position and velocity of an atom's electron are uncertain and must be described by probabilities. The atom's energy (that is, the energy of the electron in the atom) is different, however. The Schroedinger equation for an isolated atom predicts a precise energy, with no uncertainty, for each of the allowed states. Thus the atom has a precise energy when it is in state 1, and a different precise energy in state 2, and so on. These allowed energies, associated with the allowed states 1, 2, 3, and so on, are called the **energy levels** for the atom. The energy level, E_1, associated with state 1 is the lowest permitted energy, the energy level, E_2, associated with state 2 is higher than E_1, and so on.

It is helpful to exhibit the energy levels for a particular atom as a diagram such as that shown in Figure 18.4. The allowed energies E_1, E_2, and the rest, measured perhaps in joules, are marked off along an energy axis. The figure shows only a few of the many energy levels for the hydrogen atom. The energy levels are different for different types of atoms. The energy level diagram for helium, for example, would differ from the diagram for hydrogen.

Because the atom's energy is restricted to certain separate, specific values, we say that energy is **quantized.** This usage is similar to Planck's use of the word **quantum** to refer to a separate burst of radiant energy. When the atom suddenly switches from one allowed state to another, the atom's energy suddenly jumps from the one allowed energy to another. The atom performs a **quantum jump.**

We ran across energy quantization and quantum jumps in Chapter Sixteen in connection with Planck's theory of radiation. Planck's idea was that a vibrating charge particle is restricted to certain specific states of vibration. In our present terminology, these states of vibration are the allowed quantum states and the energies of these states are the allowed energies. Thus the modern quantum theory explains Planck's ideas.

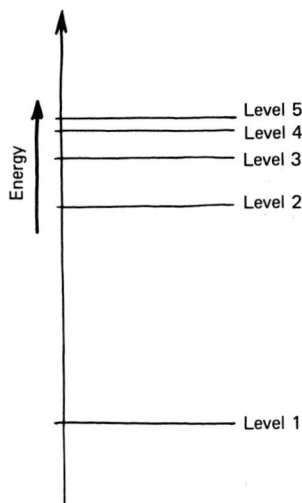

FIGURE 18.4 Energy level diagram for hydrogen.

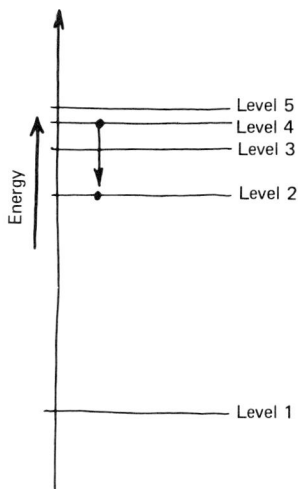

FIGURE 18.5 Energy level diagram for hydrogen, showing a quantum jump.

The Emission of Radiation

Suppose a hydrogen atom is in state 4 of Figure 18.3 and suppose it performs a spontaneous quantum jump to state 2. As you can see from the energy level diagram in Figure 18.4, the atom lost energy. Where did this energy go?

The answer, according to the quantum theory, is that it went into electromagnetic radiation. That's where photons come from.* When an orbiting electron undergoes a quantum jump to a state of lower energy, the emitted energy usually shows up in the form of a single burst of radiation, a photon. The photon is created at the instant of the jump.

Exercise 3

How fast is the photon moving once it's been created? What is its rest mass?

Exercise 4

After the quantum jump, that is, after the emission of a photon, how does the atom's mass compare with the mass it had before the quantum jump? Is it the same, has it increased, or has it decreased? *Hint:* $E = mc^2$.

Quantum jumps are shown as arrows on energy level diagrams (Fig. 18.5). The arrow from level 4 to level 2 represents the jump from state 4 to state 2.

From knowledge of the energies of level 4 and level 2, you can figure out the frequency (and hence the wavelength) of the photon emitted in this quantum jump. Here's how. The atom lost energy in the jump from level 4 to level 2. The amount of energy lost is $E_4 - E_2$, the difference between the energies of the two levels. The principle of conservation of energy tells us that this energy must go into the photon, so the photon's energy is $E_4 - E_2$. But according to the quantum theory of radiation

---

*More precisely, that's where some photons come from. Photons are also created in many processes that do not involve quantum jumps in atoms. Most visible light arises from quantum jumps in atoms.

(Chapter Sixteen), the energy of a photon is proportional to its frequency:

Energy of photon in joules $= 6.6 \times 10^{-34} \times$ frequency of photon

Because we know that the energy of the photon is $E_4 - E_2$, we can use the above relation to find the frequency of the photon. In order to actually figure out the numerical value of the frequency, in vibrations per second, we would need to know the numerical values of the energy levels E_4 and E_2, in joules. Physicists predict these numbers by solving the Schroedinger equation—something we won't try to do here.

Here are some exercises to check and to extend your understanding of this important relation between an atom's energy levels and the frequencies of the photons that that atom can emit.

Exercise 5
Which jump of the hydrogen atom would create the highest frequency photon, a jump from state 4 to state 2 or a jump from state 4 to state 3? Which of the two photons would have the longest wavelength?

Exercise 6
Which quantum jump of hydrogen would produce the highest frequency, a jump from state 4 to state 3, from state 3 to state 2, or from state 2 to state 1? (Note: As shown in Figure 18.5, the energy levels for hydrogen get closer together as they get higher.) Which of these jumps would produce the longest wavelength?

Exercise 7
Suppose hydrogen atoms had only five different allowed states (actually, they have many more), corresponding to the five energy levels of Figure 18.4. How many different types of photons (i.e., how many different frequencies) would hydrogen be able to emit?

Exercise 8
For the five-level hydrogen atom of Exercise 7, which quantum jump will produce the highest frequency? Which will produce the lowest frequency?

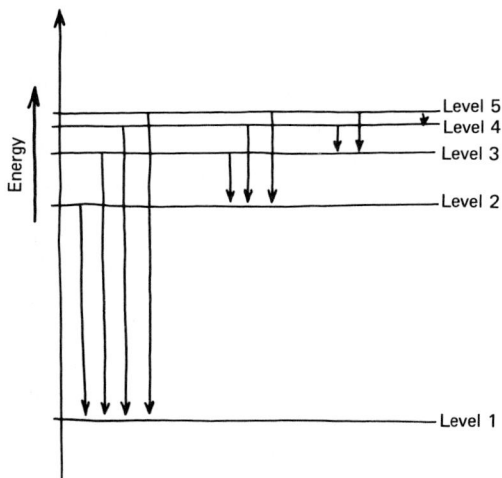

FIGURE 18.6 The ten possible downward quantum jumps between the lowest five energy levels of the hydrogen atom.

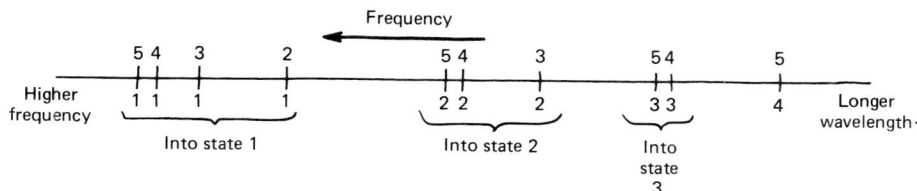

FIGURE 18.7 Graph of the frequencies emitted in the ten quantum jumps of Figure 18.6. The initial states are shown above the allowed frequency, and the final states are shown below.

Suppose you have a container full of hydrogen atoms. Most of these atoms will be in their ground state, but some fraction of the atoms will be in each of the excited states 2, 3, 4, and so on. These atoms will occasionally perform quantum jumps to lower states, emitting a photon with each jump. In other words, the box of atoms will emit electromagnetic radiation.

You've seen in Exercises 5 through 8 that our box of hydrogen atoms can emit only certain specific frequencies. These frequencies are determined by the energy differences of the quantum jumps between the allowed states. If we restrict our thinking to only the first five states (ground state plus four excited states), we can show the possible quantum jumps into lower states as arrows between the five allowed energy levels (Fig. 18.6). There are ten possible jumps, hence ten different frequencies of the emitted photons, between these five levels. All ten types of quantum jumps are going on all the time in our box of atoms*, so all ten types of photons are emitted. Figure 18.7 is a graph of the frequencies of these ten types of photons. In addition to plotting the frequency, the figure shows the initial and final states corresponding to each of the ten frequencies. If you carefully compare Figures 18.6 and 18.7, you might be able to see why these frequencies are arranged the way they are in Figure 18.7. This arrangement is a consequence of the arrangement of energy levels in Figure 18.6.

Thus we have seen how the quantum theory predicts the spectrum, the collection of emitted frequencies, of an element such as hydrogen.

Observing Atomic Spectra

The nice thing about atomic spectra is that they are easy to observe and they can be observed with great precision. These facts make spectral observation science's most important source of information about the atomic world.

Any device designed to observe the spectra of atoms or molecules is called a *spectroscope* (Fig. 18.8).

Suppose that you want to observe the spectrum of a box of hydrogen gas. Unfortunately for our story, in an ordinary box of hydrogen gas the hydrogen atoms are mostly combined into hydrogen molecules, H_2, and the spectrum of the hydrogen molecule is different and more complicated than the spectrum of the hydrogen atom. To keep things simple, we'll imagine that our gas is *atomic hydrogen,* not combined into H_2 molecules.

Because we want to observe the spectrum, we'll want the hydrogen to emit

*If the hydrogen is to continue emitting photons, we must keep putting energy into the box in order to balance the radiant energy emitted. That is we must continually "excite" the gas by putting some of the atoms into higher energy states. Later in this chapter we'll see how to excite atoms.

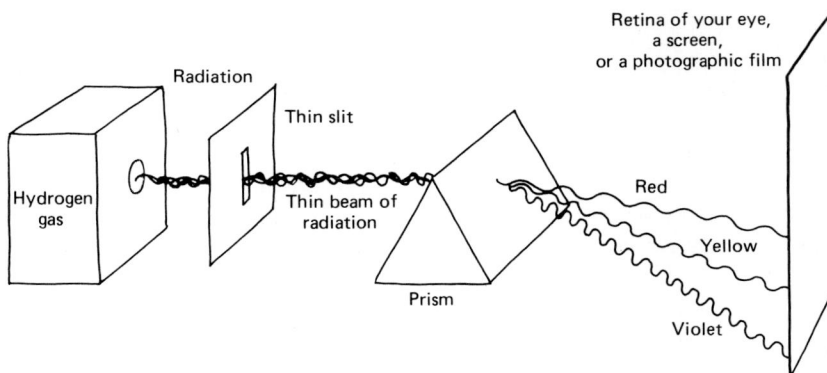

FIGURE 18.8 The spectroscope. Schematic diagram of the experimental arrangement for observing the spectrum of hydrogen.

enough radiation to be visible. This means that we need to put a large fraction of the atoms into excited states. As we'll see below, we can do this by heating gas or by sending an electric current through the gas.

The excited gas emits photons through a window in the box. The experiment works best with a narrow beam of radiation, so we'll send the emerging radiation through a thin slit.

The beam emerging from the slit contains photons of many wavelengths (many different frequencies), in fact, of all the wavelengths emitted by the gas. The crucial step in the experiment is the separation of this beam into its components, in other words, into several different beams, each having only a single wavelength. For visible radiation these components would be the different colors present in the beam.

One device widely used to separate radiation into its components is a *prism*, a triangular piece of glass or transparent plastic. Radiation bends, or refracts, as it passes from one medium into another, so the beam bends upon entering the prism and again upon leaving. The degree of bending is different for different wavelengths, so by the time the beam emerges from the prism it has split into several beams, each having only a single wavelength and each going in a different direction.

Several slit-shaped beams of radiation emerge from the far side of the prism. All we need to do now is view these beams by intercepting them with our eyes, a screen, or a photographic plate. From a measurement of the amount of bending in each component beam, it is possible to calculate the precise wavelength and frequency of that component; thus we can obtain a complete listing of all the frequencies emitted by the hydrogen gas.

Figure 18.1 shows the visible spectrum of atomic hydrogen. The most obvious feature of this spectrum is that it is not a continuous, rainbowlike band of frequencies. This is a direct consequence of the quantization of the atom's energy levels. A spectrum like that in Figure 18.1 shows each component wavelength as a narrow image of the slit, called a **spectral line.** Apparently, atomic hydrogen produces just four spectral lines in the visible part of its spectrum.

Exercise 9

Each of the four lines in Figure 18.1 corresponds to a different quantum jump. Which line corresponds to the largest jump and which corresponds to the smallest jump?

Exercise 10

The four lines in Figure 18.1 occurs at precisely the wavelengths that the quantum theory predicts for the four smallest quantum jumps into state 2 (the state just above the ground state) of the hydrogen atom. For each of the four lines, what is the initial state of the hydrogen atom before the quantum jump?

Exercise 11

Would you predict that infrared, visible, or ultraviolet is emitted when a hydrogen atom jumps from any higher state into state 1? Into state 3? (*Hint:* See Exercise 10.)

When scientists compare an observed spectrum such as that of Figure 18.1 with the spectrum predicted by the quantum theory, the results agree—that's our main reason for accepting the theory.

How to Excite an Atom

Whenever an atom emits a photon, the atom goes from a more energetic state into a less energetic state. If this was all that ever happened to atoms, it wouldn't be long before all atoms were in their ground state, emission of radiation would cease, and the lights would go out all over the universe. How do atoms move up into higher-energy states? We'll look at three ways.

The first way to excite an atom is by *absorption of radiation,* exactly the opposite of emission of radiation. Absorption of radiation occurs when an atom encounters a photon and absorbs it. The photon disappears and, according to conservation of energy, the atom must jump into a higher-energy state. This will happen only if the photon's energy exactly matches the energy difference between the two energy states of the atom, because the atom must completely absorb the photon. This in turn means that the photon's frequency must just equal one of the atom's possible emission frequencies. For example, a hydrogen atom could jump from state 1 to state 3 by absorption of a photon, provided the photon's frequency just matched the frequency that would be emitted by the atom in jumping from state 3 to 1. Thus we can excite atoms by shining electromagnetic radiation on them. For example, light passing through the air in a room will excite the atoms and molecules of the air into higher-energy states.

A second way to excite a collection of atoms is to heat it. This makes the atoms move faster and bump into each other harder, which excites electrons into higher energy levels. This is a form of *excitation by atomic collisions.*

A third way to excite a collection of atoms is to pass an electric current, usually in the form of a stream of electrons, through the atoms. For example, suppose we place hydrogen gas between two parallel metal plates and then attach one plate to the positive terminal and the other to the negative terminal of a battery. The positive and negative charges produced on the two plates create an electric field between the plates. If this field is large it will pull electrons off the negative plate. These electrons will flow through the gas to the positive plate, bumping and exciting lots of hydrogen atoms in the process. This is called *excitation by electric discharge,* because the negative plate discharges its excess electrons toward the positive plate.

More Complex Atoms and Molecules

Because the universe is made of more than hydrogen, we would like to know what the quantum theory has to say about objects such as the helium atom, which has two electrons, or the water molecule, which has ten electrons moving in the vicinity of three nuclei.

In all such cases, the idea is to solve the Schroedinger equation for the particular system of interest, whether a helium atom or a water molecule. The result for, say, the water molecule, is a set of permitted psi-waves, or allowed states, for that molecule. Each of these states corresponds to a specific precise energy of the molecule. The lowest-energy state is the ground state of the molecule, and the other states are excited states. The psi-wave for any particular state gives the probabilities of finding the electrons at various possible places. In the case of molecules, the psi-wave also gives the probabilities of finding the different nuclei in various possible configurations. The molecule can undergo quantum jumps, during which the state suddenly changes from one of the allowed psi-waves to another. Transitions to states of lower energy are accompanied by the emission of a photon. Jumps into higher-energy states are caused by excitation mechanisms such as those discussed in the preceding section.

So the situation for more complicated objects is pretty much what it is for the hydrogen atom.

Because the allowed energy levels are different for different types of atoms and molecules, the spectra are different. The spectra of atomic hydrogen, molecular hydrogen, helium, water, and so on, are all different. This difference means that spectral analysis can be used to identify different substances. For example, most of our knowledge of the chemical composition of stars, planets, and other astronomical objects, comes from analysis of their spectra. The light from these objects, after it is received by a telescope, is sent through a spectroscope, and the resulting spectral lines are studied.

Applications

There are applications of the quantum theory of the atom all around you.

The **neon tube** is an example of excitation by electric discharge. Neon vapor (in other words, a gas made of neon atoms) is enclosed in a glass tube. Electrons flow through the gas from a negatively charged piece of metal at one end toward a positively charged piece of metal at the other end. The moving electrons bump and excite the neon atoms. The atoms emit radiation as they jump back into lower-energy states. Neon happens to have many strong spectral lines in the red and orange portions of the visible spectrum, so neon-filled tubes glow red orange. Other gases yield colors.

When raindrops fall through air, the drops and the air both become charged just as your hair and comb become charged when you comb your hair (Chapter Twelve). The rising air eventually produces a buildup of positive charge on the clouds. The positive charge attracts massive numbers of electrons from the ground to the clouds, which excites the air molecules in the path of the moving electrons. The air molecules then deexcite by emitting light. This visible spark is known, of course, as **lightning.** Lightning, and in fact every other type of spark, is an example of an electric discharge, as is the neon tube. The thunder that accompanies lightning comes from rapid heating of the air in the path of the lightning stroke. The heating causes a sudden violent motion of the air molecules. This molecular motion is transmitted outward as a powerful sound wave.

A **black-light poster** is an example of excitation by absorption of radiation. A black light is a light bulb that produces large amounts of ultraviolet radiation and is painted black or violet to prevent most of its visible radiation from getting out. So only ultraviolet photons get out. Some of these photons are absorbed by the poster and excite the atoms of the poster. Many of the poster atoms then deexcite by emission of visible radiation. The poster glows mysteriously because you can't see the ultraviolet radiation that is hitting it.

The **fluorescent bulb** demonstrates several of these ideas. The long glass tube is filled with mercury vapor. Like the vapaor in the neon tube, this mercury vapor is excited by an electric current flowing through the vapor from one end to the other. Mercury atoms deexcite by emission of ultraviolet as well as visible radiation. The ultraviolet radiation strikes a powdery coating on the tube's inner surface, which excites the atoms of the coating by absorption of radiation. Finally, the atoms of the coating deexcite by emitting visible light.

Because of mercury's strong emission in the ultraviolet, many black lights are mercury-vapor discharge tubes (i.e., fluorescent bulbs minus the powdery coating) painted black.

The **northern lights,** or **aurora,** the atmospheric glow occasionally visible at night in the far northern and far southern latitudes, is an example of excitation by atomic collisions. The sun continually projects large numbers of fast-moving ions (charged particles, Chapter Twelve) outward in all directions. Many of these *solar wind* particles are trapped temporarily in Earth's magnetic field, where they form the **Van Allen radiation belt,** a broad ring of high-energy ions encircling the globe. Earth's magnetic field holds most of the Van Allen belt far above Earth's atmosphere, but in the vicinity of the poles the shape of the field allows the ions to approach Earth and enter the atmosphere. Air molecules are then excited by collisions with these fast-moving ions, and the air molecules emit light as they drop back into less excited states.

Spectral analysis can also be used to study the **chemical composition of the atmospheres of stars.** The outermost portion of most stars includes a dilute layer of cool gas (much cooler than the body of the star), known as the star's **atmosphere.** The hotter gas in the main body of a star emits radiation containing all wavelengths, so that the spectrum of starlight is a continuous band rather than a line spectrum such as that of Figure 18.1. As this starlight passes through the star's cooler atmosphere, some of the photons in the starlight are absorbed. The particular frequencies absorbed are those corresponding to the quantum jumps that the atoms of the star's atmosphere can make. Thus the starlight that gets through this atmosphere will show discrete *dark* lines. Observation of these lines tells us the composition of the star's atmosphere.

Checklist

spectrum of an element
Bohr's theory of the atom
Schroedinger's equation applied to the atom

spectroscope
prism
spectral line
excitation by absorption of radiation

psi-wave of an atom

allowed states

ground state

change of state

excited state

energy levels

energy level diagram

quantization of energy

quantum jump

emission of radiation by an atom

excitation by atomic collisions

excitation by electric discharge

neon tube

lightning and thunder

black-light poster

fluorescent bulb

northern lights

Van Allen radiation belt

composition of atmospheres of stars

Thought Questions

1. One idea encountered in science fiction is that our solar system is only an atom in a superuniverse, with the planets playing the roles of electrons and our sun as nucleus. Support or attack this theory.

2. Does it seem to you that a decent scientific theory of the atom should be able to give a comprehensible *picture* of the atom? Does the quantum theory of the atom give such a picture?

3. Recall that Democritus (Chapter Three) once said, "Nothing exists except atoms and empty space. All else is mere conjecture." Now that you have studied the modern theory of the atom, what comments do you have on Democritus's idea?

4. Ernest Rutherford, a central figure in the development of the planetary model of the atom around the turn of the century, was asked at a dinner party whether he thought electrons and atomic nuclei really existed. He answered with the reproof: "Not exist . . . not exist . . . why, I can see the little beggars in front of me as plainly as I can see that spoon!" Comments?

Further Exercises

12. Why does the burning of different materials often produce flames of different colors? The next time you stare at a fire, pick out as many colors as you can.

13. Could a round hole be used instead of a thin slit in a spectroscope? How would the spectrum appear? Why is a thin slit better than a hole? (Partial answer: The spectrum would be a collection of circles rather than lines.)

14. The neon atoms in a neon sign deexcite as they emit light. Why doesn't the sign eventually run out of excited atoms and turn off?

15. Astronomers have found that the sun's atmosphere contains iron. How might they have discovered this?

16. Name at least two types of molecules that should be identifiable in the spectrum of the northern lights.

17. How can a hydrogen atom, which has only one electron, have so many spectral lines?

Recommended Reading

Louise B. Young, editor, *The Mystery of Matter*, Oxford University Press, New York, 1965. An excellent anthology of readings in physics. Part 2 traces the development of the atomic idea from the ancient Greeks through the nineteenth century; Part 3 deals with twentieth-century developments in the theory of the atom and includes articles by Marie Curie, Erwin Schroedinger, Werner Heisenberg, Albert Einstein, and others.

19

RELATIVITY MEETS THE QUANTUM

If this is, as many people say, the scientific age, then it is important to know how science describes the basic stuff of which the universe is made. For on some level the scientific view of the universe defines the manner in which all of us view the world and the manner in which we view ourselves. Indeed, this is what we mean when we refer to our time as a scientific age.

Throughout much of history, science has been gripped by the vision of a universe made of tiny pieces, particles that are elementary in the sense that they are not made of still smaller particles. Today these ultimate building blocks of the universe are called *elementary particles.*

Exercise 1
Which of the following are certainly not elementary particles? Grain of dust, electron, hydrogen atom, water molecule, photon, proton.

In their search for the tiniest piece of the universe scientists were first led to the molecules, then to the atoms of which molecules are made, and then to the protons, neutrons, and electrons of which atoms are made. Are these the ultimate pieces? In our quest for the answer to this question we will venture into the amazing subatomic world of high-energy physics. We will find that this world is difficult to describe by means of the conventional image of a universe made of tiny pieces. We will find a world that is far from what we may have expected or, at any rate, far from the mechanistic images of Democritus and Newton.

The Restless Microworld

In order to study something as small as a proton or an electron, you have to make some extremely accurate measurements. Now, according to Heisenberg's uncertainty principle (Chapter Seventeen), if you want to measure an object's position accurately, you have to pay a price. The price is a large disturbance of the particle's speed. For example, if you use electromagnetic radiation (photons) to observe a neutron's structure, the photons must have a short wavelength and hence high frequency and high energy. These high energy photons will give a large and unpredictable jolt to the neutron. On the average, such a large jolt will set the neutron into rapid motion.

The upshot is that, if you measure the structure of a small object, you will observe it to be moving fast.

The world of the small meets the world of the fast! The Heisenberg uncertainty principle demands it. The subatomic world is restless; it won't sit still, cannot sit still, for our observations. As we go to smaller and smaller levels of observation we find the objects of our observation to be moving faster and faster. The microscopic world is a chaos of motion and energy, the chaos increasing at smaller levels.

Thus the search for general principles of the elementary particles must involve not only the quantum theory but also the theory dealing with fast-moving objects, namely, Einstein's relatively. Quantum theory and relativity must come together in a **relativistic quantum theory.** Since the search involves high energies (because of the high speeds involved), experimental work in this field is called **high-energy physics.**

Elementary particle physics is truly a child of our century, offspring of the two grand twentieth-century visions of nature: quantum theory and relativity.

Dirac's Idea

Einstein developed his theory of relativity during the years 1905 to 1915. Planck, Bohr, Schroedinger, Heisenberg, and others (including Einstein) created the quantum theory between 1900 and 1930. Thus it was natural that, in the 1920s, physicists were thinking about how these two theories could be combined.

One of the most successful of these theorists was a young Englishman named Paul Dirac (Fig. 19.1). In 1929, Dirac invented a theory concocted from the quantum theory, relativity, and electromagnetism. In its present-day form, this theory of **quantum electrodynamics** is the most successful scientific theory ever created in that it makes the most precise experimentally verified predictions made by any theory. One feature of this invention was the prediction of an unexpected phenomenon known as **pair creation.**

Pair creation is simple, really. The phenomenon consists of a photon spontaneously turning into a pair of particles of matter (Fig. 19.2).* Radiation turns into matter. The pair creation process doesn't produce something out of nothing—conservation of energy won't allow that—but it creates matter where no matter existed. In 1929, when Dirac predicted pair creation, nobody had seen anything like this happen, and there was no reason, other than Dirac's theorizing, to think that anything like it would happen.

In 1932, 3 years after Dirac's prediction, pair creation was observed in the laboratory.

A phenomenon that produced matter from nonmatter was surprising enough, but pair creation held yet another surprise. One of the particles created was the familiar electron; the other was a type of subatomic object that had not been observed before. It had the mass and charge of an electron, but the charge was positive instead of negative. This "positive electron" is known as the **positron.**

Exercise 2

In terms of the rest mass of an electron (or positron), what is the least energy a photon must have in order to create a positron-electron pair? *Hint:* Remember Einstein's mass-energy relation.

---

*Figure 19.2 is simplified. Pair creation can only occur in the vicinity of another object, such as a nucleus, not shown in the figure. The nucleus doesn't participate directly in the creation process and is unchanged (except for a change in its motion) by the process.

FIGURE 19.1 Paul Dirac. (AIP Neils Bohr Library.)

Particle-Antiparticle

The subatomic level is a nonpermanent and uncertain world. Its uncertainty arises from Heisenberg's uncertainty principle, which tells us that *every* situation holds the potential for a variety of outcomes. Its nonpermanence arises from the fact that microscopic particles, and even matter itself, can pass instantaneously in and out of existence. As examples photons are created or destroyed when atoms undergo

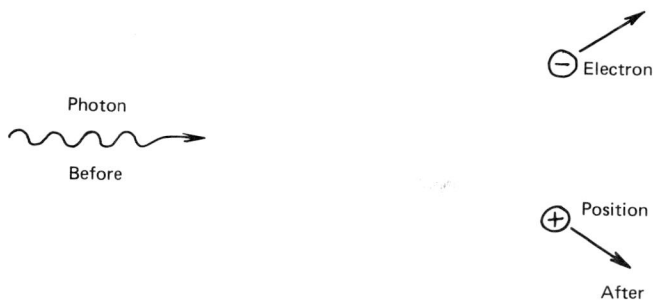

FIGURE 19.2 Creation of a positron-electron pair.

quantum jumps to lower or higher energy states, and electron-positron pairs are created from high-energy photons.

These two features, nonpermanence and uncertainty, come together in the contemporary theory of pair creation. According to this theory, any photon having sufficient energy is potentially a positron-electron pair. That is, any photon whose energy is greater than $2mc^2$ (see Exercise 2) could turn out to be an electron-positron pair instead of a photon. But this doesn't always occur; it occurs only with a certain likelihood. There is no way to predict when this object will be observed to be an electron-positron pair and when it will be observed to be a photon. Quantum uncertainties forbid such a prediction. We should therefore think of a high-energy photon as a tiny particle of radiation *and* a combined electron-positron pair (Fig. 19.3). Perhaps it should be called a photon/electron-positron because both possibilities are present.

Every object in our ordinary surroundings is some definite thing—a baseball, a newspaper, a car, a dust particle. But it *seems* that subatomic objects (if we call them objects) are not things in this sense. It does no good to ask what a photon/electron-positron really *is* at any given moment because both possibilities are always there. Which of the two possibilities will actually be realized upon observation depends on the throw of the cosmic dice.

In low-energy quantum theory (Chapter Seventeen), the uncertainty principle led us to question the nature of such subatomic objects as electrons and photons because such objects cannot simultaneously have a definite position and velocity. But at least we could say that the object in question was, for example, a photon. In the high-energy subatomic world, the uncertainty principle forbids us from even naming the object we are dealing with. A high-energy photon isn't necessarily a photon at all—it might be an electron-positron pair instead.

Thought Question 1

Would you say that such objects are real? Should we call them objects?

This isn't the end of the particle-antiparticle story. It turns out that a photon can also change into a proton-*antiproton* pair. The antiproton is not found in ordinary matter (composed of electrons, protons, and neutrons). It has the mass of a proton and is like the proton in other respects, but its charge is negative instead of positive.

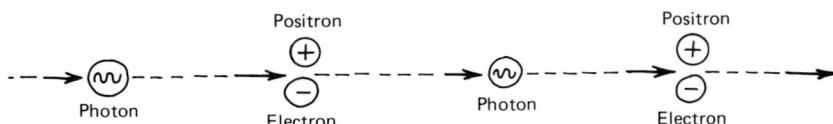

FIGURE 19.3 An attempt to visualize a photon as potentially a particle of radiation and also potentially an electron-positron pair.

Exercise 3

Explain how to calculate the minimum amount of energy needed to create a proton-antiproton pair. (*Hint:* See Exercise 2).

Exercise 4

Compared to the photons that create electron-positron pairs, would you expect photons that create proton-antiproton pairs to have higher, lower, or equal frequencies?

Every photon with sufficiently high energy is potentially an electron-positron pair and potentially a proton-antiproton pair. Maybe it should be called a photon/electron-positron/proton-antiproton. Furthermore, there are many other pairs of this sort. All possibilities are ordinarily not realized, or actualized, because the photon energy needed to create these pairs is quite high. Photons for production of even the lowest-energy pair, the electron and positron, must be in the gamma-ray range.

We've seen that the electron and proton each have their antiparticles: the positron and the antiproton. The **antiparticle** corresponding to any particular type of particle is similar to the original particle but with some property, such as charge, reversed. The positron is the electron's antiparticle because it is just like the electron only with a positive instead of a negative charge.

The relativistic quantum theory asserts that for *every* type of particle there is a corresponding antiparticle. We have seen two examples of such particle-antiparticle pairs. A third example is the neutron-**antineutron** pair. The antineutron differs from the neutron in certain of its magnetic properties. You might wonder if there is an antiphoton. According to the theory, the photon and the antiphoton are identical, that is the photon is its own antiparticle. There is only one type of photon.

Particle-antiparticle pairs can flash into existence, born perhaps of a photon, and they can also flash out of existence, giving birth to a photon. In fact, this **annihilation** process occurs even more easily than the creation process, because once you have the particle and its antiparticle there is no minimum energy requirement for annihilation. Whenever the particle and antiparticle are close together, it is likely that they will annihilate each other and turn into photons.

These creation and annihalation processes have been occurring on a routine basis in high-energy physics laboratories since 1932. They occur with great frequency in stars. They even occur all around us, in Earth's atmosphere, as a.result of high-energy photons and other particles coming in from outer space. Matter (i.e., stuff that has mass even when it is at rest) can be created and it can be destroyed. It is energy, not matter, that must always balance out, that is neither created nor destroyed. Energy seems to be the basic stuff of the universe.

We find ordinary protons, electrons, and neutrons all around us, but we find very few of their antiparticles. Nevertheless, these antiparticles are quite real, they have ordinary positive masses, and they are no more mysterious than protons, electrons, and neutrons. They are in all ways analogous to particles of ordinary matter. It is possible, for example, to make antihydrogen atoms by getting a positron to orbit an antiproton. The spectrum of an antihydrogen atom would be identical to the spectrum of an ordinary hydrogen atom.

There is very little of such **antimatter** on Earth. If Earth contained large amounts of antimatter, the matter would annihilate with the antimatter and you wouldn't be here to read this story.

Although no one knows how much antimatter is in the rest of the universe, we believe that the universe is made primarily of matter. Our sun and solar system are made almost entirely of matter, but it is conceivable that some other stars, even some entire galaxies, are made primarily of antimatter. We wouldn't know if an isolated galaxy were made of antimatter, because the radiation from an antigalaxy would be identical to the radiation from a normal galaxy.

Exercise 5
Would you like to travel to an antigalaxy? Why?

Exercise 6
What would we expect to detect, on Earth, from a region of space in which large-scale matter-antimatter annihilation is taking place?

Tracking the Subatomic World

The way to observe the high-energy subatomic world is to start with a high-energy particle, let it smash into another particle, and see what happens. Nature provides us, free of charge, with lots of high-energy particles in the form of *cosmic rays.* These are particles that enter Earth's atmosphere from space. Many of them come from explosions in stars. A lot of high-energy physics has been done using cosmic rays as the source of fast particles. Scientists like to control their experimental conditions, though, so they have built *particle accelerators* that produce just the energies and the number of particles desired.

Figure 19.4 shows the huge accelerator recently constructed at the Fermi National Acceleratory Laboratory (Fermilab) near Chicago. Inside the evacuated main ring, protons circulate 200,000 times in just a few seconds, a distance farther than that to the moon and back. They pick up speed with every turn, eventually reaching more than 99.999 percent of lightspeed. At this speed, their inertial mass is some 400 times the mass of a proton at rest. When released from the ring, the protons shoot like a stone from a sling toward one of the three different end-point facilities shown in the map above the figure.

Scientists have recently proposed an even larger accelerator at Fermilab. The new machine, financed by the U.S. government, would pack ten times the energy of the present machine. It would be 3.25 miles in diameter and would extend to the perimeter of the 6,800-acre Fermilab site. Many physicists think that it is impractical for individual nations to build such large accelerators and that the next generation of accelerators, if such accelerators are to be built at all, must be international ventures. Indeed, some machines are already multinational. Perhaps the curiosity that drives

FIGURE 19.4 (a) Fermi National Accelerator Laboratory. The main ring is 6.4 km in circumference. The scale of the main ring may be judged from the 16-story main lab building. (b) Map of Fermilab.

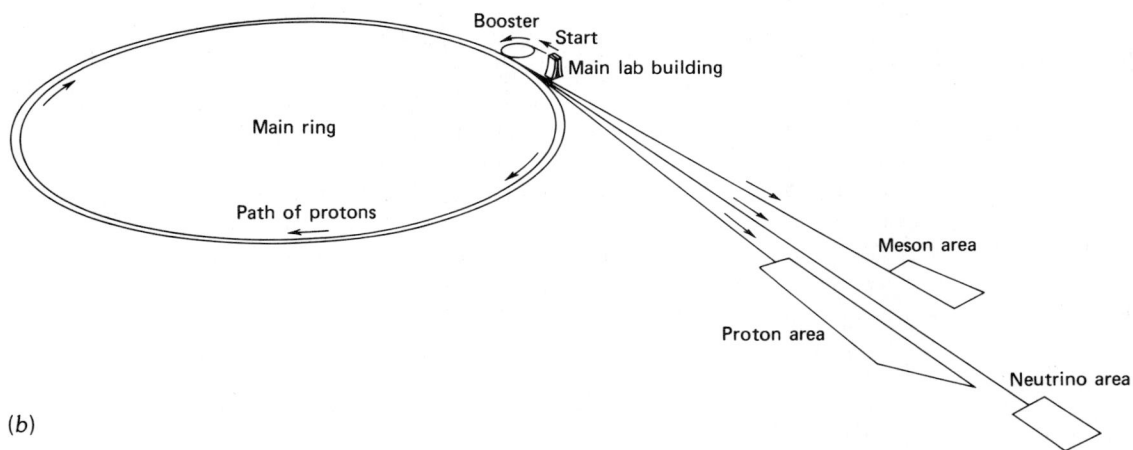

us to search for the unity behind natural phenomena will one day help unite the nations.

The fast-moving particles from the accelerator enter a device called a **bubble chamber,** a tank of liquid on the verge of boiling. Any charged particle passing through the liquid makes the liquid boil along the particle's path. The tracks made by the incoming particle and by the debris created as that particle smashes into other particles, which smash into other particles, and so on, are then visible. Technicians study photographs of these tracks for signs of interesting interactions. Figure 19.5, a typical bubble-chamber photograph, shows the creation of three electron-positron pairs. The photons that created these pairs left no tracks because photons are not charged particles; the paths of the electrons and positrons are curved because a magnetic field is present in the bubble chamber.

Some physicists, struck by the beautiful and fantastic churning of the subatomic world, have expressed their fascination metaphorically. In this spirit, Figure 19.6 superimposes a photo of the "dance" of the elementary particles on a photograph of the Hindu god Shiva, king of dancers. The dance of Shiva symbolizes the cosmic cycles of creation and destruction and the human rhythm of birth and death. The wild, yet graceful, gestures of Shiva are reflected in the restless beauty of the microworld captured in a bubble-chamber photograph. This photograph is based on a similar photograph in *The Tao of Physics* (Ref. 2).

Quarks and Leptons: The Ultimate Pieces?

Three quarks for Muster Mark

James Joyce, *Finnegans Wake*

Thanks to particle accelerators and cosmic rays we now know that nature makes many subatomic particles in addition to the photon, proton, electron, neutron, and their antiparticles. The world's high-energy physics laboratories have discovered an embarassingly large "zoo" of new particles.

These new particles differ in qualities such as electric charge and mass and in other, recently discovered qualities with such names as spin, isospin, and strangeness. Perhaps we should say that such qualities have been invented because, like Ptolemy's circles within circles, like Copernicus's sun-centered circles, like Kepler's ellipses, these qualities are just ideas that physicists use to help organize their thinking. Such organization is usually called understanding. For example, the concept of electric charge was invented as a way of organizing (or understanding) certain facts about the electromagnetic interaction. Certain types of particles, such as electrons, protons, and muons (one of the new particles), feel the electric force. Others, such as photons, neutrons, and neutrinos (another one of the new particles), do not. This is simply a basic feature of nature, not explainable on the basis of any present theory. To help organize this feature in our minds, we say that electrons, protons, and muons are charged and that photons, neutrons, and neutrinos are not charged. The word *charge* doesn't really explain anything, it simply helps physicists catalogue a certain feature that they have experienced in the natural world.

Thought Question 2

In what way is mass simply an invented quality of objects? In what way does this concept help us organize our thinking about the natural world?

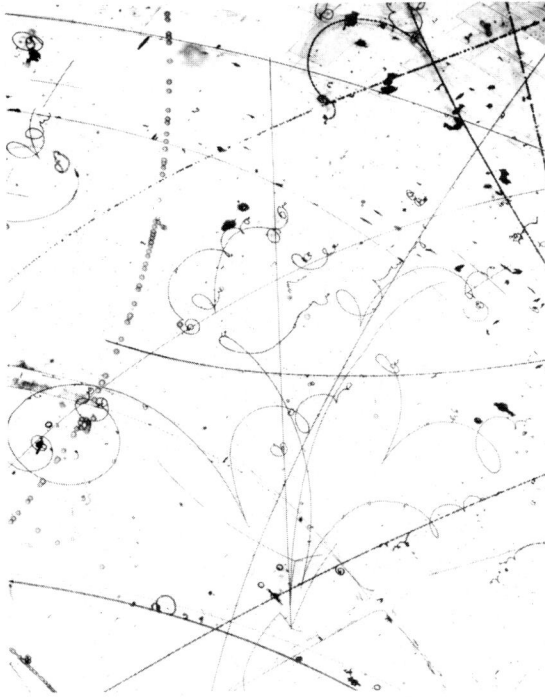

FIGURE 19.5 The creation of matter. The three arrows in the map below show the points at which three electron-positron pairs are created from three high-energy photons.

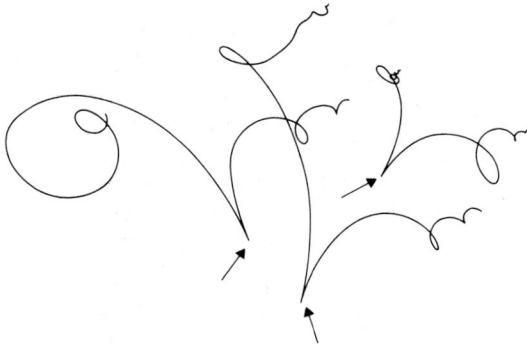

   Hundreds of different subatomic particles have been discovered. Now, if you believe that there is any kind of underlying simplicity in nature, it is difficult to believe that the universe is made of so many different kinds of things. Perhaps all these subatomic particles are themselves made of just a few still smaller particles, and perhaps these still smaller particles are the truly elementary particles. Then beneath

FIGURE 19.6 The cosmic dance—the world of the elementary particles and Shiva, king of dancers. (Nelson Gallery, Kansas City, Mo. and Fermilab.)

the molecular level, beneath the atomic level, beneath even the subatomic level of protons and neutrons, would lie the really fundamental level.

Enter the quark.

In their search for the particles behind the particles behind the particles, physicists added one more level of smallness during the 1960s. They suggested that many of the subatomic particles are not elementary but are instead composed of a small number of truly elementary particles. Lifting a word from James Joyce's novel *Finnegans Wake,* Murray Gell-Mann, one of the inventors of the new theory, dubbed these particles **quarks.** Sceptics pointed out that in German the whimsical new word means "cream cheese" or "nonsense." But the theory caught on. Quarks seem destined to be with us for awhile.

There seem to be six different kinds of quarks, along with six antiquarks. Prior to 1975 experimental evidence existed for only three types of quarks, known as the up, the down, and the strange quarks (don't try to attach any particular significance to the names). The quark theory suggests that quarks come in pairs and that the up and the down quarks form one such pair. So a fourth quark, a quark to accompany the strange quark, was predicted. In 1975 two independent groups of high-energy physicists discovered a new particle in their bubble chambers. One of the constituents of the new particle seemed to be the predicted fourth quark. The fourth quark, now reasonably well confirmed, is called the charmed quark.

The picture was further complicated in 1977 when another new particle was found at high energies, a particle that seemed to contain a fifth quark. The fifth quark has been dubbed the bottom quark. Because the theory implies that quarks come in pairs, the discovery of the bottom quark means that a sixth quark should exist. The as-yet-undiscovered (as of this writing) sixth quark will be, naturally, the top quark.

So maybe there are six quarks. Plus their antiquarks.

According to the quark theory, a proton is made of two up quarks and one down quark and a neutron is made of one up quark and two down quarks. Other combinations of quarks yield many of the other members of the particle zoo. However, not all particles are made of quarks; as far as is known, electrons and neutrinos are truly elementary.

After Gell-Mann and others dreamed up the quark theory, the experimentalists went to work searching for individual quarks in their bubble chambers. Even after several years, however, no individual quarks had been found.

Now for a lesson in the methods of science. Most people, after searching in every conceivable place for some as-yet-undiscovered object, such as, say, a unicorn, might conclude that the object doesn't exist. But physicists found the quark idea so helpful in organizing their thinking about the subatomic world, so pleasing to the mind (recall Copernicus, Chapter Two), that rather than relinquish the quark concept they decreed that quarks cannot be separated from each other and so are never found in isolation. They are found, it seems, only in groups of two or three or more, for example, in the form of a proton (two ups and a down) or a neutron (one up and two downs).

But how then to account for the fact that quarks are never found alone? A new force is needed, a force between quarks that is so strong that quarks cannot be separated from other quarks. This assumed force between quarks is called, for obscure reasons, the **color force.** Unlike every other known force, the color force between two quarks gets stronger as the quarks get farther apart.

We have had cause at several points in this book to ask whether a certain thing really exist. Do Copernicus's circular orbits really exist? Kepler's elliptical orbits? The electromagnetic field? Electrons? Psi-waves? And now we have the quarks, teetering between existence and nonexistence. What is the status of such creatures? Do they exist physically, or do only the more directly observable particles, such as protons and neutrons, exist? (Note that all of this is quite apart from the question of whether, in view of the Heisenberg uncertainty principle, protons and neutrons should be said to exist!) Is the quark idea just a convenient fiction? What does *exist* mean?

305

Relativity meets the
Quantum

Thought Question 3
What do you think?

If this all sounds a little crazy to you, don't be discouraged. It sounds a little crazy to a lot of people. Contemporary physicists do not let this deter them. In fact, the situation in high-energy physics caused one physicist to remark, on attending a lecture announcing a new concept, "We all know that this new theory is crazy. The question is, whether it is crazy enough to be true."

Current theory has come up with just one other category of truly elementary particles in addition to the quarks. These other particles are called **leptons.** Unlike quarks, leptons can be found in isolation. In fact, the familiar electron is a lepton. Leptons, unlike quarks, do not feel the color force; this is the defining feature of leptons, just as the defining feature of uncharged particles is that they don't feel the electric force. It seems that there are just six types of leptons (although this has not been entirely confirmed by experiment): the electron, the muon, the tau, and three types of neutrinos.

So perhaps the universe is made of six kinds of quarks and six kinds of leptons, plus their twelve corresponding antiparticles. Perhaps higher-energy accelerators will add more particles to the list. If the list of quarks and leptons gets too long, perhaps physicists will begin thinking about still another level of smallness below the quarks and leptons. Or perhaps some unimagined new theory will emerge.

Thought Question 4
The main tool in our search for knowledge of the subatomic world is the high-energy accelerator. These machines are expensive, however. The accelerator pictured in Figure 19.4 cost about half a billion dollars. For comparison, this is about one-tenth of 1 percent of the annual U.S. federal budget. Are such machines worth the money?

An Ultimate Cosmic Force?

Unification has always been one of the main themes of science. Kepler's three principles unified the motions of all the planets in one larger scheme. Newton's laws unified the planetary motions with such earthly events as the fall of an apple, thus unifying Earth and the heavens.

The four fundamental forces (Chapter Eight) provide another unification. All the forces in our environment, from the tension in a rope to the pull that holds Earth in its track around the sun, can be understood in terms of only four forces: gravitational, weak, electromagnetic, and strong. Perhaps these four forces can be further unified! Such a unification of basic forces has happened before. Little more than a century ago, the electric and the magnetic forces were thought to be independent; nineteenth-century physicists unified these two forces in a single electromagnetic force. Today electric and magnetic forces are seen as different aspects of the same force, namely, the force acting between charges.

The theories of the four fundamental forces are presently in a state of flux. The color force that acts between quarks has arisen as a new fundamental force. This new force is apparently responsible for the strong nuclear force. Other new theories have attempted to unify the weak force with the electromagnetic force. One of these theories predicted the existence of a certain new type of particle. In 1973 physicists found evidence for the predicted particle.

So there is both theoretical and experimental support for a further unification of the fundamental forces, for the idea that the weak and electromagnetic forces are just different manifestations of a single electro-weak-magnetic force. Perhaps the color force will eventually be seen as just one aspect of an underlying electro-weak-magneto-color force, which will explain all the weak, electromagnetic, and strong interactions. Perhaps someday the gravitational force will be combined with the others, so that all forces will be seen as aspects of a single fundamental force, which might then be called the grav-electro-weak-magneto-color force.

But if this ultimate unification is achieved, there will be no need to give it an extended title. We will simply call it: The Force.

The Eternal Flow: A Personal View

We have come a long way since the atomism of Democritus and the mechanism of Newton. Contemporary high-energy physics has stretched traditional beliefs about the physical world to the breaking point. It has become nearly impossible to understand physics in terms of particles moving mechanically in empty space. The particles of modern physics don't have specific positions and velocities, they are sometimes particles and sometimes waves, they spring unpredictably in and out of existence, and some of them cannot even be individually isolated. It is a considerable distortion of the language to call such entities particles. In addition, the space of modern physics is far from empty. It is filled with electromagnetic fields, gravitational fields, psi-waves, and potential creation and annihilation. It is stretching an image a long way to perceive all this in terms of atoms and empty space.

Perhaps a new image, a new paradigm,* is in order.

In this clutter of subatomic phenomena, can we identify any basic unchanging principles? Of the physicists who think about such matters, many would answer that the fundamental principle is *change itself*. Some might call it motion, energy, the cycle of creation and annihilation, or, perhaps, the eternal flow. It seems that nothing, no thing, no object, is eternal; all things change, and this ceaseless change applies even more strongly to the world of atoms than it does to the world of you and me. In fact, this idea of ceaseless change is, roughly, the concept that is captured in the principle of conservation of energy. For, according to this principle, although external features may be radically transformed, energy (motion, flow, the capacity to create) is eternal.

It is difficult to express this idea in words. We can easily visualize atoms moving in space just as we can easily visualize automobiles moving along a highway. In our ordinary ways of thinking, objects are prior to motion. We can visualize objects that don't have motion, but we have trouble visualizing motion without objects. Modern

---

*According to the influential philosopher-historian Thomas Kuhn, a science is primarily characterized at a particular stage of its development by a paradigm, which might be a basic set of beliefs, characterizing the structure of a particular aspect of the natural world.

physics seems to be telling us that this object-oriented manner of thinking is not appropriate. We should, instead, make motion the primary concept. The image of material objects moving in space must be replaced by an image of motion or energy creating material objects.

Thus we seem to live in a universe of motion rather than a universe of things.

A second principle that, in the opinion of many physicists, arises from subatomic physics, is the notion of *connections*. Like many of the significant ideas of contemporary physics, this idea is closely related to the uncertainty principle. Physicists have usually learned about the natural world by isolating a small part of it and studying that part, a process known as experimentation. Experimentation in this sense breaks down when we approach the atomic world. According to the uncertainty principle, small particles cannot be isolated from the observer because the observer always affects them in essential ways, and this problem becomes more severe as we investigate smaller and smaller objects. An object, such as an electron, is now a particle, now a wave, depending on the experimental conditions under which it is observed. The properties of the electron have meaning only in relation to the processes of measurement. In this situation, it seems absurd to say that we are observing the electron. It is much more natural to say that we are observing a certain specific experimental arrangement, a certain electron source, screen, and slit. We cannot observe the electron itself, we can observe only the electron plus environment.

Thought Question 5

The word *observe* seems to be getting in our way. Would the word *experience* be better? For instance, should we say that experimental physicists observe electrons or that they experience electrons?

In addition to the connection between the observer and the object of observation, there are connections between the objects of the atomic and subatomic world themselves. An excited atom creates a photon, a high-energy photon creates an entire hierarchy of possible particle-antiparticle pairs, a particle-antiparticle pair annihilates to create a photon or possibly a different pair, and so on. All these creation and annihilation events are unpredictable and can occur at any time. But again we are stretching the language. When we say that a photon creates an electron-positron pair, what we really mean is that at any specific time when we observe this object, we might find it to be a photon or an electron-positron pair. We must think of this object as *both* a photon and an electron-positron pair. The two partake of each other at all times; they are two aspects of the same underlying reality, connected in the most fundamental sense.

Thus it seems that the two foremost achievements of twentieth-century physics, the relativity theory and the quantum theory, force us to radically revise a fundamental scientific paradigm. No longer do isolated, unchanging atoms move silently in the void. Today's physics views the cosmos not as a universe of objects, but rather as a dynamic web of relations in which time and change are the essential elements and within which there is a basic unity of all things and events.

Checklist

elementary particle	pair annihilation
high-energy physics	cosmic rays
Paul Dirac	particle accelerator
quantum electrodynamics	bubble chamber
pair creation	quark
positron	lepton
antiproton	color force
antiparticle	unification of forces

Further Thought Questions

6. Find a suitable reference, such as an encyclopedia yearbook, a *Scientific American* article, or one suggested by your instructor, and report on recent progress in elementary particle theory.

7. Is it sensible to speak of the parts of an object when these parts cannot conceivably be separated from each other?

8. What significant developments do you suppose came from the unification of the electric and the magnetic forces in the nineteenth century?

9. Do you think that the unification of two or more of the four fundamental forces would be a significant achievement? That is, would it ever make much difference to anyone except a few high-energy physicists?

10. Has science dematerialized matter?

11. Are relationships more real than things?

12. The ancient Greek philosopher Plato thought that abstract ideas were the ultimate reality. Was he right?

Further Exercises

7. Is any net electric charge (i.e., overall charge after subtracting negative charge from positive charge) created in any of the following events? Emission of a photon by an atom; removal of an electron from an atom (ionization) as a result of collision with another atom; combination of oxygen with hydrogen to form water; creation of an electron-positron pair; annihilation of a proton-antiproton pair. (Note that a *conservation of charge* principle appears to be operating here.)

8. According to the quark theory, which of the following are elementary particles? Oxygen molecule (O_2); oxygen atom; electron; proton; neutron; up quark; charmed quark; positron; antiproton.

9. Even without using particle accelerators physicists often find such antiparticles as positrons and antiprotons on Earth. Any such particle exists on Earth only for an instant before being annihilated by ordinary matter. Where might such particles come from?

10. What type of force do you suppose bends a proton's path into a circle inside a particle accelerator? (*Hint:* In what direction, relative to the direction of motion of the proton, would the force have to act in order to change the direction of the motion?)

11. A certain radiation source produces gamma radiation whose photons have sufficient energy for electron-positron pair production. The quantum theory predicts that, in a certain experiment, one of these photons has a 20 percent chance of being found as an electron-positron pair and an 80 percent chance of being found as a photon. The experiment is performed ten times. About how many times will the experimenter observe an electron-positron pair? About how many times will he or she observe a photon? Are these answers exact or only approximate? Why?

Recommended Reading

1. Nigel Calder, *The Key to the Universe*, Viking Press, New York, 1977. A well-written but somewhat difficult guide to the world of high-energy physics.
2. Fritjof Capra, *The Tao of Physics*, Shambhala Publications, Berkeley, Calif., 1975. Capra, a high-energy physicist, explores the parallels between modern physics and Eastern mysticism. The book includes excellent accounts, suitable for nonscientists, of high-energy physics. Figure 19.6 is from this book.

20

RADIOACTIVITY: ONE WAY TO CHANGE A NUCLEUS

Humankind's understanding and control of the atomic nucleus have changed the shape of the twentieth century. International relations are dominated by nuclear weapons, national energy policies are influenced by nuclear energy. Nuclear physics affects our lives in these and many other ways.

The next four chapters will survey the physics and social implications of the atomic nucleus.

Nuclear Forces

The basic features of atomic structure are determined by the properties of the electromagnetic force and the strong nuclear force (Chapter Eight). Let's review some of these properties.

Exercise 1
Name the four fundamental forces. Which of these forces is strongest? Which force is weakest? Which forces have a long range and which have a short range? (Answers below.)

Arranged from strongest to weakest, the four fundamental forces are: strong, electromagnetic, weak, and gravitational. Two of these, the strong and the weak, have ranges extending to only about 10^{-15} m; in other words, two objects that are more than 10^{-15} m apart cannot feel each other's strong or weak force. The other two forces, the electromagnetic and the gravitational, have unlimited range, although they become weaker as the distance increases.

Another fact about these forces that is important in determining the structure of the atom is that the electromagnetic force acts only between the proton and the electron, whereas the strong force acts only between the proton and the neutron—neutrons are not charged particles, and electrons do not feel the strong force.

The "glue" that holds the nucleus so compactly together is the strong force that, at distances of about 10^{-15} m, is strongly attractive. The nucleus must be small, roughly 10^{-15} m across, because two nuclear particles that are farther apart won't be attracted to each other by the strong force. Most nuclei are tightly bound, hard to break apart, because the nuclear force is so strong. At short distances, the strong nuclear force is much stronger than the electromagnetic force, which explains why several protons can be held together in the nucleus despite the electromagnetic repulsion between them. You won't usually find electrons in the nucleus, because electrons don't feel the strong nuclear force that binds the protons and neutrons into the nucleus. Instead, you'll find electrons at a large distance (10^{-10} m—100,000 times the size of the nucleus!) from the nucleus, where they are held by the attractive electromagnetic force between the protons and the electrons. The reason the electrons are so far from the nucleus is that the electromagnetic force is so weak compared to the strong force.

The gravitational force acting between individual subatomic particles is so much weaker than the other three forces that it plays no role in atomic structure. The weak force is short ranged, so it plays a role only within the nucleus. It is much weaker than the strong force, so the major features of nuclear structure are determined by the strong force. However, the weak force is responsible for an important phenomenon known as beta decay (discussed below).

It is much easier to alter an atom's electron structure than to alter its nuclear structure, because the strong force is much bigger than the electromagnetic force. So the energies involved in **nuclear reactions** (alterations of atomic nuclei) are much bigger than the energies involved in **chemical reactions** (alterations of electron structure).

The universe would be a very different place if the four fundamental forces had different properties from those which they actually have. If the strong force weren't so strong, nuclei wouldn't hold together and matter as we know it could not exist; if the electromagnetic force had only a short range, electrons would not be held in orbits around nuclei and, again, matter could not exist. In either case, intelligent life in anything resembling its present form could not exist. Perhaps there would be nobody here to comprehend the universe. Perhaps there would be only isolated subatomic particles drifting randomly—a very dull universe. Why does it happen to work out that the four fundamental forces have the strengths and ranges that they have, properties that make the universe such an interesting and varied place? The high-energy physicists (Chapter Nineteen) are trying to find out.

Thought Question 1

Does a universe totally devoid of even the possibility of life and intelligence seem conceivable to you? Such a universe would spin out its days, for all time past and future, without even being comprehended, without ever being observed. To the scientist, things that cannot, even in principle, be observed should not be regarded as real. What is your opinion? Could there possibly be such a nonobserved universe?

Nuclear Reactions

Medieval alchemists mixed chemistry with magic in their attempts to transform, or transmute, materials such as iron and lead into gold. In modern terminology, the alchemists were trying to change one element into another.

Exercise 2

Is it conceivable that atoms of one element could be converted into atoms of another element? What would be involved in such a change? (Answers below.)

Exercise 3

What would you have to do to an atom of lead to make it into an atom of gold? (*Hint:* See Fig. 3.1).

The alchemists weren't so far off. One element *can* be changed into another, but chemical methods won't do the job and neither will magic. The energies involved in a chemical reaction (i.e., recombination of atoms by alteration of their electron structure) are far too small to produce the changes in nuclear structure needed to transmute the elements. Although science has finally achieved the alchemists' dream of transmuting the elements, transmuting lead into gold would cost much more than the gold would be worth. However, substances more valuable than gold can be made this way.

Any alteration of a nucleus is called a **nuclear reaction.** There are three main types of nuclear reactions: **fission, fusion,** and **radioactive decay.** We'll study radioactive decay in this chapter, and fission and fusion in the next chapter.

As you know, the name of an element tells us the number of protons in its nucleus, that is, its **atomic number** (Chapter Three). In nuclear reactions, the number of neutrons in the nucleus is also likely to be important. For example, the element carbon comes in three different forms in nature. All three forms have the same number of protons because they are all forms of the same element, but they have different numbers of neutrons. Some carbon atoms have six neutrons, some have seven, and some have eight. These three types of carbon are known as different **isotopes** of carbon. That is, two nuclei of a given element are said to be different isotopes if they have different numbers of neutrons.

We label the different isotopes by giving the name of the element and the total number of nuclear particles (protons plus neutrons). Because the mass of a nucleus is roughly proportional to the total number of nuclear particles it contains, this number is called the **mass number.** We will represent a particular isotope by giving the name or abbreviation of the element followed by the isotope's mass number. For instance, carbon 12 (C-12) means the isotope of carbon having 12 nuclear particles.* It has atomic number 6 and mass number 12. The naturally occurring isotopes of carbon are C-12, C-13, and C-14. They each have atomic number 6 but different mass numbers.

Exercise 4

The istotopes H-1, H-2 (often called deuterium), and H-3 (often called tritium) play an important role in nuclear technology. How many protons and how many neutrons does each isotope contain? (Recall that H is the symbol for hydrogen.)

Different isotopes of a given element all have the same number of electrons, provided they are not ionized, because they all have the same number of protons. For example, C-12 and C-13 both have six orbital electrons. Because chemical reactions involve only the electron structure of the atoms participating in the reaction, different isotopes of a given element must behave the same way in chemical reactions. Thus chemical reactions cannot be used to separate C-12 and C-13 in a mixture of the two isotopes, because the two isotopes enter the same way into each reaction. However, different isotopes have different nuclear structures, so they behave differently in nuclear reactions. For example, the two naturally occurring isotopes of uranium, U-238 and U-235, behave differently in the nuclear reactions going on inside a nuclear reactor.

*Most authors use the notation $_6C^{12}$, which shows both the atomic number and the mass number. We'll use C-12 for simplicity.

Radioactive Decay

In 1898, Marie and Pierre Curie accomplished the monumental task of chemically separating a previously unknown element from a ton of tarry black pitchblende. The Curies obtained only 200 mg (less than 0.01 oz) of the new element, which they named radium. Radium has fascinating properties, including the ability to emit highly penetrating rays and to glow spontaneously in the dark. For her discovery of radium and other accomplishments, Marie Curie (Fig. 20.1) became one of the few people in history to receive two Nobel prizes in science: the 1903 prize for physics and the 1911 prize for chemistry. She died of leukemia in 1934. The references listed at the end of this chapter contain biographies of Marie Curie and of other women scientists.

Many other substances have been found that spontaneously emit penetrating rays. These substances are said to be *radioactive.* This behavior occurs in the nucleus of certain types of atoms. A radioactive nucleus is simply one that, either because the balance of protons and neutrons is unstable or because the nucleus is too large, can't quite hold itself together. Such a nucleus will spontaneously and unpredictably emit a small particle in an effort to reach a more stable structure. This spontaneous emission

FIGURE 20.1 (*a*) Marie Curie. (Figure 20.1 *b-c* appear on following pages.)

FIGURE 20.1 (*b*) Marie Curie in her laboratory in 1912, the year after she received her second Nobel Prize.

process by a nucleus is called *radioactive decay.* Streams of these particles emerging from radium nuclei form the penetrating rays observed by the Curies. These particles emerge with high speeds because of the great energy associated with the strong nuclear force.

Two types of particles can be emitted during radioactive decay. The first type consists of two protons bound to two neutrons. This particle is identical with the nucleus of a helium atom and, in fact, after this particle is emitted it usually picks up a couple of electrons from its surroundings to become a normal helium atom. This particle, which we could define as a helium nucleus emitted from a larger nucleus during radioactive decay, is called an *alpha particle.* Because a nucleus loses two protons and two neutrons when it emits an alpha particle, the mass number of the nucleus must decrease by four, and its atomic number must decrease by two.

Exercise 5
Radium 226 decays by emission of an alpha particle. When a radium 226 nucleus decays, what is the name, atomic number, and mass number of the remaining nucleus? (*Hint:* See Fig. 3.1).

The other kind of particle emitted during radioactive decay is an electron. This is rather surprising because the atomic nucleus doesn't contain any electrons, it contains only protons and neutrons. So what is an electron doing emerging from a nucleus? The answer is that any neutron has a certain probability of turning into

(c)

FIGURE 20.1 (c) Marie Curie in 1934. (AIP Neils Bohr Library.)

a proton plus electron, somewhat as a high-energy photon might turn into an electron-positron pair (see Chapter Nineteen). When a neutron turns into a proton plus electron, the electron might be ejected from the nucleus. Any electron emitted in this manner from a nucleus is called a *beta particle.*

Exercise 6
What happens to the atomic number and the mass number of a nucleus that decays by emission of a beta particle? (answer below.)

When a beta particle is emitted one neutron in the nucleus changes into a proton, so the atomic number of the nucleus increases by one, whereas the mass number is unchanged.

Exercise 7
As a result of an unstable balance of protons and neutrons, carbon 14 is radioactive. It decays by emitting a beta particle. What nucleus (name, atomic number, mass number) results?

Exercise 8
U-238 is an alpha emitter. Give the name, atomic number, and mass number of the resulting nucleus.

Exercise 9

You found in Exercise 8 that U-238 decays into Th-234. But Th-234 is also radioactive. It is a beta emitter. Give the name, atomic number, and mass number of the resulting nucleus.

Radioactive decay can also cause a nucleus to emit electromagnetic radiation, much as an atom emits radiation when anything upsets its electron structure (Chapter Eighteen). But the primary force operating in the nucleus, the strong force, is much stronger than the electromagnetic force operating among the electrons. So the photons of nuclear radiation have much larger energies than the visible, ultraviolet, and X-ray photons associated with changes in atomic structure. This high energy radiation emitted by nuclei during radioactive decay and during other nuclear reactions is called *gamma radiation,* and the associated photons are *gamma photons.*

Half-life

As we saw in Chapter Seventeen, atomic and subatomic phenomena are unpredictable. As a direct consequence of these quantum mechanical uncertainties, the time at which a radioactive nucleus will decay is unpredictable. We can predict that any particular U-238 nucleus will eventually emit an alpha particle, but we cannot say when it will emit its alpha particle. It might occur during the next second, or it might not occur for the next billion years!

Recall however, that the quantum theory can predict probabilities of various events even though it cannot predict with certainty. Thus if we begin with a large number of U-238 nuclei, quantum theory can predict roughly what fraction of them will have decayed in any given period of time, even though the theory cannot predict which nuclei will decay. It's just like flipping coins. If you flip one coin once it is impossible to predict the outcome, but if you flip 1000 coins you can predict roughly what fraction of them will turn up heads even though you cannot predict which individual coins will come up heads.

Thus it is possible to predict how long it will be before 10 percent of a large collection of U-238 nuclei will decay or before 20 percent will decay, and so on. For example, it is known that the time for half of a large collection of U-238 nuclei to decay is 4.5 billion years. This time is called the half-life of U-238. The *half-life* of any radioactive substance is the time required for half of the nuclei in a sample of this substance to decay.

As another example, the half-life of C-14 is 5000 years. If you begin with a million C-14 nuclei, roughly 500,000 of them will decay during the next 5000 years.

Exercise 10

Recall (Exercise 7) that C-14 is a beta emitter. Suppose you have a kilogram of pure C-14. (a) How much C-14 will remain after 5000 years and what will have become of the rest of the material? (b) How much C-14 will remain after 10,000 years? (c) After 15,000 years? (d) After 20,000 years? (Answer below.)

Figure 20.2 is a graph of the results of Exercise 10. The graph shows the mass of C-14 remaining after various times. After 5000 years, ½ kg remains (the other ½

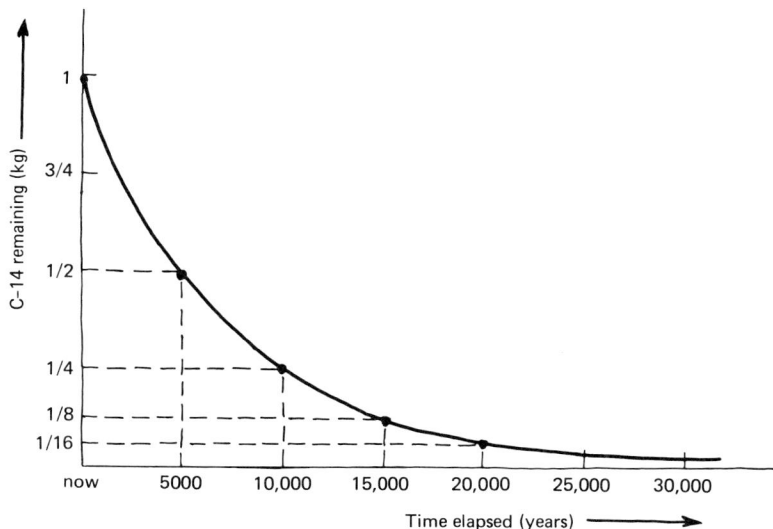

FIGURE 20.2 Decay curve for C-14.

kg has turned into N-14). After another 5000 years, half of this ½ kg will have decayed because the half-life concept applies to the ½ kg just as it applied to the entire 1 kg. Thus after a total of 10,000 years, only ¼ kg remains. After 15,000 years, half of that, or ⅛ kg, remains. After 20,000 years, 1/16 kg remains. Connecting these points by a smooth curve, we get the ***decay curve*** of Figure 20.2.

Exercise 11
If you start today with 100 kg of pure U-238, how much U-238 will remain after 4.5 billion years? After 9 billion years? After 18 billion years (4 half-lives)?

Even though each radioactive substance has a different decay time, the decay curve for each will look like that in Figure 20.2, but with the numbers along the two axes altered. Half the original amount will decay in one half-life, half of that will decay in the next half-life so that only one-quarter remains, and so on.

Applications of Radioactive Isotopes

Every element heavier than bismuth (atomic number 83, slightly heavier than lead, which is number 82) is radioactive. There are no nonradioactive, or **stable,** isotopes of these heavier elements. Their nuclei are just so big that even the strong force cannot hold them together permanently. Most of the elements lighter than bismuth come in the form of one or more stable isotopes, along with one or more ***radioactive isotopes.*** For example, carbon comes in the form of the stable isotopes C-12 and C-13 and the radioactive isotope C-14. Some radioactive isotopes, for instance C-14, occur naturally on Earth. Others are manufactured artificially in nuclear reactors.

Radioactive isotopes find a wide range of technological applications because they have the same electron structure and thus the same chemistry as the nonradioactive atoms of the same element. For instance, a molecule of CO_2 (carbon dioxide) that had a C-14 atom instead of the more common C-12 atom would behave chemically just like any other CO_2 molecule, but would be radioactive.

Roughly half of all cancer patients receive **radiation therapy.** Treatment is based on the fact that rapidly dividing cancer cells are especially vulnerable to the high-energy alphas, betas, and gammas emitted by a radioactive substance, so radiation destroys these cells more readily than it destroys healthy cells. The radiation may be administered externally using cobalt 60 (see the introductory paragraphs of Chapter One). The radioactive isotope Co-60 is manufactured from the naturally occurring stable isotope Co-59 by placing the Co-59 in a nuclear reactor, where it is bombarded by neutrons. Occasionally, a Co-59 nucleus will absorb one of the neutrons to become Co-60. Co-60 is a convenient source of gamma photons. For therapy, a quantity of Co-60 is placed outside the patient's body, which is shielded in such a way that only the cancerous area is exposed to gamma radiation.

Therapists often implant radium-filled needles, small pellets containing radon gas (radon is one of the heavier-than-bismuth radioactive elements), or wires containing manufactured radioactive isotopes, directly into a tumor, or a patient might swallow small amounts of an appropriate radioactive isotope. For example, the radioactive isotope iodine 131 behaves chemically just like its stable relative I-127 so it collects in the thyroid just like normal iodine, and it is used in treating cancer of the thyroid.

Radioactive substances are easily detected using any device that is sensitive to high-energy alphas, betas, and gammas. Hence these substances find important applications as **tracers** in industry, agriculture, and medicine. For example, engineers can locate gas leaks in an underground pipe by using radioactive isotopes to make the gas radioactive and then using a radiation detector to find the leaking gas. By "labeling" fertilizer with a radioactive isotope, agricultural researchers learn where the fertilizer goes after it is placed in the soil. The path followed by herbicides and pesticides are studied in the same way. In medicine, physicians use radioactive labels to study the path in the human body followed by any particular chemical.

The Dating Game

How old are the Dead Sea Scrolls? When did agriculture originate? How old is **Homo sapiens**? How old is life on Earth? How old is Earth itself? We have learned the answers only recently, using **radioactive dating.**

Suppose you have a particular object and you want to know how old it is. Suppose that your object contains at least one radioactive substance. If you use a radiation detector to measure the amount of the radioactive substance in the object today, and if you have some way of knowing the amount present when the object was originally formed, you can calculate the age of the object from the known half-life of the radioactive substance. For instance, if only one-half of the original amount remains, the object is 1 half-life old. This is the idea behind radioactive dating.

Let's look at a specific type of dating: **carbon 14 dating.**

A certain small percentage of the carbon atoms in the atmosphere are radioactive C-14 atoms rather than the stable isotopes C-12 or C-13. During photosynthesis (Chapter Three) all plants consume CO_2 from the atmosphere, thus incorporating atmospheric carbon into their structure, so a certain small fraction of the carbon atoms in all living plants is radioactive C-14. This fraction can be measured in plants living today. It is thought that plants living in the past had approximately the same fraction of C-14 when they were alive. C-14 has a half-life of 5000 years and decays by beta emission into the stable isotope N-14. Once a plant dies, it stops consuming atmospheric carbon, so its allotment of radioactive carbon begins to decay. Archaeologists determine the date of the plant's death by using a radiation detector to measure the remaining C-14 and by comparing the measured amount with the amount known to be in the plant when it died.

Exercise 12

Suppose you measure the beta radiation from an old ax handle and find the ax handle to be only half as radioactive as a living plant. How old is the ax handle? What if it was one-sixteenth as radioactive as a living plant?

The C-14 method won't work for nonbiological objects, such as rocks, because they don't contain C-14, and it won't work for objects older than about 10 half-lives, or 50,000 years, because hardly any C-14 remains and the measurement is extremely difficult to make.

Where does C-14 in the atmosphere come from? Why hasn't it all decayed away by now, since the atmosphere is much older than 5000 years? The answers begin in the distant stars, with **supernova explosions,** which can occur during the gravitational collapse (Chapter Eight) of the more massive stars when a cataclysmic explosion rips the entire star apart and throws most of it into outer space. The small remaining core continues collapsing into one of the compact states described in Chapter Eight. The most recent supernova explosions observable with the naked eye from Earth were in the years 1054, 1572, and 1604.

The energy output and the brightness of these exploding stars is enormous. A supernova explosion could rival the brightness of the full moon. As you might expect, such an explosion sends a multitude of subatomic particles, mostly protons or helium nuclei, hurling into space at high speeds. Some of these eventually enter Earth's atmosphere. Astronomers think that most **cosmic rays,** the high-energy particles continually showering down on Earth, originate in this way.

An entering cosmic-ray particle is likely to collide with the atoms of the upper atmosphere, and, when it does, lots of new particles are formed. Among these secondary particles are many high-energy neutrons. If one of these neutrons happens to strike a nucleus of the common atmospheric element nitrogen, a nuclear reaction produces the isotope C-14. And *that's* where atmospheric C-14 comes from.

Thus many of the museum dates on ancient artifacts come to you courtesy of supernova explosions.

There are several other radioactive-dating methods, based on other radioactive isotopes. These methods are independent of each other, as they are based on entirely different radioactive decay processes. Thus the agreement between the dates obtained by the various methods bolsters one's confidence in radioactive dating.

Radioactive dating has traced human culture back to its origins 10,000 years ago. Our species, *Homo sapiens,* has been traced back about 100,000 years. Humanlike creatures have been dated at over 3 million years ago. Evidence of life has been firmly established back to about 2.3 billion years, but fossils found in 1979 seem to have pushed this figure to 3.5 billion years. The oldest rocks on our planet solidified some 4.5 billion years ago.

To get a feeling for these dates, imagine that the entire 4.5-billion-year history of Earth is compressed into a 12-hour period, say from noon to midnight. On this time scale, life appears early, around 3 P.M., but humanlike creatures do not appear until 20 *seconds* before midnight. *Homo sapiens* dashes onto the scene at 1 second before midnight, human culture originates at one-tenth of a second before midnight, and the entire scientific-technological age of the past 3 centuries begins at .003 seconds before midnight!

Viewed in perspective, humankind has been on this earthly stage for only a moment and has produced all the trappings of technology during the bat of an eyelash.

Biological Effects of Radiation

The high-energy alpha, beta, and gamma particles emitted during radioactive decay can destroy living tissue. These particles do their damage as they move through a living cell by knocking electrons out of the atoms in their path, thereby ionizing (Chapter Twelve) those atoms and producing a radical change in the chemistry of the cell. As nature has spent several billion years perfecting the cell, any change is likely to be for the worse.

These disruptions can kill the cell or, worse yet, alter it in such a way that the alteration is passed on when the cell divides into two cells; this can produce a cancer. The harmful effects of radiation were not recognized until the 1920s, when the government formulated the first limitations on radiation exposure. Prior to that time, scientists, physicians, and industrial workers often received massive radiation doses. Many developed malignancies, sometimes decades after the exposure had ceased.

The biological effect produced by high-energy subatomic particles is measured in units called **millirems,** a measure of radiation-produced biological damage. Thus a person receiving 200 millirems has twice as much cellular damage as a person receiving 100 millirems. The number of millirems received is usually called the **radiation dose,** even though much of the effect is caused by particles of matter (alphas and betas), rather than by electromagnetic radiation (gamma photons). All sources of biologically harmful high-energy particles are included in the radiation dose: not only particles emitted during radioactive decay, but also X-ray photons, cosmic-ray particles, and so on. For example, during the 1940s and 1950s, the X-ray photons emitted by X-ray shoe-fitting machines in stores contributed to the radiation dose. Unfortunately, these machines were especially popular with children—I know, because I was one of them!

The average radiation dose for persons living in the United States is about 180 millirems per year. Table 20.1 shows the sources of this radiation dose.

Note that about 100 millirems of the 180 millirem average dose is from natural sources, sources that the human race has always been exposed to. Medical diagnoses, especially dental X rays, account for almost all the remaining 80 millirems. Nuclear-power reactors contribute a negligible amount. In fact, I have included nuclear power separately in the list only because of the public interest in this radiation source and to emphasize that its average contribution is small. Many other low-level sources contribute more than nuclear power; their combined effect is listed in the table under "miscellaneous." Fallout from weapons tests caused 4 millirems in 1970, despite the fact that the United States and the Soviet Union have not exploded nuclear weapons

Table 20.1 Sources of radiation in the United States

Source	Dose per Person (millirems per year)

Cosmic rays, U.S. average
    Sea level, 41 millirems
    Denver (5000 ft), 70 millirems
    Leadville, Colo. (10,000 ft), 160 millirems
	44

Rocks and soil, U.S. average
    Atlantic coastal plains, 23 millirems
    Rocky Mountains, 90 millirems
	40

Internal consumption of naturally occurring radioisotopes	18
Total from natural sources, U.S. average	102
Fallout from weapons tests (1970)	4
Nuclear power	0.003
Medical diagnostic	72
Miscellaneous artificial sources (e.g., television screens)	4
Total from artificial sources, U.S. average	80
Total from all sources, U.S. average	182

Source: Based on data provided by the U.S. Environmental Protection Agency and by the National Academy of Sciences Committee on Biological Effects of Ionizing Radiation.

in the atmosphere since 1963. Most of today's fallout comes from continued atmospheric testing by China. If the United States and the Soviet Union had continued their massive testing in the atmosphere, the contribution from fallout would have been much larger, perhaps 50 to 100 millirems per year.

Thought Question 2

How do you feel about the small contribution that China makes to your average radiation dose? Should the United States do anything about this? How do you think other countries in the world felt about the much more massive U.S. and Soviet testing during the 1950s? Should they have done anything about it?

The above doses are small compared to the amounts known to cause ill effects. Whole body doses under 25,000 millirems have no directly demonstrable effect. Doses above 100,000 millirems produce extensive damage to the blood-forming tissues. Doses over 500,000 millirems, if delivered during a short time, cause death within a few weeks.

There is considerable controversy about the effects of low doses of radiation, doses comparable to the 100 millirems per year received from natural sources. Many experts feel that harmful effects are present at these levels even though it is difficult to demonstrate them conclusively. The National Academy of Sciences estimates that if the artificially caused dose were increased from 80 millirems to 500 millirems per year, averaged across the U.S. population, an additional 3,000 to 15,000 cancer deaths would occur annually. But these additional deaths would be difficult to observe or demonstrate conclusively because of the long incubation period for cancer and because the effects would be hard to discern among the 300,000 cancer deaths that already occur every year.

The National Academy of Sciences estimates that medical diagnostic X rays cause 1500 to 3000 cancer deaths annually. The average yearly dose from these X rays is increasing, yet it is estimated that medical exposures could be greatly reduced without any loss of the benefits of diagnostic X rays. Clearly the medical and dental professions have a responsibility to avoid unneeded X rays and to use well-shielded machines. Although the benefits of X rays are often worth the small risk, diagnostic X rays should not necessarily be equated with good medical practice in evaluating health care.

Thought Question 3

It is difficult to balance benefits and risks in matters involving human lives. We permit nonessential uses of the automobile even though the automobile is responsible for 50,000 deaths a year. Should we do anything about the 1500 to 3000 cancer deaths caused annually by diagnostic X rays? Should we, for example, prohibit their use in dental offices? Keep in mind that diagnostic X rays also save many lives.

Fallout means the radioactive isotopes produced during a nuclear explosion. Most of these isotopes are carried into the atmosphere on dust particles. These dust particles then gradually settle to the ground. This falling out of radioactive dust is most severe during the first few hours after the blast and in the first few hundred miles downwind. The dust contains large numbers of radioactive isotopes that cause damage when they decay hours or years after the explosion.

The radioactive isotope strontium 90 is one of the important components of fallout. Strontium, atom number 38, is normally a stable element. Because its electron structure is similar to that of calcium, the chemistry of strontium is similar to the chemistry of calcium. Now, where does calcium ordinarily go in your body? To your bones. That's where strontium goes, including any strontium 90 that you may have swallowed. And, since red blood cells are manufactured in the bone marrow, it isn't surprising that the leukemia rate in Hiroshima and Nagasaki after the nuclear bombing was 15 times the normal rate or that Marie Curie died of leukemia. Leukemia is a cancer of the blood.

With a 30-year half-life, strontium 90 is a long-lived radioactive isotope. If you get Sr-90 on your skin and wash it off within a few hours, it won't do much damage because only a small fraction of the Sr-90 nuclei will have decayed during that time. If you ingest Sr-90, however, it will go to your bones and remain there for years, working its damage the entire time. Thus some radioactive isotopes are thousands of times more harmful when they are swallowed.

Exercise 13

Iodine 131 is another important radioactive isotope in fallout. It is taken up by plants, which are eaten by cows; it becomes concentrated in the cows' milk; the milk is drunk by humans; and the I-131 becomes concentrated in the humans' thyroid glands. I-131 has a half-life of only 8 days. Roughly how many days after a nuclear bombing would you have to wait for I-131 radioactivity to decrease to 10 percent of its original value? For it to decrease to 1 percent of its original value?

High-energy subatomic particles can cause biological mutations and thus affect evolution. A *mutation* is any alteration of a DNA molecule (the molecule that carries genetic information) in an egg cell or a sperm cell. High-energy alpha, beta, or gamma particles can produce such alterations. If an altered egg or sperm cell gets together with a cell of the opposite sex, the resulting offspring will carry the alteration in every cell of his or her body, including his or her sex cells. Thus such an alteration is transmitted to future generations.

Any increase in the average radiation dose, for instance by X rays or fallout, causes some increase in the mutation rate. It is not known how much the average dose must be increased before the increased mutation rate becomes significant. The survivors of the nuclear bombing of Hiroshima and Nagasaki and their offspring showed no observable increase in their mutation rate despite the fact that the doses were sufficient to cause a disastrous increase in the cancer rate. Of course, this should not be taken to mean that the bombing caused no mutations; it only means that the number of genetic defects was so low as to be undetectable.

Checklist

four fundamental forces	stable isotope
nuclear reaction	radioactive isotope
chemical reaction	radiation therapy
three types of nuclear reactions	radioactive tracers
atomic number	radioactive dating
isotope	C-14 dating
mass number	supernova explosion
Marie Curie	cosmic rays
radioactive nucleus	millirems
radioactive decay	radiation dose
alpha, beta, and gamma particles	sources of U.S. radiation dose
half-life	fallout
decay curve	mutation

Further Thought Questions

4. Thanks to radioactive dating, we now know when human culture originated, when the human race evolved, and when the planet Earth solidified. Who cares?

5. Some people find a conflict between their religious beliefs and the historical dates determined by radioactive dating. For example, some people feel that the Bible places Adam and Eve at about 5000 B.C. What is your opinion?

6. Does it seem to you that it is important for us to realize that, relative to the long history of Earth and relative to the history of the human race, science and technology have developed during the bat of an eyelash? Discuss.

Further Exercises

14. How many protons and how many neutrons are there in each of the isotopes U-238 and U-235 of uranium? (*Hint:* See Fig. 3.1).

15. Let's consider an extremely small amount of a radioactive substance. Suppose we have precisely eight C-14 nuclei (half-life 5000 years). Can you predict precisely how many will remain after 5000 years? Why not? Roughly how many will remain? How many will remain after 10,000 years? After 15,000 years?

16. For the eight C-14 nuclei of the preceding exercise, how many C-14 nuclei do you suppose will remain after 50,000 years? Can you make a rough estimate as to how long it will take for the entire sample of eight nuclei to decay?

17. Strontium 90 is a beta emitter with a half-life of about 30 years. What nucleus results from its decay? How long would you have to remain in your fallout shelter in order for the Sr-90 radioactivity level to be reduced to roughly 10 percent of its original value? Would you wait that long?

Recommended Reading

1. H. A. Boorse and L. Motz, *The World of the Atom,* Basic Books, New York, 1966. Contains excerpts from the writings of the Curies, Rutherford, and others as well as biographies.

2. Anne M. Briscoe and Sheila M. Pfafflin, editors, *Expanding the Role of Women in the Sciences,* New York Academy of Sciences, New York, 1979. These 33 papers provide a comprehensive review of the status of women scientists in the United States.

3. Bernard Cohen, *The Heart of the Atom,* Science Study Series, Anchor Doubleday, Garden City, N.Y., 1967.

4. H. J. Mozan, *Woman in Science,* reprinted by MIT Press, Cambridge, Mass., 1974. Although it was written in 1913, this classic remains a valuable source book for women's contributions to science. It includes chapters on women in mathematics, in physics, and in chemistry as well as Marie Curie's early contributions.

5. Olga S. Opfell, *The Lady Laureates,* Scarecrow Press, Metuchen, N.J., 1978. The lives and achievements of the 17 women who have won the Nobel Prize.

6. Robert Reid, *Marie Curie,* New American Library, New York, 1978. Biography of a woman scientist during an era when male supremacy was unquestioned.

21

TWO MORE WAYS TO CHANGE A NUCLEUS

Radioactive decay releases nuclear energy. In other words, it changes some nuclear energy into other forms—into kinetic energy of the remaining nucleus, kinetic energy of the emitted alpha or beta particle, and radiant energy of any emitted photons.

For better or for worse, nature has arranged two other ways of releasing the energy tied up in the strong nuclear force. Nuclear energy is released when two sufficiently light nuclei are put together into a single unit and also when a single sufficiently massive nucleus is split into two lighter nuclei. The first type of reaction is called *fusion;* the second is called *fission.*

Fusion

Let's make an energy analysis of the nucleus. Imagine taking a helium 4 nucleus apart into its two individual protons and its two neutrons (Fig. 21.1). Because the nucleus is bound together by the strong nuclear force, you must do work to pull it apart. In fact, you must do a great deal of work because of the strength of the force that holds the nucleus together. According to the work-energy principle (Chapter Nine), because a great deal of work was done on these protons and neutrons, they end up with a great deal more energy than they had to begin with. Thus the separated system in Figure 21.1b has significantly more energy than the bound system in Figure 21.1a.

Now turn the process around (Fig. 21.2). Suppose you manage to get two isolated protons and two isolated neutrons to stick together. This would be hard to do because they are so small and because the electric force between protons tends to push them apart. But once any two of these four particles are close together, close enough that they are within range of each other's strong nuclear force, this strong force will grab the two particles and hold them tightly together. We've seen in the previous paragraph that the separated configuration has more energy than the bound configuration. Because this energy difference is caused by strong nuclear force, it is called *nuclear energy*.

The process of putting a nucleus together from smaller parts, illustrated by Figure 21.2, is called *nuclear fusion.* The upshot of the preceding two paragraphs is that in Figure 21.2 there is significantly less nuclear energy "after" than there is "before." According to conservation of energy, this energy cannot just vanish. In fact, it shows up as kinetic energy of the resulting helium nucleus and radiant energy of the gamma photons emitted during this nuclear reaction. Thus the fusion of isolated protons and nuetrons converts nuclear energy to other forms.

Think of fusion this way: You expend a little energy getting the two protons and two neutrons of Figure 21.2a close together, but once they are close enough, the strong nuclear force can assert itself and finish the job for you. This force grabs the four particles and smashes them rapidly and tightly together, releasing a great deal

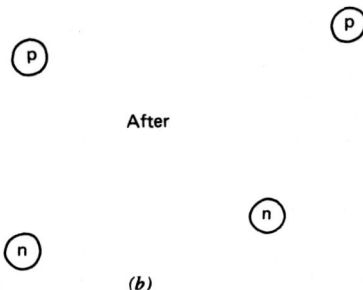

FIGURE 21.1 Taking a nucleus apart. The *p*'s are protons, the *n*'s
are neutrons.

of nuclear energy in the process. Thus gamma photons emerge and the resulting
helium nucleus gains kinetic energy.

The isolated protons in Figure 21.2*a* can be thought of as the nuclei of ordinary
hydrogen atoms, H-1. So the nuclear reaction in Figure 21.2 could be represented
by

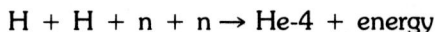

$$H + H + n + n \rightarrow He\text{-}4 + energy$$

The word *energy* on the right-hand side of the reaction is supposed to mean that
some of the original nuclear energy shows up as other forms of energy after the
reaction.

Suppose we had a large number of protons and neutrons and that we could fuse
the entire collection to form He-4 nuclei. A large amount of nuclear energy would be
converted into radiant energy of emitted photons and kinetic energy of the He-4
nuclei. From our larger-scale point of view, the kinetic energy of this collection of He-

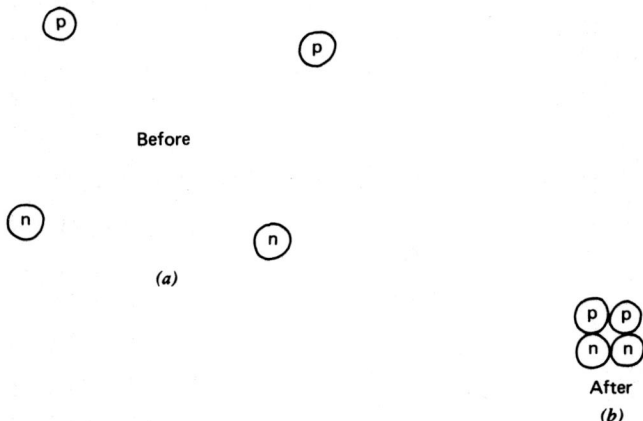

FIGURE 21.2 Putting nuclei together—nuclear fusion.

4 nuclei is just thermal energy. In other words, the fusion reaction makes the system hot and radiant.

This is why the sun and the H-bomb (hydrogen bomb) are hot and radiant—they get their energy by fusing hydrogen into helium. However, the reaction shown in Figure 21.2 is an oversimplification of the reactions that actually occur in these two applications. Several different types of fusion reactions actually occur in the sun. One of the most important of these is

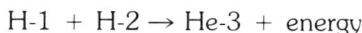

$$\text{H-1} + \text{H-2} \rightarrow \text{He-3} + \text{energy}$$

Exercise 1

How many nuclear particles (protons and neutrons) are represented on the left side of this reaction? How many are represented on the right side? How many neutrons are on the left side? How many neutrons are on the right side?

Nearly all stars derive their radiated energy from the fusion of hydrogen into helium, which eventually uses up the hydrogen initially present in the star. Our sun started its life some 5 billion years ago as a sphere of hydrogen gas. Hydrogen-helium fusion has kept the sun hot and radiant and will continue to do so for another 5 billion years. By that time, most of the hydrogen will be used up. After a few final outbursts (which would destroy any remaining life on Earth), the sun will shrink down into a white dwarf star (Chapter Eight).

It's not easy to get hydrogen to fuse into helium. In the H-1 + H-2 reaction shown above, an H-1 nucleus (a proton) must be brought very close to an H-2 nucleus (proton plus neutron) in order for the short-ranged nuclear force to grab the particles, and they must be brought together hard in order to overcome the repulsive electric force between them. The easiest way to bring the particles together like this is by heating a gas consisting of H-1 and H-2 atoms. At high temperatures, the atoms move rapidly and collide vigorously with each other. The powerful collisions strip many atoms of their electrons and bring some nuclei close enough together that they fuse. The temperatures needed to achieve this are sizable—10 million to 100 million degrees Celsius!

So the fusion energy machine at the sun's center must be as hot as 10 million degrees Celsius or the sun wouldn't shine. The sun's surface, however, is much cooler—a mere 6000° C. The hydrogen bomb's central temperature is some 100 million° C.

According to Einstein's mass-energy relation (Chapter Sixteen), whenever a system gains or loses energy it also gains or loses mass, because the total energy and total mass are always proportional to each other. For the small energies involved in chemical reactions the mass changes are so small that they can't be measured; but the energies involved in nuclear reactions are large and the corresponding mass changes are measureable.

Exercise 2

Remembering that nuclear energy is converted to other forms when hydrogen fuses to make helium, would you expect a stationary He-3 nucleus to be more massive, less massive, or equally massive, when compared with the combined mass of an H-1 nucleus plus an H-2 nucleus? (Answer below.)

An He-3 nucleus has less nuclear energy than the combined H-1 and H-2 nuclei, so it has less mass. In fact, the nuclear energy released in the fusion reaction is so great that the total mass decreases by nearly 1 percent. It's rather remarkable—the He-3 nucleus contains the same two protons and single neutron as the H-1 plus H-2 nuclei, yet it's mass is 1 percent less. Our ordinary experience tells us that when two things are stuck together to make a bigger thing, the mass of the bigger thing equals the total mass of the two smaller things. Once again, ordinary experience has led us astray. Many experiments have confirmed that the mass of He-3 actually is 1 percent less than the total mass of H-1 and H-2. When the "sticking together" involves forces as large as the strong nuclear force, Einstein's mass-energy relation begins to assert itself.

Exercise 3

Suppose we started with 2 kg of hydrogen and managed to fuse all of it to form helium. How many kilograms of helium would we have? What if we had started with 1000 kg of hydrogen?

From the fact that 1 percent of the mass is lost in hydrogen-to-helium fusion, we can figure out how much nuclear energy is converted to other forms of energy. For instance, fusion of 1 kg of hydrogen (Exercise 3) would produce 0.99 kg of helium. The mass decrease is 0.01 kg, or 10 g, or about half an ounce. According to Einstein, the amount of nuclear energy converted to other forms is mc^2, where the mass m is 0.01 kg. Recalling that c^2 is 9×10^{16} in metric units, the energy released is $0.01 \times 9 \times 10^{16}$, or 9×10^{14} J. That's enough energy to power a large, 1000-MW nuclear generating plant for 10 days!

The sun is losing mass at the rate of 5 million tons per second as a result of nuclear fusion! You can see from the preceding paragraph that this represents an enormous release of nuclear energy. Nevertheless, this mass loss is trivial compared to the total 2×10^{30} kg mass of the sun. As shown by the above discussion, if the entire mass of the sun were to undergo hydrogen-to-helium fusion, the sun's mass would decrease by only 1 percent.

Hydrogen-to-helium fusion is not the only possible fusion reaction. There are many other nuclear reactions in which lighter nuclei combine to form a heavier nucleus, with a release of nuclear energy. The temperature required for the other fusion reactions is even higher than the hydrogen-to-helium fusion temperature. Stop for a moment and see if you can guess the reason for this.

(Time out for guessing.)

The reason is that hydrogen has only one proton in its nucleus whereas all other elements have two or more protons, so the electric repulsion is larger between other nuclei and higher temperatures are needed to overcome this repulsion.

The following exercises will show how the chemical elements were born. Use the table of elements (Fig. 3.1) in answering these.

Exercise 4

Once the hydrogen-to-helium reaction has built up significant quantities of helium in the center of a star, He-4 nuclei begin to fuse with other He-4 nuclei. What is the atomic number, mass number, and name of the nucleus formed in this reaction?

Exercise 5
In the more massive stars, He-4 combines with Be-8. What nucleus does this produce?

Exercise 6
Once C-12 is produced in the more massive stars, these C-12 nuclei begin to fuse with H-1 nuclei. What is produced?

Exercise 7
The N-13 nucleus produced in the preceding reaction is a radioactive isotope. It decays by the unusual process of spontaneous *positron* (positive electron or positive beta particle) emission. What nucleus results from this decay?

Exercise 8
The C-13 nuclei produced by the preceding process combine with H-1 nuclei in stars. What nucleus is produced?

If you worked through these exercises, you learned how many of the elements have been created. It is thought that the early universe (for the birth of the universe consult Chapter Twenty-Four) consisted entirely of hydrogen and helium. Many of the heavier elements were then produced in the centers of stars by fusion reactions.

Limits on Fusion

There are limits on the ability of the fusion reaction to produce heavier nuclei from lighter ones. We've seen that a collection of nuclei will fuse only as long as the temperature remains high. In the absence of some external source of thermal energy, the temperature can remain high only if the reaction produces thermal energy. The reaction will then sustain itself.

It turns out that not all fusion reactions produce thermal energy. If the fusing nuclei are very massive, the reaction instead absorbs thermal energy. That is, thermal energy is converted into nuclear energy.

To be more specific, fusion produces thermal energy only when two nuclei combine to form a nucleus whose atomic number is 26 or less. For example, the fusion of manganese (number 25) with hydrogen to form iron (number 26) produces thermal energy, so this reaction will sustain itself. But the fusion of iron with hydrogen to form cobalt (number 27) absorbs thermal energy. This conversion of thermal energy to nuclear energy cools the collection of cobalt and hydrogen nuclei, so this reaction will not sustain itself: an outside energy source is required to keep the reaction going. Thus iron (atomic number 26) is the limit of self-sustaining fusion reactions.

Exercise 9
Compare the mass of the cobalt nucleus to the total mass of iron plus hydrogen.

We have seen that stars manufacture heavier elements out of hydrogen and helium by nuclear fusion in their high-temperature centers. This is how all the elements from lithium (atomic number 3) through iron were created. The high temperatures that prevail at the centers of stars cause hydrogen and helium to fuse into heavier nuclei, a process that produces thermal energy, which causes further fusion. The entire process is self-sustaining so long as the star isn't creating elements heavier than iron.

But there are some 66 naturally occurring elements that are heavier than iron. Where did they come from? An external energy source is needed for their production. A likely candidate for such an energy source is the supernova explosion of a giant star (Chapter Twenty). It is generally thought that many of the nuclei heavier than iron were formed by nuclear fusion in the high-energy interiors of exploding stars. The thermal energy required for these reactions came from the supernova.

Consider: Your body is made mostly of hydrogen (atom number 1), carbon (number 6), nitrogen (number 7), and oxygen (number 8), with significant contributions from phosphorus (number 15) and sulfur (number 16). The nuclei of all the hydrogen atoms were made when the universe was born. The carbon, nitrogen, and the rest were made by fusion in stars that burned in far reaches of the galaxy billions of years ago. Your body also contains trace amounts of biologically crucial heavier elements such as cobalt (number 27), copper (number 29), zinc (number 30), and iodine (number 53). The nuclei of these atoms were born in the distant fires of ancient supernova explosions.

We are all made of star-stuff.

Fission

We have seen that fusion reactions between nuclei heavier than iron absorb thermal energy. This implies that the opposite reaction, namely the splitting of a heavier-than-iron nucleus to form two lighter nuclei, produces thermal energy. This is the idea behind **nuclear fission,** or the splitting of a heavy nucleus into two large fragments to produce thermal and radiant energy.

A heavy nucleus will not fission spontaneously; it must be hit with a particle. The particle that works best is the neutron.

Exercise 10

Why would we expect the neutron to work better than the proton? Why would we expect the neutron to work better than the electron?

Some nuclei split more readily than others. Two widely used fissionable nuclei are uranium 235 and plutonium 239.

Let's consider U-235. When a neutron strikes a U-235 nucleus, the heavy nucleus breaks up. The breakup can occur in many possible ways, but it always produces two large fragments plus a few neutrons. For example, a typical pair of large fragments produced when U-235 fissions is krypton 91 and barium 142.

Exercise 11

How many neutrons must have been produced in this reaction? (Answer below.)

In Exercise 11 you should add up the number of nuclear particles (protons and neutrons) in the two fragments. This number is 233, the sum of the mass number, 91, of krypton and the mass number, 142, of barium. The mass number of U-235 is two more than this. So two nuclear particles are left over in the reaction. These particles cannot be protons, because the number of protons in U-235 equals the number in Kr-91 plus Ba-142, as you can see by consulting the table of the elements in Figure 3.1. So the two extra particles must be neutrons.

This typical nuclear fission reaction can be represented by

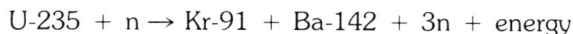

$$\text{U-235} + n \rightarrow \text{Kr-91} + \text{Ba-142} + 3n + \text{energy}$$

The n stands for neutron and 3n stands for 3 neutrons, the two produced by the reaction plus the one that caused the reaction. Figure 21.3 pictures this reaction.

Most of the fragments produced by fission of a heavy nucleus are radioactive isotopes because they do not have the balance of protons and neutrons needed for stability. For example, Kr-91 and Ba-142 both have far too many neutrons to be stable. The heaviest stable isotopes of krypton and barium are Kr-86 and Ba-138. So these fragments, or **fission products,** are radioactive.

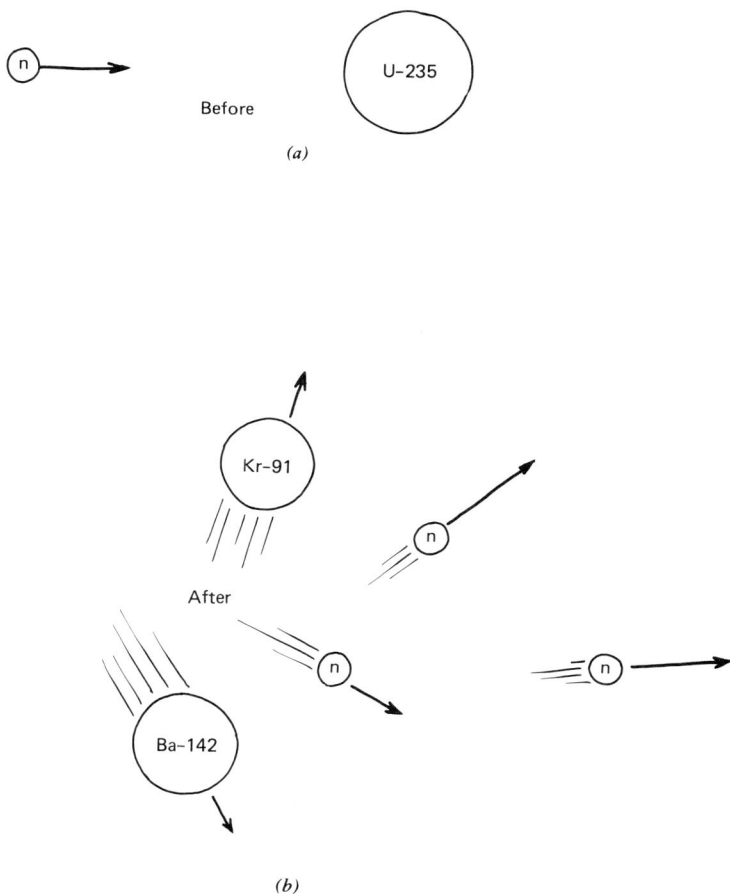

FIGURE 21.3 Breaking a nucleus apart—nuclear fission.

As we have seen, the fission of a heavy nucleus converts nuclear energy to other forms of energy. These other forms are thermal energy of the fission products and radiant energy of the gamma photons produced in the reaction.

The Discovery of Fission

Nuclear fission was discovered in Berlin in 1938. It was a significant time and place in the history of the world.

The neutron had been discovered just 6 years earlier by the British physicist James Chadwick. The new particle was found to be an effective promoter of the nuclear reactions that interested scientists at that time. We have seen (Chapter Twenty) that bombardment of atmospheric nitrogen by neutrons produces the radioactive isotope C-14. From 1934 on, physicists produced many new isotopes by neutron bombardment of many of the elements.

This neutron bombardment of the elements must have produced many fission reactions, but nobody noticed them, at least not until 1938. The trouble was, nobody was looking for them. What scientists were looking for, and what usually occurred, was absorption of the neutron by a nucleus. This absorption made the nucleus unstable, and the nucleus then decayed radioactively.

When Italy's Enrico Fermi (Fig. 21.4) was asked why he missed the discovery of fission he replied, "It was a thin piece of aluminum that stopped us all from seeing what took place." Fermi, among others, had been bombarding uranium atoms with neutrons, but he had always placed a foil in front of the detectors monitoring the reactions. The purpose of the foil was to filter out the lower-energy particles produced when uranium atoms absorbed a neutron and then decayed radioactively. It happened that the foil also absorbed all traces of the fission reactions that occasionally occurred in the uranium.

During this period, two Swiss physicists accidentally left the foil off their detector. Their equipment immediately began recording dramatic evidence of nuclear fission. But the two scientists were not looking for such evidence, and they didn't recognize it. Instead, they thought their equipment was malfunctioning. They replaced the apparently faulty detector with a new one—with the foil in place.

In retrospect, Fermi was glad he had missed this discovery. An earlier discovery would have given the world 4 additional years to prepare nuclear weapons for World War II.

In Berlin, Otto Hahn, Lise Meitner, and Fritz Strassman had been working on radioactive isotopes and neutron bombardment. Because she was Jewish, Meitner was dismissed from her post in 1938. While she was in exile in Sweden, Hahn and Strassman discovered that neutron bombardment of uranium produced barium, lanthanum, and cerium (atomic numbers 56, 57, and 58). It is difficult to see how nuclei as light as this could be produced by radioactive decay of neutron-bombarded uranium (atomic number 92), because uranium is so very much heavier than these three elements. It was Lise Meitner, residing at the time in Sweden, who showed that the experimental results demonstrated the fission of the uranium nucleus and that barium, lanthanum, and cerium were among the fission products.

It was one of the ironies of science and of world affairs that as the human race slipped toward global war, nuclear fission was discovered in Berlin, the hub of Hitler's

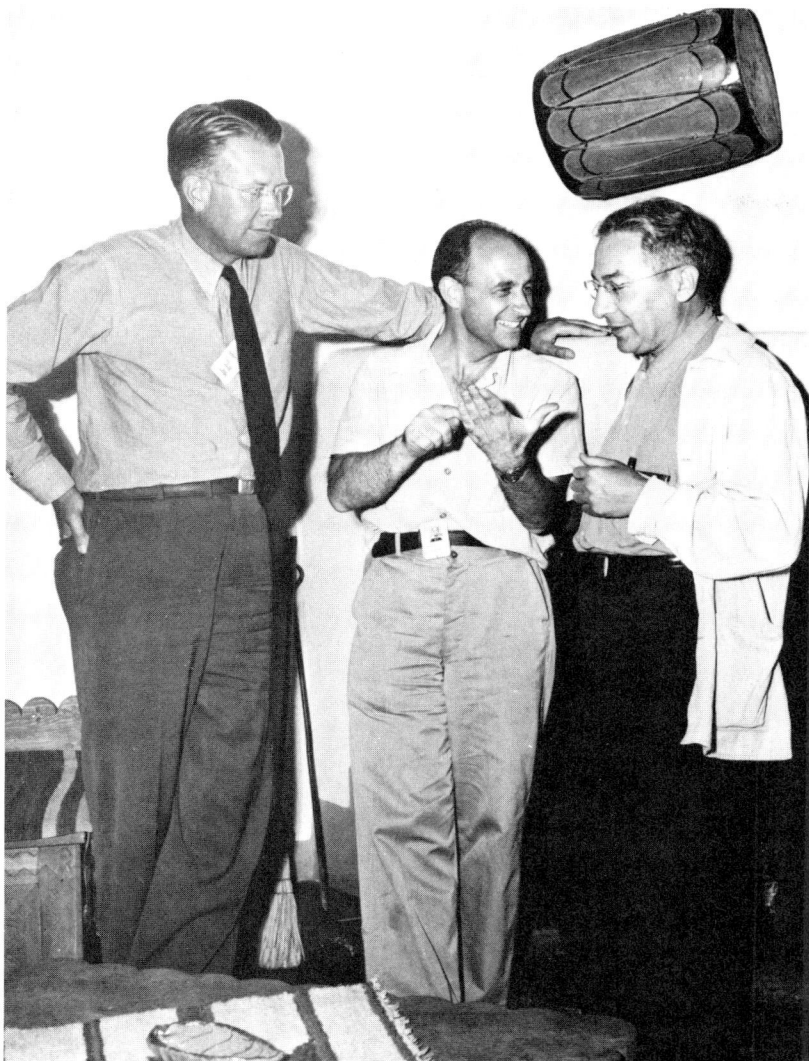

FIGURE 21.4 Enrico Fermi, center, is conversing with physicist I. I. Rabi, while Ernest Lawrence, inventor of the cyclotron particle accelerator, looks on.

planned empire, and interpreted theoretically by a Jewish exile from anti-Semitic Germany.

Thought Question 1

New discoveries in basic science are usually considered a benefit for the human race. Apparently, Fermi thought that the discovery of fission in the 1930s was *not* beneficial, at least not at that time. Can you think of other basic discoveries that have been of dubious benefit to the human race? Now list a few that have been of undoubted benefit.

The Chain Reaction

We have seen that neutrons can fission a U-235 nucleus and that this reaction produces thermal and radiant energy and releases neutrons.

It's not hard to guess where these facts lead. The excess neutrons can be used to fission other U-235 nuclei, and the excess neutrons from the fissions of those other nuclei can be used to fission still other nuclei, and so forth. This process is called a *chain reaction* (Fig. 21.5).

The reaction can build up in a hurry. Each U-235 fission produces on the average about three neutrons. So once a single nucleus is fissioned, perhaps by firing a neutron into a sample of U-235, there will be 3 excess neutrons. If each of these neutrons fissions one U-235 nucleus, there will then be 9 excess neutrons. If these each cause one fission, there will be 27 neutrons. Then 91, 243, and so forth. It won't be long before all the U-235 will fission.

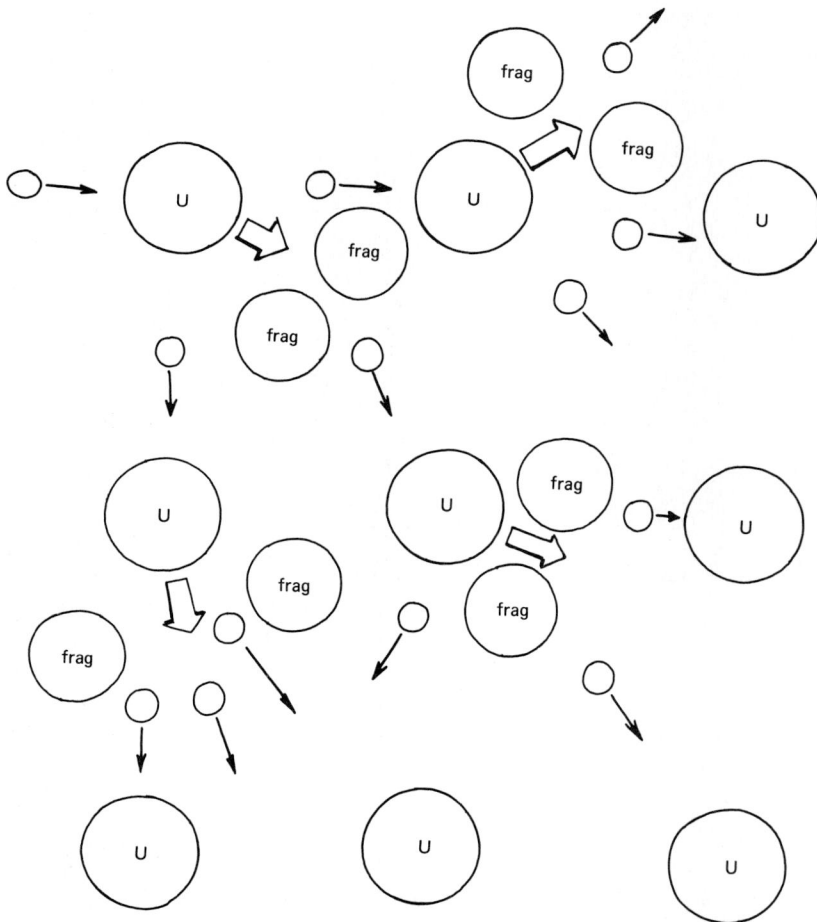

FIGURE 21.5 The chain reaction. *U* means a U-235 nucleus; *frag* means a fission fragment, such as the Kr-91 and Ba-142 of Figure 21.3. The small circles are neutrons.

That is how nuclear reactors get their energy. And that is how A-bombs get their energy.

Checklist

nuclear energy

fusion

creation of elements in stars

self-sustaining nuclear reaction

creation of elements in a supernova

the limit on fusion

fission

fission products

chain reaction

Further Thought Questions

2. The early workers in fission did not know whether a chain reaction was physically possible; it all depended on the average number of neutrons released in a single fission reaction that went on to cause another fission. If this number was less than 1, the chain reaction could not be self-sustaining and would soon die out with little release of energy. As it turned out, this number was larger than 1 for the isotopes U-235 and Pu-239. Would we have been better off if chain reactions had not been possible?

3. For the past several centuries one of the guiding principles of Western civilization and, in particular, of western science has been that it is better to know than not to know. And yet Fermi was glad he had missed the discovery of fission in 1934. Would it be better if we didn't know about nuclear physics? Suppose that a law prohibiting the world's scientists from studying the nucleus had been in effect since 1938 and that this law could be enforced. Would this be a good thing?

Further Exercises

12. Why is nuclear fusion a poor prospect for powering automobiles?

13. Which process would release energy from lead—fission or fusion? Which would release energy from carbon? From iron? (Partial answer: neither fission nor fusion would release energy from iron).

14. The heaviest stable isotopes of krypton and barium are Kr-86 and Ba-138. How many excess neutrons are there in the fission products Kr-91 and Ba-142?

15. Radioactive materials are always slightly warmer than their surroundings. Why?

16. Use the table of the elements, Figure 3.1, to find another pair of elements other than barium and krypton that might be formed in the fission of uranium. Find a pair that might be formed in the fission of plutonium.

17. List some similarities between combustion (Chapter Three) and fission. List some differences.

Recommended Reading

1. Laura Fermi, *Atoms in the Family: My Life with Enrico Fermi,* Phoenix Books, University of Chicago Press, 1961. Laura Fermi's delightful biography of her famous husband, who directed the first self-sustaining chain reaction.

2. Louise B. Young, editor, *The Mystery of Matter,* Oxford University Press, New York, 1965. Part 4 contains an eyewitness account of the human race's first large-scale release of nuclear energy. This anthology contains several other articles on the development of nuclear energy and nuclear weapons.

22

WARFARE IN THE NUCLEAR AGE

In this chapter and the next we will study two of the most significant technologies spawned by twentieth-century physics: nuclear weapons and nuclear reactors.

Nuclear weapons are one of the unfortunate realities of modern times. Even if they are never used, their very existence affects all of us. In the interest of learning about the real world around us, in the interest of learning to coexist with nuclear weapons, and in the interest of ultimately ridding ourselves of this menace, it behooves us all to be informed about this technology.

Although it is unconventional to include a chapter on modern warfare in a physics text, I feel that no physics-related topic is of greater significance than the development of nuclear weapons. Like nearly every important question, nuclear war is an inherently interdisciplinary topic. Thus if our discussion is to be meaningful, it must touch on military, political, and moral matters.

As a vehicle for the discussion of nuclear weapons, I have chosen to present a brief history of the international nuclear weapons race. This competition, dominated by the United States and the U.S.S.R., is part of a larger worldwide buildup in armaments of all kinds.

The buildup has been expensive. The total annual world military expenditure of over $600 billion is about twice the entire national income of the poorer half of the human race.

The buildup has produced dramatic inventories of explosive power. The world's stockpile of nuclear weapons is currently equivalent to some four tons of TNT per head of the world's population. The 20 megatons of explosive power carried by a single Strategic Air Command (SAC) bomber is greater than the total firepower unleashed by all sides in all wars in the history of the human race.

Lest there be any doubt about the seriousness of our topic: The Department of Defense estimated in 1977 that in a single, brief, all-out nuclear exchange between the United States and the Soviet Union, over 150 million Americans would be killed. That's three out of every four. This estimate is for the first 30 days alone, and thus does not account for subsequent deaths among the injured due, for example, to the long-term effects of radiation, inadequate health care facilities, and a shortage of physicians. Civil defense measures might reduce the toll to 110 million lives; a more lengthy exchange would increase the toll. See Reference 8 for more details.

Students often want to know whether a nuclear war could conceivably destroy all human life on Earth. The answer is that, yes, it is conceivable, but unlikely. As one possible scenario, there is little doubt that a few cobalt bombs (a cobalt jacket wrapped around a large H-bomb, designed to produce intense radioactive fallout) detonated in the Northern and Southern Hemispheres would destroy the human race.*

*See *Unless Peace Comes*, edited by Nigel Calder, Viking Press, New York, 1968. The above statement is taken from the article by Dr. David Inglis, a senior physicist at Argonne National Laboratory and a former chairman of the Federation of American Scientists.

Fortunately, such a doomsday machine would be of little use to anyone not bent on complete annihilation.

In our journey through the arms race we will study the physics and technology of the major weapons as we come to them. This historical approach will help bring out the moral and political dimensions of the problem.

1938–1949: The Fission Bomb, or A-Bomb

As we saw in the preceding chapter, nuclear fission was discovered by Hahn, Strassman, and Meitner in 1938. In 1939 there was a flood of research papers on the new phenomenon along with the suggestion by Fermi and others that the chain reaction (Chapter Twenty-One) might be feasible. As the clouds of world war gathered, the military implications of nuclear fission were becoming all too obvious. Prompted by the fear of a possible German fission bomb, Einstein urged President Roosevelt to proceed with development of the weapon (Chapter Fourteen).

During 1939 and 1940 several U.S. laboratories conducted small research projects to test the possibilities of a chain reaction. In 1941 the U.S. fission bomb project, known as the Manhattan Project, began in earnest. It was the largest industrial project in human history. As part of this effort, a research group under Enrico Fermi at the University of Chicago initiated the first self-sustaining chain reaction in 1942. It was humankind's first large-scale release of nuclear energy (the scene is recounted in Chapter One).

The surrender of Germany in May 1945 removed the fear that had prompted the Manhattan Project. Realizing that the Japanese were in no position to develop nuclear weapons, some Manhattan Project scientists now opposed further bomb research. By this time however, the project had developed a momentum of its own, and the rush to produce a usable weapon continued. The world's first nuclear weapons were developed at Los Alamos, New Mexico, under the inspired direction of physicist J. Robert Oppenheimer (Fig. 22.1). Los Alamos is still the site of one of the two major weapons laboratories in the United States. A plutonium fission device, the world's first A-bomb, exploded in the New Mexican desert on July 16, 1945.

On August 6, 1945, a uranium fission bomb exploded over Hiroshima, Japan (Fig. 22.2). Three days later a plutonium fission bomb exploded over Nagasaki, Japan. On August 15, Japan surrendered and the world war ended.

Thought Question 1

The United States used these bombs against a nation that did not possess nuclear weapons. The bombs and their aftereffects killed more than 200,000 persons, nearly all civilians, and the aftereffects continue—2200 persons died during 1980 from diseases attributed to the bomb, and Japan has designated nearly 400,000 persons as suffering from the effects of the two bombs, entitling them to medical benefits. But if the bombs had not been used, an invasion of the islands of Japan might have been required to end the war. Such an invasion would have cost more than 200,000 civilian lives and many more than 200,000 American and Japanese soldiers' lives. Should we have used our fission bombs?

The Manhattan Project illustrates two principles that motivate the arms race. The first principle is ***the role of fear and uncertainty***. The United States developed nuclear

FIGURE 22.1 (a) J. Robert Oppenheimer. (b) Oppenheimer (left),
Fermi (center), and Lawrence (see
Fig. 21.4). (c) Oppenheimer in New
Mexico while selecting the site for
the world's first A-bomb test, 1944.
(d) Fathers of the new physics and
of the nuclear age. (AIP Neils Bohr
Library.)

(c)

(d)

FIGURE 22.2 (a) The bomb over Hiroshima, and (b, facing page) its aftermath.

weapons because of the fear of Germany and because of uncertainty about Germany's weapons projects. It was natural that the United States would interpret Germany's intentions and capabilities in the worst possible light. As it turned out, Germany's intentions were just as we had feared—there was in fact a German nuclear weapons project. But the war-torn and bureaucracy-ridden German nation was entirely incapable of actually producing such a weapon; not only was there no German bomb, there wasn't even an operational nuclear reactor. Nevertheless, the fear of a German bomb was sufficient to stimulate the development of a bomb in the United States.

A second principle that motivates the arms race is that weapons projects, once initiated, develop a momentum of their own. There was no pause in the Manhattan Project when Germany surrendered. It was clear by that time that a fission bomb was possible and that it could be completed within a short time. Politics and technology combined to produce a ***technological imperative*** to complete the project.

Fission bombs, often called atomic bombs, are based on the chain reaction of either the uranium isotope U-235 or the plutonium isotope Pu-239.

The U-235 needed for a uranium bomb is not easy to come by. Natural uranium is a mixture of over 99 percent U-238 and less than 1 percent U-235. Because U-238 won't chain-react, it acts as a damper on the reaction of the U-235. Uranium for a bomb must contain at least 20 percent U-235, and high-grade bomb material is nearly pure U-235. It isn't easy to separate U-235 from U-238. Two isotopes of the same element behave identically in all chemical reactions, so no chemical process can separate them, and the separation must be based instead on the small mass difference of the two isotopes.

Exercise 1

By what percentage does the mass of the U-238 nucleus exceed the mass of the U-235 nucleus?

U-235 separation, or *uranium enrichment*, was carried out during World War II at a plant at Oak Ridge, Tennessee, still the site of an enrichment plant and nuclear laboratory. The method used was *gaseous diffusion*. In this process, a gaseous

(b)

substance (uranium hexafluoride, UF_6), whose molecules contain the uranium atoms, is allowed to work its way through, or **diffuse** through, a series of porous barriers. Lighter molecules move faster and diffuse more easily than heavier molecules. Thus, after diffusion through a barrier, the gas is slightly enriched in molecules containing U-235. After thousands of such diffusion processes, nearly pure U-235 is obtained.

The enrichment of uranium remains technologically difficult and enormously expensive. The diffusion method is still used, but the newer **gas centrifuge** method is cheaper. In this process, uranium hexafluoride gas is whirled rapidly in a drum. The U-238, more massive than the U-235, has more inertia and so resists deflection into a tight circle. That is, the U-235 moves in a smaller circle than the U-238. The U-238 "gravitates" to the outside, and gas rich in U-235 is extracted from the center.

Scientists are currently investigating yet a third separation technique, based on the use of the high-precision electromagnetic radiation frequencies available from lasers. Since the two isotopes have the same numbers of protons and electrons, the shapes of their electron orbits (more precisely, the shapes of their psi-waves) are nearly identical. However, the slightly different masses of their nuclei produce a slight difference in their electron orbits and thus a slight difference in their spectra (Chapter Eighteen). Precise lasers can capitalize on this difference in the spectra of U-235 and U-238 to separate the two isotopes. This **laser separation** method may soon make the enrichment process a lot simpler and cheaper.

Thought Question 2
Would this development be a good thing?

The bomb requires a certain minimum amount of uranium, enough to sustain the chain reaction. If the sample of U-235 is too small, most of the excess neutrons coming from each fission will simply escape through the surface of the sample without hitting any nuclei. In a larger sample, most of the excess neutrons must travel farther through the sample before reaching the surface, so their likelihood of hitting a nucleus is greater. Thus there is a certain **critical mass** of U-235 above which a chain reaction will sustain itself and below which the reaction will die out. This critical mass turns out to be about 15 kg and about as big as a grapefruit. In the world's first uranium bomb,

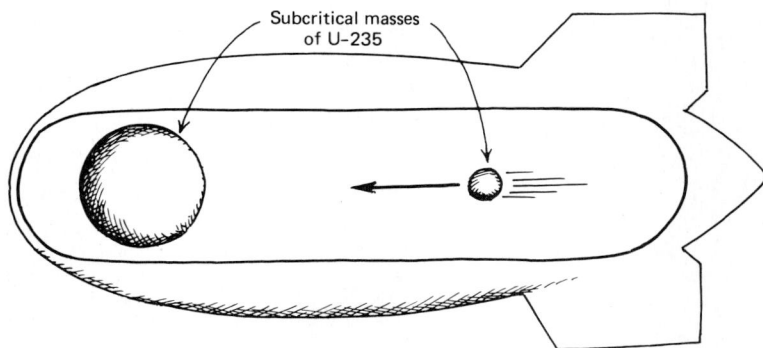

FIGURE 22.3 Principle of the uranium A-bomb.

this material was stored in two subcritical pieces. These two pieces were brought rapidly together to initiate the chain reaction in the skies over Hiroshima (Fig. 22.3).

The fission of one U-235 nucleus releases about 10 million times as much energy as the chemical reaction of a single TNT molecule. The few pounds of uranium in the Hiroshima bomb released nuclear energy equal to the chemical energy released by some 15 thousand *tons* of TNT. In today's world, such a 15 **kiloton** bomb is a fairly small nuclear weapon.

Plutonium, the second A-bomb explosive, doesn't occur naturally on Earth. Despite this fact, the chain-reacting isotope Pu-239 is easier to obtain than U-235— the main requirement is a nuclear reactor. Nearly every reactor is fueled with uranium, but this uranium needn't be highly enriched in U-235. It can be natural uranium, a commodity much easier to obtain than the pure U-235 needed for the uranium bomb.

Reactors produce plutonium as a by-product of the fission process. Many of the excess neutrons produced by the chain-reacting U-235 nuclei strike U-238 nuclei rather than U-235 nuclei. Instead of fissioning, the U-238 nucleus absorbs the neutron and then radioactively decays by beta emission.

Exercise 2

What is the mass number of the uranium after absorbing the neutron? What is the atomic number, the mass number, and the name of the isotope after beta emission? (Answer below.)

This isn't the end of the process. The Np-239 (atomic number 93) produced from the U-238 nucleus is radioactive and decays by beta emission with a half-life of about 2 days. The result? Pu-239. The plutonium stays around for a long time. Its half-life is 24,000 years.

Exercise 3

The plutonium produced in a reactor is intimately mixed with U-235 and U-238, so a separation process is required to remove it for use in a Pu-239 bomb. Is this process likely to be easier or more difficult than the process required to obtain pure U-235? Why?

To summarize an important point: Any country with a reactor can produce Pu-239 for a plutonium bomb. On the other hand, production of a uranium bomb requires the difficult and expensive separation of pure U-235 from natural uranium.

The test device detonated in the New Mexican desert in July 1945 was based on plutonium. The plutonium for that device, and for the bomb exploded over Nagasaki, came from reactors in Hanford, Washington. There are now several military reactors at Hanford that produce plutonium for nuclear weapons.

The 1945 test device and the Nagasaki bomb were based on an inward-explosion, or **implosion**, technique (Fig. 22.4). In this approach, conventional explosives surround a subcritical mass of Pu-239. When the explosives are triggered simultaneously, they drive the Pu-239 inward until it is about half its normal volume, or, in other words, twice its normal density. At this higher density the fissionable nuclei are closer together and a stray neutron is more likely to run into a nucleus, thus

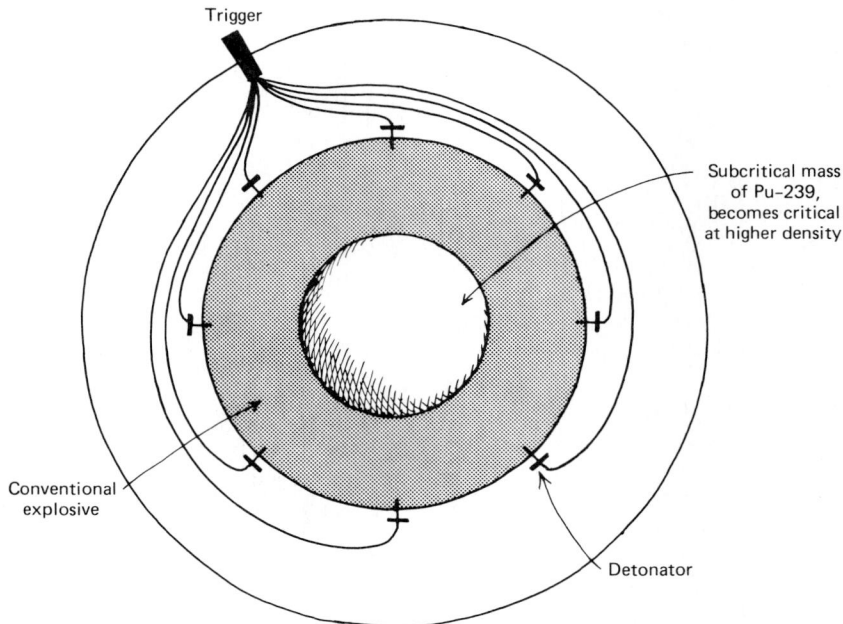

FIGURE 22.4 Principle of the plutonium A-bomb.

the critical mass needed for a chain reaction is less than at normal densities. Thus in the implosion technique the conditions for a chain reaction are attained by compressing the Pu-239 to a high density.

The United States was the sole possessor of nuclear weapons from 1945 until 1949. In 1949 the first Soviet fission bomb exploded.

1949–1955: The Fusion Bomb, or H-Bomb

The Soviet bomb in 1949 and the outbreak of the Korean War in 1950 instilled fear in the U.S. political and scientific community, fear similar to that which motivated the earlier development of the fission bomb. The result was a vehement but secret debate among a few scientists and politicians as to whether the United States should proceed with the development of an even more devastating weapon, the fusion bomb. Edward Teller, a physicist and European refugee, led the faction supporting a fusion bomb program as the appropriate U.S. reply to the Soviet bomb and to the political realities illustrated by the Korean War. J. Robert Oppenheimer, former director of the Los Alamos laboratory that developed the fission bomb, was prominent in the faction that opposed the fusion bomb on technical, political, and moral grounds. The Teller faction eventually persuaded President Truman to initiate the project in 1950.

Thought Question 3

The fusion bomb debate illustrates one of the dilemmas that the arms race poses for democratic societies. The decision to initiate the fusion bomb was one of the more important U.S. decisions of the 1950s. Yet it was carried out without public discussion, because we feared that an open debate would reveal weapons secrets to the Soviet Union. Should the public have participated in this debate? Is there some way that the public could participate more fully in such decisions without unduly compromising U.S. security? Might such public discussion ultimately make us more secure?

In what is now seen by many as one of the saddest chapters in the history of U.S. science, the government removed Oppenheimer's security clearance in 1954 because of controversies related in part to his reservations about the development of fusion weapons (Ref. 10). The father of the atomic bomb never again worked for the government.

The United States detonated a fusion test device at Eniwetok atoll in the Pacific Ocean in 1951; the U.S.S.R. detonated a similar device in 1953. The United States tested the first true fusion bomb at Bikini Island in 1954. The power of the sun was unleashed on Earth.

The *fusion bomb,* or *hydrogen bomb,* is based on the fusion of the hydrogen isotopes H-2 and H-3. The reaction is initiated by high temperatures (Chapter Twenty-One). The isotope H-3 is radioactive, with a half-life of 12 years, so it can't be stored inside the bomb because it would soon decay away. The high temperature is produced by a fission bomb, and the H-3 is produced from lithium 6 (Fig. 22.5). The fission "trigger" provides the high temperature needed for fusion, and the neutrons released by the fission chain reaction interact with the Li-6 to form H-3 and helium:

$$\text{Li-6} + n \rightarrow \text{H-3} + \text{He-4}$$

The H-3 then fuses with the H-2:

$$\text{H-3} + \text{H-2} \rightarrow \text{He-4} + n + \text{energy}$$

This weapon is called a thermonuclear weapon, because the fusion reaction sustains itself at temperatures like those at the center of the sun.

Since it requires both a fission chain reaction and a fusion thermonuclear reaction, the H-bomb is a much more technologically advanced device than the A-bomb.

Exercise 4

Underdeveloped country X has recently received a nuclear reactor from developed country Y. Which of the following weapons would be the most likely to be produced by country X: uranium A-bomb, plutonium A-bomb, H-bomb?

Some military laboratories are looking into the use of laser beams to produce the temperatures needed for fusion bombs. If successful, this approach will make thermonuclear weapons much easier and cheaper to build. The justification given for this project is that the radioactive fallout (Chapter Twenty) from the fusion bomb is caused mainly by the fission trigger, so the new triggering device will produce a fallout-free, "clean" bomb. Such a bomb could destroy a military target without showering the surroundings with radioactive debris.

Thought Question 4

Should the United States be working on this device? Keep in mind that the U.S.S.R. may be working on it.

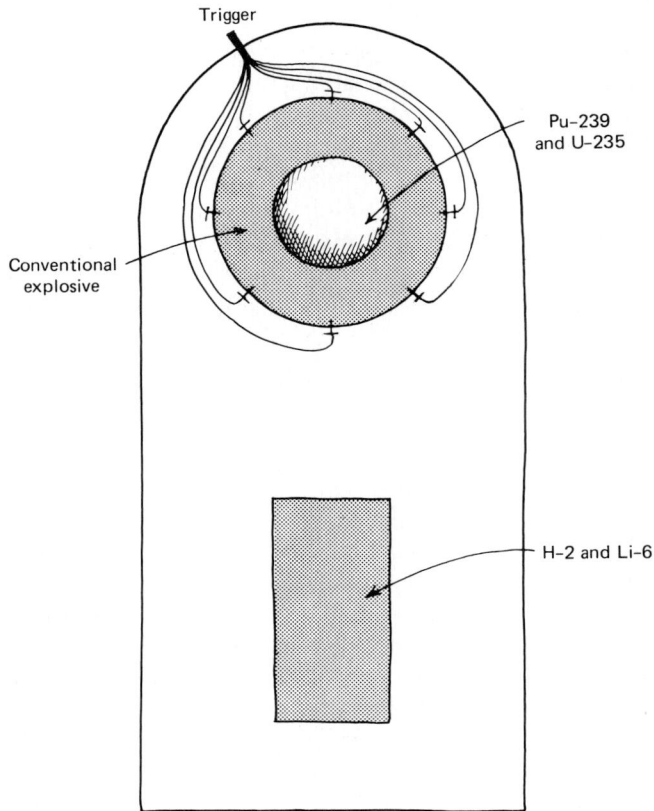

FIGURE 22.5 Principle of the H-bomb.

Fusion weapons and fission weapons differ in three important ways. First, the energy released per pound of material is some ten times greater in fusion explosions. Second, fusion weapon materials are much cheaper. Thus a fusion bomb may be doubled or tripled in explosive power with little increase in cost, giving "more bang for the buck." (Note the implications for poor countries if an inexpensive laser triggering device becomes available). Third, whereas fission bombs are limited in size by the possibility of a spontaneous chain reaction in a sufficiently large mass of fissionable material, fusion bombs have no such natural limitation.

Because of all these factors, the explosive power of the fusion bomb ranges from a fraction of a megaton to several megatons. A **megaton** is a million tons of TNT equivalent and is 1000 times as large as a kiloton. Fission bombs are in the kiloton class, whereas fusion bombs are in the megaton class. A megaton makes an impressive explosion. For comparison, a total 2,700,000 tons of explosives, or 2.7 megatons, were dropped on Germany during World War II. SAC bombers and some Soviet missiles carry bombs as large as 20 megatons. Such a weapon, exploded over a typical large city, for example, Detroit (population 4 million), would dig a crater 1.5 miles in diameter and 200 yards deep, knock down brick houses at 10 miles from ground zero, produce severe burns at 40 miles, kill half of the population, and injure most of the remainder. Lethal fallout would drift downwind in a path 20 miles wide and 200 miles long, extending across Lake Erie and covering the cities of Cleveland, Akron, Youngstown, and Pittsburgh. The more distant areas, such as Pittsburgh, could be safely reinhabited in 2 years, while areas closer to Detroit would be safe only after 10 years. People would probably have to go on living in these cities despite the high radiation levels; many of them would eventually contract radiation-induced cancer. For further details on nuclear weapons effects, see Reference 8.

An even more lethal weapon has been concocted by surrounding a fusion bomb with a shell of natural uranium (99 percent U-238). Under the conditions obtaining in the fusion reaction, even the U-238 can be brought to fission. The result is a more powerful explosion and, more importantly, a much larger fallout of radioactive isotopes (Chapter Twenty). The intense fallout makes this three-stage fission-fusion-fission weapon more deadly against a spread out, rural population.

1955–1965: Missiles and Deterrence

The missile age arrived in the mid-1950s. In rapid succession the United States developed six types of large ballistic missiles and four types of cruise missiles.

Ballistic means "thrown," like a ball. A ***ballistic missile*** is powered by rocket engines (Chapter Six) for only the first few miles of its flight, just as a ball is powered (accelerated) by your arm only until the ball leaves your hand. The engines shut off when the rocket has reached a speed of about 25,000 km/hr, a speed sufficient to carry it to its target without a further boost from the engines. It follows a curved path, similar to the curved path of a ball, slowing down because of gravity as it rises and speeding up as it comes down, landing on target if it was aimed properly when the engines shut off. A typical intercontinental flight takes about 30 minutes. These missiles are quite accurate. Minuteman III, the primary U.S. land-based missile, has a probable error of 200 m at a distance of 12,000 km. Future developments will make these missiles even more accurate.

A ***cruise missile*** is just a pilotless jet airplane. It is powered all the way from launch to target, so it needn't achieve the speed or height of ballistic missiles. The V-1 "buzz bombs" used by Germany during World War II were cruise missiles; the later V-2 weapons (also used by Germany) were ballistic missiles.

Today U.S. strategic power is based on three distinct types of weapons. This so-called ***strategic triad*** is composed of over 1000 land-based ***Intercontinental Ballistic Missiles*** (ICBMs), over 600 ***Sea-Launched Ballistic Missiles*** (SLBMs), and 300 ***strategic bombers.*** Each of these forces is defended against surprise destruction in a different way. The ICBMs are mostly Minuteman missiles stationed in blast-resistant, or ***hardened,*** silos in the northern Midwest. The SLBMs are Polaris and Poseidon missiles stationed on Polaris and Trident submarines hidden in the seas. The bombers are B-52s, which can be airborne within minutes.

One of the central realities of the missile age is ***deterrence,*** or ***Mutually Assured Destruction*** (MAD). The idea is that no matter how hard the other side (U.S.S.R. or United States, depending on where you are) hits your side in a first strike, you have enough power left to cause unacceptable destruction to the other side in a second strike. This fact deters the other side from trying a first strike. Thus both sides should refrain from initiating nuclear war because both sides fear the retaliatory capability of the other side.

An important feature of deterrence is that the second strike, or retaliatory blow, would be not only at military targets but also at the other side's civilian and industrial targets. There are at least two reasons for this. After the other side's first strike, their military targets would consist mostly of empty missile silos. Furthermore, the threat of retaliation against vulnerable civilian and industrial targets should have a highly deterring effect on anyone considering a first strike.

In a nutshell, peace is maintained by a balance of terror. It's like the balance between two scorpions who are deterred from stinging each other by the fear that the other would, before dying, sting in return.

It is essential to the success of deterrence that neither side possesses a **first strike capability**—the ability to remove the other side's retaliatory capacity in a single massive blow. If either side possessed a first strike capability, the world would be an even more dangerous place than it is already. The trigger fingers of both sides would be itchy, because each side would constantly fear that the other side was preparing a massive first strike.

Thought Question 5

If you were the U.S. chief of staff, would you attempt to achieve a first strike capability? What if you were the U.S.S.R. chief of staff?

Luckily, it is at present impossible for either side to achieve anything approaching a first strike capability. Such a capability requires the ability to destroy essentially all of the other side's protected (or hardened) missile silos, essentially all of their nearly invulnerable submarines, and essentially all of their strategic bombers. For illustration, suppose the U.S.S.R. initiated a first strike and miraculously destroyed all U.S. forces with the exception of just one Polaris submarine.* This submarine is equipped with 16 Poseidon missiles; each missile carries ten bombs, or warheads, and each warhead is independently targeted for its own destination (see MIRV, below); each warhead carries a 40-kiloton fusion bomb. So the retaliation from even a single submarine would amount to a deluge of 160 separately targeted nuclear bombs, each of them packing an explosive power equal to nearly three Hiroshima bombs. This onslaught could reduce the centers of the 160 largest Soviet cities to ashes.

The United States has 41 such submarines, most of them hidden in the depths of the ocean and invulnerable to surprise attack.

The United States and the U.S.S.R. possess far more strategic weapons than there are cities in either country. By 1980 the United States had the capacity to hit 9200 strategic targets, and the U.S.S.R. could hit 6000.† The Congressional Budget Office predicts that, by 1985, the United States will be able to hit 14,000 strategic targets, while the U.S.S.R. will be able to hit 8000. Nonstrategic nuclear weapons like the U.S. Pershing missiles stationed in Europe and the Soviet Backfire bombers raise these figures even higher.

Earth has flirted with nuclear holocaust several times. The closest brush probably occurred in 1962 during the Cuban Missile Crisis. As the secretary of state, Dean Rusk, said at the time, "We were eyeball to eyeball." Some U.S. military leaders

---

*The new Trident submarine is similar, except that each submarine carries 24 missiles, each missile can carry 17 warheads (hence 408 warheads on each submarine), and each warhead carries the power of five Hiroshima bombs.

†Reported in *Bulletin of the Atomic Scientists,* June 1980, p. 28.

considered using nuclear weapons during the Vietnam War. General William C. Westmorland, field commander in Vietnam during most of that war and army chief of staff from 1968 to 1972, wrote in his memoirs that he thought it a "mistake" not to consider the threat or use of "a few small tactical nuclear weapons," although "no one could say with certainty" that it would have ended the war quickly. According to a 1976 report in *Time* magazine, Israel assembled several atomic bombs and readied them for use against the Syrians and Egyptians during the Yom Kippur War of 1973. The bombs were rushed off to waiting air force units. Before any triggers were set, however, the battle on both fronts turned in Israel's favor and the bombs were no longer needed.

There is a lesson here: Despite the horrors of mutually assured destruction, nuclear war can occur. It is not some remote possibility that we can afford to neglect. On the other hand, there is a lesson in the fact that nuclear war has not occurred during these 40 years of the nuclear age: Nuclear war won't necessarily occur anytime soon. Perhaps, if we can muddle through for a few more decades, we will all agree to put away our bombs.

1965–1975: ABM, MIRV, SALT I

About 1960 thes U.S.S.R. began deployment of an *Antiballistic Missile* (ABM) system around Moscow.

An ABM is a missile designed to shoot down another missile. The ABM system eventually developed by the United States consisted of a long-range missile carrying a large fusion warhead, a fast short-range missile carrying a fusion warhead, radar for spotting and tracking incoming missiles, and a computer for directing the battle. The long-range missile intercepts incoming missiles while they are still above Earth's atmosphere, and the short-range missile intercepts missiles in the atmosphere should the long-range missile miss.

There are three primary "kill" interactions: 1. *Blast.* The blast from any explosion is an intense pressure wave in air, essentially an intense sound wave. Thus this effect occurs only within Earth's atmosphere. 2. *Neutrons.* The fusion of H-2 and H-3 to form He-4 releases one excess neutron for each individual fusion reaction. Neutrons, because they are uncharged, can penetrate matter of great depths. Thus the neutrons released by the exploding ABM can penetrate the incoming warhead and destroy its delicate electronics or perhaps produce some fission in the warhead. 3. *Electromagnetic radiation.* Gamma rays and X rays from the exploding ABM can heat and vaporize the outer portions of the incoming warhead and thereby render it ineffective.

The ABM is a complicated and uncertain weapon. Hitting a missile with another missile is a little like hitting a bullet with another bullet. The ABM system can be defeated by a variety of methods, such as decoys, "precursor" nuclear bursts to confuse the ABM's radar, and destruction of the radar system. Nevertheless, the system just might work. So from the U.S. point of view the safe thing to do in the early 1960s was to assume that the Soviet system deployed around Moscow would work and to plan accordingly. Thus, the United States embarked on a program to develop an ABM. It wasn't clear at the time whether the Moscow system was the beginning of an ABM devployment throughout the entire U.S.S.R. or whether the defense of Moscow alone was contemplated. With the benefit of hindsight, it now appears that only Moscow was to be protected. At any rate, the Soviets made no other deployments during the next 10 years, after which the United States and the Soviet Union signed a treaty banning further ABM deployment.

The United States embarked not only on an ABM system, but also on a second large weapons program in response to the Soviet ABM system. Again with the benefit of hindsight, it might appear that the United States overreacted, but at the time a massive response seemed safest.

Thought Question 6
What response would you have recommended in the 1960s?

The second new U.S. program was the *Multiple Independently-targetable Reentry Vehicle* (MIRV).

The MIRV was originally designed as the surest way of penetrating a Soviet ABM system. One early suggestion was to let each U.S. booster rocket release, along with the fusion warhead, several empty warheads to act as decoys. But it was feared that the Soviet system just might be able to distinguish the real from the decoy warheads (fear and uncertainty, once more). So the United States replaced the single warhead in each missile with several smaller warheads. At first this was only a shotgun technique in which the several warheads were scattered randomly around a single target, but technologists soon found ways to separately aim and target each of the warheads from a single booster.

MIRV was born.

The payload of each U.S. Minuteman missile (our main land-based ICBM) consists of three 160-kiloton independently targetable warheads. Each Poseidon missile (our main SLBM) contains ten 40-kiloton independently-targetable warheads.

MIRV and ABM are both *destabilizing weapons,* that is, they both make a first strike seem a little more plausible, a little less suicidal, and so make the world more dangerous. This is easy to understand in the case of MIRV. Let's suppose that both sides are armed with equal numbers of missiles and that each missile carries ten warheads. Then the attacking country can destroy ten enemy warheads with each hit on a missile silo, and since the attacking country has ten times as many warheads as the other country has silos, the attacking country has a high probability of knocking out most of the opposing land-based missiles.

The ABM, if deployed throughout the United States or U.S.S.R., would also be destabilizing, even though it is a purely defensive weapon. Suppose that country A initiates a first strike on country B, destroying 90 percent of country B's retaliatory capability. If country A has no defense, then country B's remaining retaliatory capability could, even though reduced by 90 percent, still obliterate most of country A's civilian population. But if country A had a widespread ABM system, it is conceivable that country A could defend itself against the remaining 10 percent of country B's strategic force. A first strike by country A is then more plausible.

The combination of these two destabilizing weapons would be especially dangerous. This was the situation the world was drifting toward in the late 1960s, as both sides developed both of the new weapons.

Thought Question 7
Some weapons systems have a more stabilizing effect than others. For example, within the context of deterrence, the submarine-launched ballistic missile has a

stabilizing effect because it is at present nearly invulnerable and packs a large retaliatory capacity. Unarmed spy satellites are generally stabilizing because they reduce one side's uncertainty about the other side. Can you think of other examples of stabilizing innovations in the arms race?

Then a ray of hope appeared. The first round of the U.S.-U.S.S.R. Strategic Arms Limitation Talks (SALT I) came to a close in 1972 with the signing of an agreement banning the ABM.* Many observers judge this the most important arms agreement in history. It eliminated an entire weapons system and removed the dangerous ABM-MIRV combination.

But unfortunately, MIRV, initially conceived as the U.S. response to the Soviet ABM, was already out of the bag. Today the MIRV missile force of either side could wipe out much of the land-based ICBM force of the other side. A first strike by either side thus looks a little more plausible than it used to. Each side is currently looking for a "safe" response to this problem, a way to protect the land-based ICBM systems. This response, perhaps in the form of mobile ICBMs (see below), may be the next major step in the arms race.

Clearly, the history of ABM and MIRV contains many lessons.

Thought Question 8

Should the United States seek a way to protect its ICBMs against the threat posed by Soviet MIRVs? Should the U.S.S.R. similarly protect its ICBMs against U.S. MIRVs?

1975–Present: Cruise, Neutron Bomb, MX, . . .

An entire new round of U.S. nuclear weapons is currently under discussion, development, or deployment: a new bomber to replace the B-52; the Trident submarine to replace the Polaris submarine; a mobile ICBM to replace the Minuteman ICBM; and entirely new weapons, such as the cruise missile and the neutron bomb.

The new *cruise missile* is, like the earlier cruise missiles, a pilotless jet airplane that carries a bomb. The major difference between the new and the old cruise missiles lies in the guidance system. A map of the flight path is programmed into the missile's computer. Sensors monitor the terrain between the moving missile and compare it to this map. Any deviations from the programmed path are corrected during flight. This missile will fly at 500 mph at just 100 feet above the ground (to elude radar and ground defense) for distances up to 1700 miles and deliver a 200-kiloton fusion warhead to within 100 feet of its target. A product of recent developments in miniature electronics, this pilotless bomb that can see terrain and read maps is one of the foremost technological developments of the decade.

The cruise missile can be released from airplanes flying outside the borders of the target country, or from ships, submarines, or land. In favor of the new weapon it is argued that it is a needed response to the recent vulnerability of the strategic bomber. This vulnerability is caused by the new Soviet look-down radar, which can

---

*More precisely, each side was allowed to have ABM protection for a single limited area. The United States chose not to install the allowed system because it would cost more than it was worth and, as we have seen, it would probably be ineffective anyway.

spot low-flying airplanes from above (the United States deployed look-down radar years ago). It is also argued that the cruise missile will upgrade our aging B-52 force more effectively and less expensively than would the proposed new B-1 bomber. In fact, this is why President Carter, in 1977, approved the cruise missile and canceled the B-1 bomber. Against the cruise missile it is argued that this weapon is a new step in the arms race; that if it is deployed the Soviets will respond by matching it; that it will complicate future arms agreements because it is difficult for one side to verify the number of cruise missiles possessed by the other side; and that, like all large weapons programs, it will be expensive.

The **neutron bomb** is in a different category from the other weapons in this chapter. It is a **tactical nuclear weapon** designed for local battlefield use, as opposed to the **strategic nuclear weapons** that operate across continents and are designed to destroy entire cities or ICBM sites.

The neutron bomb is basically just a small fusion bomb. But it is a fusion bomb with a difference. The neutron bomb utilizes the penetrating neutrons released by the nuclear reaction of H-2 and H-3. The bomb produces an intense shower of neutrons out to a radius of perhaps 1200 m ($\frac{3}{4}$ mile). At the same time, the fusion reaction is arranged to produce only a small blast effect, relative to other nuclear weapons. Its blast is reported to be equivalent to 1 kiloton of TNT, or about 10 percent of the Hiroshima blast (still sizable). This blast destroys structures out to a radius of 200 m.

Exercise 5
Since neutron destruction extends six times as far as blast, how much more *area* is covered by neutrons as compared with blast?

So the neutron bomb is unique in two respects: It is smaller than other nuclear weapons, and it destroys mainly by neutron damage.

Unlike blast and heat, neutrons don't damage structures. In fact, they pass along unimpeded through cement, steel, and the like. Their penetrating power results from the fact that they are uncharged and so are not deflected by electric forces caused by the electrons and protons in matter. On the other hand, like alpha and beta radiation, high-energy neutrons do great damage to biological cells. So this bomb kills people at up to 1200 m but destroys structures only out to 200 m. The victims die of radiation sickness within a few hours or a few days. Since few radioactive isotopes are produced in the small blast, little radiation lingers and the area is "clean" in a short time.

The neutron bomb could be placed in a small, short-range missile or in an 8-in. artillery shell. It is designed primarily as a defense against a tank attack in Europe. Tanks are resistant to the heat and blast of a conventional or nuclear explosion but are easily penetrated by high-energy neutrons.

The significance of this weapon lies not so much in its neutron-radiation feature, but rather in its small size, especially its small blast effects. It is much less destructive of nearby civilians and buildings than are other nuclear weapons, and hence it is more likely actually to be used on the battlefield.

Supporters of the neutron bomb argue that it is more humane than other nuclear weapons because it limits civilian damage; that it is a clean weapon that produces little radioactive debris; that it is a credible defense against a tank attack in Europe because its small size suits it to the battlefield; and that this credibility will help forestall any actual tank attack in Europe. Opponents argue that the neutron bomb makes nuclear war more thinkable, and hence more probable; that Soviet policy is to respond to any nuclear attack with a massive nuclear counterattack, so that use of the neutron bomb would quickly escalate to all-out nuclear war; and that a tank attack can be countered more safely with the "smart," guided nonnuclear shells now available.

The MIRVing of the U.S. and Soviet missile forces, along with the increasing accuracy of both forces, is beginning to threaten the survivability of the land-based ICBM forces of both sides. Note that the submarine and bomber forces are not threatened by this new development.

The obvious reply to the new threat is for both sides to agree to just do away with their land-based ICBMs and to rely on SLBMs and bombers to deter nuclear war. Unfortunately, such agreement has not been forthcoming. A second possible reply is to do nothing, because even the surviving small fraction of ICBMs could wreak havoc on the other side and thus deter an attack, not to mention the fact that the entire SLBM force and the entire bomber force would survive.

A third possible response is to build a **Multiple Aim Point** (MAP) missile system. This is a system of land-based missiles that are secretly moved around from time to time so that the other side doesn't know where they are. The missile itself would be much more technologically advanced than the current Minuteman missile. The new missile, known as Missile X (MX), might contain ten independently-targetable 200-kiloton fusion warheads. The most significant feature of the proposed system, aside from its enormous cost, is not its mobility or its enlarged power, but rather its increased accuracy. The missile's **terminal guidance** (see below) system will make it accurate to within 15 m, compared with the 200-m accuracy of the current Minuteman missile.

The guidance systems of new missiles are becoming increasingly significant in the arms race, and increasingly dangerous. With poor accuracy, a first strike will not destroy many of the other side's land-based missiles. Thus the older missiles are useful only in a second, retaliatory strike. But the 15-m accuracy of MX is sufficient to destroy even a hardened missile site. These more accurate missiles make a first strike seen more plausible to both sides. Another way of putting this problem is that it is difficult to imagine that the new, highly accurate missiles exist purely to deter nuclear war. Instead, the new weapons seem designed to fight a nuclear war should one break out. Such a situation is less stable than a situation in which missiles exist only to deter war.

Proponents of the MAP-MX system argue that the new system is the proper response to the threat posed by the more accurate Soviet missiles and that the new system will be a more effective deterrent to nuclear attack than is the more vulnerable Minuteman system. Opponents argue that our submarines and bombers are a sufficient deterrent and that the increased accuracy makes the new system a first strike weapon that destabilizes the arms race.

The increased accuracy of today's weapons has opened a new era in the arms race, an era in which deterrence has been partially replaced by **counterforce.** A counterforce strategy is one that concentrates on destroying an opposing country's missiles, submarines, and bombers. This doctrine has played a role in U.S. strategic thinking at least since the early 1970s, but was announced as official policy only in 1980. The possible danger in this policy is that counterforce weapons could be used in a first strike to remove the other side's retaliatory capability. Thus if country A develops counterforce weapons capable of destroying many of country B's nuclear forces, then country B may fear that country A is actually preparing a surprise first strike. Trigger fingers on both sides would become more itchy. If U.S.-Soviet tensions increase during some future confrontation, as they did for example during the Cuban missile crisis, both countries will fear a first strike. Being fearful of such a surprise attack, either country may see some advantage in launching a first strike of its own to remove some of the other side's weapons and to reduce its own casualties. Briefly, a counterforce strategy makes nuclear war fighting a more plausible alternative.

Proponents of the counterforce strategy argue that the United States must be able to respond to a wide range of Soviet threats and, hence, must have the ability to strike specific nuclear installations, and that a policy of aiming at missile bases is more humane than a strategy of aiming at cities. Opponents of this strategy argue that our present ability to destroy military and industrial targets is sufficient to deter any conceivable Soviet threat, that a counterforce strategy makes deterrence less effective and thus destabilizes the arms race, and that the death and destruction would be enormous even if the missiles were miraculously restricted to purely military targets.

The new, highly accurate guidance systems are probably the most significant weapons development of the past decade. These systems are a result of advances in electronics, computerization, and other technologies. Once a missile's main rocket engines burn out a few minutes after launch, the missile is supposed to be headed in the proper direction at the proper speed to bring it down roughly on target. In addition, all missiles contain an *inertial guidance system* to determine inaccuracies in the actual path and small rocket engines to make corrections during flight. Inertial guidance is also used in the space program and to tell military and commercial aircraft where they are regardless of weather and changes in heading and speed. Typically, the system consists of three high-precision instruments known as gyroscopes in mutually perpendicular fixed directions. Each gyroscope can measure acceleration in one direction. Computing machinery adds these accelerations during the flight and so "knows" the system's position relative to that at launch.

Technological evolution has refined inertial guidance to the point that a missile thrown halfway around Earth can be brought down within 200 m of its target. Nevertheless, a new refinement, *terminal guidance,* is currently being added. Terminal guidance is similar to the cruise missile guidance system. It uses sensing devices (radar, infrared, lasers) to see the terrain. Comparison of this data with a map encoded in the missile's electronic brain can bring the missile to within a few yards of its target.

The historical background of the MAP-MX system makes an excellent lesson in the dynamics of the arms race. This system is a response to the new vulnerability of the Minuteman system; but this vulnerability was a result of the development of MIRV and improved guidance; MIRV, in turn, was a result of the uncertainty and fear surrounding the installation of a Soviet ABM system around Moscow, and improved guidance is a perfect example of the technological imperative that drives science and technology to produce ever improved instruments of destruction.

In the words of former secretary of defense Robert McNamara,

There is a kind of mad momentum intrinsic to the development of all nuclear weaponry. If a system works— and works well— there is a strong pressure from all directions to produce and deploy the weapon out of all proportion to the prudent level required.

Thought Question 9

Is there anything that you personally can do to reduce the likelihood that these weapons will be used? Could you help simply by being informed? Could you help by intelligent voting or other forms of political activity? Would it help for you simply to be *peaceful:* to deal with people and the environment in a thoughtful and peaceful way? Or is there nothing that you can do?

Checklist

Manhattan Project	cruise missile
J. Robert Oppenheimer	strategic triad
Hiroshima, Nagasaki	ICBM
role of fear and uncertainty	SLBM
technological imperative	SAC
fission bomb	deterrence
uranium bomb	first strike capability
plutonium bomb	ABM
enrichment	MIRV
gaseous diffusion	stabilizing, destabilizing
gas centrifuge	SALT I
laser separation	neutron bomb
critical mass	MAP
kiloton, megaton	MX
plutonium production	counterforce
fusion bomb	inertial guidance
ballistic missile	terminal guidance

Further Thought Questions

10. Give examples of the role of fear and uncertainty in the arms race.

11. Give examples of the technological imperative in the arms race and in civilian technology.

12. Does the United States response to the Soviet ABM in the 1960s illustrate the role of fear and uncertainty in fueling the arms race? If so, how?

13. How can we reduce fear and uncertainty between nations?

14. Today's SLBMs are too inaccurate to be effective against hardened missile sites and are thus targeted against larger civilian and military sites. Would a more accurate SLBM, capable of knocking out hardened missiles, have a stabilizing or a destabilizing effect on the arms race? Should the United States develop such a missile? If you were a Soviet military adviser, would you recommend that the U.S.S.R. develop such a missile?

15. Should the United States deploy the new cruise missile? Should it deploy the neutron bomb? Should it deploy the MAP-MX system? Why or why not?

Further Exercises

6. The fissionable material for an A-bomb is a sphere of uranium or plutonium. Would the critical mass be less, more, or the same if the shape was an elongated egg shape? Why? *Hint:* Which shape has the greater surface area?

7. Inertial guidance systems measure accelerations directly. Is there any device, totally contained within a missile as is this one, that could directly measure velocities? Why? *Hint:* Remember the theory of relativity.

8. U-235 releases an average of 2.5 neutrons per fission, and Pu-239 releases an average of 2.7 neutrons per fission. On the basis of this information, which of these elements would be expected to have the smaller critical mass?

9. There is a limit to the amount of bomb-grade fission fuel that can be stored in one single lump. Why? Why is there no limit to the amount of fusion fuel that can be stored in a single lump?

10. Why does an H-bomb produce radioactive fallout?

11. India exploded an A-bomb in 1974. The explosive material for this bomb came from a nonmilitary reactor supplied by Canada. What type of bomb was it?

12. Germany recently sold an enrichment plant to Brazil in connection with that country's nuclear power program. What type of nuclear bomb could be built with the help of this plant?

13. The largest H-bomb ever tested produced a blast estimated at 38 megatons. How many 15-kiloton Hiroshima bombs would be required to produce a blast this large?

14. Is the megaton a unit of force, a unit of energy, or a unit of power? That is, could it be converted to newtons, to joules, or to watts? (Answer below.)

15. The megaton is a unit of energy, equal to 10^{15} cal or 3×10^{15} ft lb or 4×10^{15} J. To what height could this much energy lift the 200 million citizens of the United States? Assume that an average weight of 150 lb per person. (Answer: 100,000 ft or 19 miles.)

Recommended Reading

1. *The Bulletin of the Atomic Scientists,* a monthly journal, deals with science-related social questions, especially the arms race and nuclear power. It is not a technical journal and should be readable by anyone who has read *Physics and Human Affairs.*

2. Nigel Calder, *Nuclear Nightmares,* Penguin Books, New York, 1981. An investigation into possible wars and into our society's military-industrial predicament.

3. Stephane Groueff, *Manhattan Project,* Bantam Books, New York, 1968. A complete history of one of the largest technological projects in the history of the human race.

4. Michihiko Hachiya, *Hiroshima Diary,* translated and edited by Warner Wells, The University of North Carolina Press, Chapel Hill, N.C., 1955. The journal of a Japanese physician, August 6–September 30, 1945.

5. *Hiroshima-Nagasaki, August, 1945,* Museum of Modern Art, New York, 1969. This 16-minute documentary film of the bombing and its aftermath is shown continuously at the nuclear war memorial in Hiroshima. Excellent for classroom use, church groups, and the like.

6. Robert Jungk, *Brighter Than a Thousand Suns*, Penguin Books, London, 1958. An account of the research and development of the atomic bomb.

7. D. Schroeer, *Physics and its Fifth Dimension: Society*, Addison-Wesley Publishing Company, Reading, Mass., 1972. Chapters 17, 18, 20, and 25 deal briefly with the Manhattan Project, the H-bomb, and the ABM.

8. United States Office of Technology Assessment, *The Effects of Nuclear War*, Allanheld, Osmun and Company, Montclair, N.J., 1979. An authoritative study of the consequences of nuclear war under several possible scenarios.

9. *The War Game*, Films Incorporated, Wilmette, Illinois, 1965. Reviewers have said that this "may be the most important film ever made. It should be screened everywhere on earth." This 47-minute documentary-style film shows how World War III would appear from a suburb of London. Originally produced for BBC Television, this film won the Academy Award for documentary of the year in 1966. The film remains topical. Recommended for classrooms and discussion groups.

10. Herbert York, *The Advisors: Oppenheimer, Teller and the Superbomb*, W. H. Freeman and Company, San Francisco, 1976. An inside look at the great debate that led to the decision to build the first H-bomb.

11. Louise B. Young, editor, *The Mystery of Matter*, Oxford University Press, New York, 1965. Part 4 of this excellent anthology deals with the Manhattan Project. It contains the Einstein letter to Roosevelt, eyewitness accounts of the first atomic reactor, and other articles by participants in the project.

23

ENERGY IN THE NUCLEAR AGE

Nuclear knowledge has made humankind powerful, "full of power" in the strict physical sense of being able to do a lot of work in a short time. This power can wreak destruction, or it can produce useful energy.

Any device that transforms nuclear energy into other forms of energy in a controlled fashion is called a **nuclear reactor.**

Nuclear reactors can be based on either fission or fusion, just like nuclear bombs. All **fission reactors** derive their energy from either U-235, Pu-239 (the two A-bomb isotopes), or U-233.

The idea of the **fusion reactor** is fascinating: control the fusion reaction, energy source for the sun and for the H-bomb, to obtain electric power from the nucleus of nature's most abundant element, hydrogen. The idea sounds simple, but it is difficult in practice. Supporters of fusion power estimate that even with the help of a proposed multibillion dollar crash program, it will be at least the year 2000 before the first demonstration fusion reactor is ready for operation. It would then be at least 2025 before fusion reactors are commercially available, and even further in the future before fusion power began to make a significant contribution to our energy needs. Less optimistic observers argue that the technical problems are so great that it is not clear whether fusion power will ever provide a practical source of energy.

Exercise 1
Roughly what temperature would you expect would be required for the fusion fuel to undergo the fusion reaction?

There is not space in this book for a complete discussion of alternative strategies for electric power generation. This chapter focuses on just one approach, nuclear fission, because it is a widely used source of electric power today, because it could become much more widely used in the future, because it is a controversial issue (Fig. 23.1), and because it is closely related to basic physics. Although there is a brief comparison of nuclear power with fossil power (especially coal), this chapter passes over the many suggested mid-term options that might provide some significant amounts of electricity or its equivalent by the year 2000. These options include wind, solar-thermal, solar-sea, photovoltaics, geothermal, biomass, and small hydroelectric as well as the many schemes that have been proposed to increase energy efficiency (Chapter Ten) and hence provide needed services with less electricity. Many of these options were briefly discussed in Chapters Ten and Eleven. This chapter also omits the far-term options, such as fusion, which might provide electricity after the year 2025.

Proliferating A-plants
Called Threat in Study

NEW YORK (AP) — A 66-country study says the increasing number of nuclear power plants in industrialized and developing c...will result in a dramatic increase in available sup... -grade nuclear material, The
New York Tin...

Searchers Fail to Turn Up
Uranium Missing at A-plant

ERWIN, Tenn. (UPI) — Inspectors using hand-held radiation detectors probed every nook and cranny of a shuttered nuclear plant Saturday, searching for a ...issing batch of enriched ura...uld be fashioned into

Tories Prefer Nukes

Britain's Secretary of State for Energy, David Howell, told an audience in Washington, D.C., on 2 October that the new Conservative government has no qualms about developing nuclear power as a source of energy and plans to expand the nuclear program inherited from the Labor government. Speaking at a luncheon given by the Women's Economic Roundtable,

Reactor Called 'T...

To END TH...

India Affirms Peace, Right to Build A-bon...

Beyond the light water reactor

The Philippines Po...
What to Do Abo...
Out If ...
May Take Years to Clean
TMI Reactor, NRC Is Told

NRC rejects sending uranium to India

Nuclear Pow...
Will the Lig...
...ckma...
A-power Plants,
...lled Peril
...raudulent
Nuclear
Waste
Controversy

Carter to Allow
Nuclear Fuel
To Go to India
Overrules NRC,
Ignites Criticism
From Congressmen

Scientists Call
TMI Venting
Risky Operation

FIGURE 23.1 Nuclear power in the news.

Conventional Reactors

The most important peaceful application of nuclear energy is the production of electric energy in nuclear power plants. The fission and fusion processes convert nuclear energy into thermal energy, and the thermal energy is used to make steam to turn a turbine to turn a generator to make electric energy, just as in a fossil-fueled power plant. Thus nuclear plants have much in common with fossil-fueled plants, the main difference being the energy source.

Nearly every reactor in the world today is based on the fission of U-235. The main exceptions are a few experimental and demonstration breeder reactors.

There are four main ingredients in a uranium reactor: **fissionable fuel,** a **coolant** to remove thermal energy from the fuel and transfer it to the turbine, a **moderator** to slow down the neutrons, and a **control system** to control the reaction rate.

The fuel in most U.S. reactors is uranium enriched to about 3 percent U-235, although some reactors are fueled with natural uranium (1 percent U-235). A typical reactor contains some 50 tons of this slightly enriched uranium. The fuel is packed into thousands of long, thin **fuel rods** and placed into the center, or **core,** of the reactor.

Exercise 2
Could this slightly enriched uranium be used to make a bomb?

A fluid circulates between the fuel rods to remove the thermal energy generated by fission in the rods. This fluid is called the *coolant.* It must remove thermal energy rapidly and continuously, or the rods heat up and might melt. In most plants water is the coolant that removes thermal energy from the core and transfers it to the turbine. In many plants the water is heated at such a high pressure that it does not boil (high pressures prevent the formation of the gas-filled bubbles characteristic of boiling). A second water circuit, at ordinary pressure, is then used to make steam to drive the turbines (Fig. 23.2). Compare Figure 23.2 with the general power plant diagram shown in Figure 11.7.

Fission releases high-energy neutrons. In fact, these neutrons are so energetic that they need to be slowed down before they can be captured efficiently by U-235 nuclei. The chain reaction will die out without a *moderator* to slow down, or moderate, the neutrons. Either carbon, ordinary water, or **heavy water** (H_2O in which the hydrogen atoms are the isotope H-2) may be used for this purpose. The water circulating through the core of a pressurized water reactor (Fig. 23.2) serves the dual purpose of removing thermal energy and moderating the fast neutrons.

Matters must be delicately arranged in the core to ensure that the chain reaction is maintained at a steady rate. It is essential that, on the average, each fission event emits precisely one neutron, which subsequently causes another fission. Anything less, and the reaction fizzles out; anything more, and the reaction builds up and may get too hot. *Control rods,* containing a good absorber of neutrons, are inserted in the reactor vessel to maintain the reaction at the desired level. The reaction rate can be increased or decreased by manipulating these rods. The chain reaction can be stopped by fully inserting the control rods. This is done deliberately when the reactor is shut down for repairs or refueling, and it is done automatically in case the reaction rate becomes dangerously high.

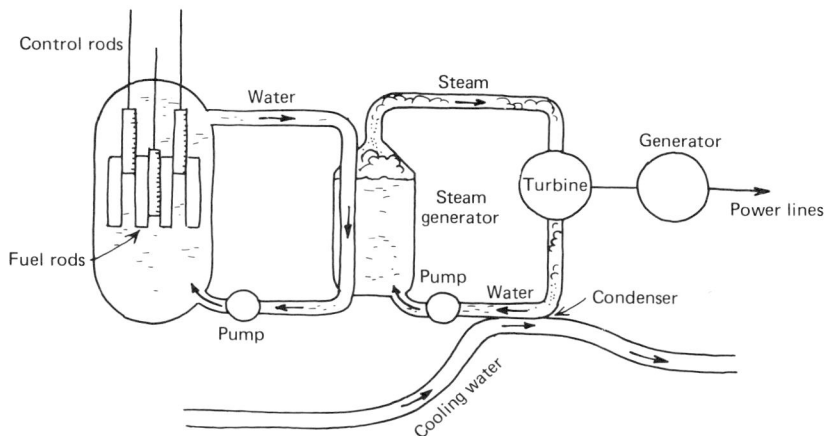

FIGURE 23.2 Schematic diagram of a pressurized water reactor. (From *Energy: An Introduction to Physics* by Robert H. Romer, W. H. Freeman and Company, copyright 1976.)

The energetic neutrons in the core are extremely penetrating and produce radioactive isotopes when they run into the atoms in their path, so a neutron-absorbing shield surrounds the core to protect the operating personnel and to prevent the area outside the core from becoming radioactive. For similar reasons, the water circulating through the core is kept in a closed cycle.

The uranium reactor cannot be a long-term solution to the energy dilemma. The problem is that U-235 is in short supply. Although there is a lot of uranium in the ground, less than 1 percent of it is the fissionable isotope U-235. There is uncertainty about the extent of uranium resources and about how long these resources will last. These uncertainties loom large in the nuclear power debate. It seems that if present projections about the expansion of nuclear power are correct, and if certain assumptions are made as to the economic feasibility of mining various grades of uranium ore, then conventional reactors will use up all the economically recoverable U-235 in some 40 or 50 years.

But U-235 is the only fission fuel found in nature. The other fuels, Pu-239 and U-233, must be artificially created. Since U-235 is projected to run out only shortly after the oil runs out, it might appear that fission energy is a technological dead end.

Enter the breeder reactor.

Breeder Reactors: A Long-term Solution?

If we're running out of the only naturally occurring fission fuel and if we want to continue using fission reactors, then we'll have to find a way to create new fuel artificially. Recall (Chapter Twenty-Two) that the Pu-239 for plutonium bombs is created in uranium reactors. Why not use this Pu-239 in a plutonium reactor rather than in a bomb? This is the idea behind the ***breeder reactor.***

There are two artificially created fission fuels, Pu-239 and U-233. The first is produced when a U-238 nucleus absorbs a neutron. As described in Chapter Twenty-Two, the resulting isotope undergoes two beta decays to become Pu-239.

The second artificially created fission fuel, U-233, is produced in a similar way, from naturally occurring thorium-232. One of the nice things about this process is that thorium is even more abundant than uranium in Earth's crust.

Exercise 3
With the help of the table of the elements (Fig. 3.1), make a guess as to the sequence of events by which U-233 is produced from Th-232.

These two schemes are known as uranium-plutonium breeding and thorium-uranium breeding. We'll look at the uranium-plutonium scheme because it is the center of most breeder reactor research and debate.

Figure 23.3 shows the basic plan. The core contains chain-reacting fission fuel. The fuel can be U-235, but after breeder reactors have been in operation for a few years, the fuel would probably be previously manufactured Pu-239. A *blanket* of natural uranium (mostly U-238) surrounds the core. The chain reaction in the core produces neutrons, which keep the reaction going (to produce power) and which also find their way into the blanket (to produce plutonium). To maintain the reaction, an average of one neutron from each fission must cause another fission. If, in addition, a second neutron is absorbed by a U-238 nucleus in the blanket, then for every nucleus used up in the core, a Pu-239 nucleus is created in the blanket. And if, on the average, the uranium in the blanket absorbs more than one additional neutron, the device produces more fuel than it uses. The fuel produced is eventually removed from the blanket and processed into usable fuel elements. It's neat: the chain reaction in the core creates fuel while it produces power.

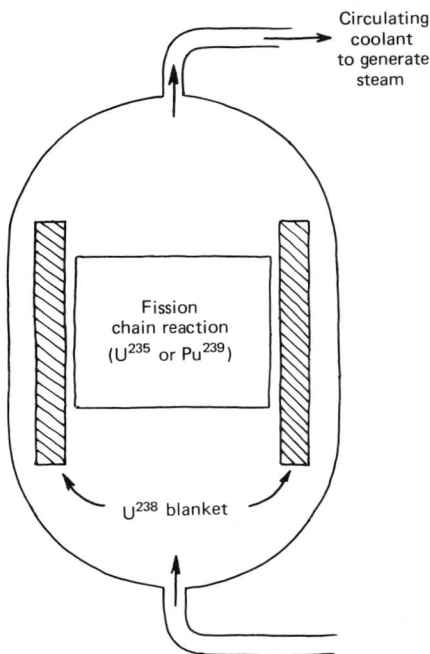

FIGURE 23.3 Central core of a breeder reactor. (From *Energy: An Introduction to Physics* by Robert H. Romer, W. H. Freeman and Company, copyright © 1976.)

We seem to have invented a perpetual-motion machine: a device that produces more fuel than it uses. But perpetual-motion machines are supposed to be outlawed by thermodynamic principles such as conservation of energy. In fact, the breeder reactor is not a perpetual-motion machine, for it can be kept going only with outside assistance, by feeding it with U-238. Breeding simply represents an ingenious way of unlocking the energy in the U-238 nucleus.

However, breeders are in some ways nearly as useful as perpetual-motion machines, because they make available large supplies of otherwise useless U-238. We can make a quick estimate of the amount. Since U-235 constitutes about 1 percent of the uranium in the ground, uranium resources can be stretched by a factor of 100 through the production of fuel from U-238 in the breeder reactor. If U-235 resources will last some 40 or 50 years at projected rates of use, then U-238 resources should last on the order of 100 times as long. Like the sun, and like fusion fuel, uranium used in breeder reactors is an essentially inexhaustible source of energy.

We've seen that in order to produce more fuel than it consumes, a breeder reactor must use, on the average, more than two of the neutrons produced in each fission event. One of these neutrons causes another fission event, one is absorbed by a U-238 nucleus in the blanket to maintain the net fuel supply at a constant level, and any other neutrons could be absorbed in the blanket to increase the net fuel supply. Now, the average number of neutrons produced when a nucleus fissions is only 2.7. Because some of these are lost in various ways, an average of about 2.2 useful neutrons per fission event is the best we can hope for. A carefully designed plant, capable of achieving 2.2 useful neutrons per fission event, should be able to double its original fuel inventory in a few years.

Breeder reactors must use fast neutrons rather than the slowed-down or moderated neutrons used in conventional reactors. Whereas slow neutrons are more efficient in promoting fission, the resulting fission reaction produces, on the average, significantly fewer than 2.7 neutrons. It takes fast neutrons to produce that figure of 2.7. To compensate for the inefficiency of these fast neutrons in promoting fission, breeder-reactor fuel rods must be packed close together. Thus the safety problems are different from those in conventional reactors.

It is conceivable that a breeder reactor core could explode like a poorly designed A-bomb. A nuclear explosion cannot occur in a conventional reactor, because the low-enriched U-235 fuel is not bomb grade and is not packed sufficiently close together. The power-producing core of a breeder reactor, though, is typically a cylinder just a foot or two across and a foot high. Compare that with the 10,000 tons of coal required every day by a coal-fired plant. The enormous difference in volume is a reflection of the enormous difference in the energies released in nuclear and chemical reactions.

Because fast neutrons are desired, a moderator to slow neutrons is not needed or wanted. So the coolant circulating through the core should be something other than water, because water slows neutrons. The coolant should also have good heat-transfer properties and a high boiling point so as to remain in liquid form at high temperatures. Liquid sodium meets these requirements.

Sodium, however, has its problems: It reacts violently with air and with water, so it must be well isolated. It becomes highly radioactive under neutron bombardment, so it must be well shielded. The presence of even trace amounts of sodium oxygen

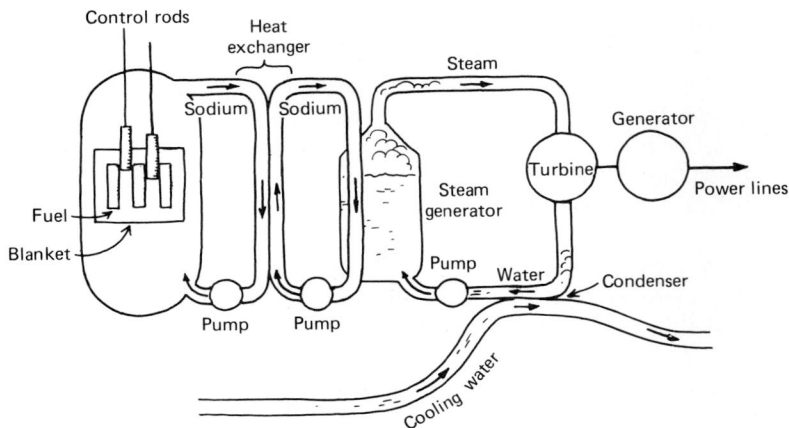

FIGURE 23.4 Schematic diagram of a liquid sodium breeder reactor. (From *Energy: An Introduction to Physics* by Robert H. Romer, W. H. Freeman and Company, copyright 1976.)

compounds (oxides of sodium) makes sodium very corrosive, which would spell trouble for the stainless steel containment pipes. And because sodium absorbs neutrons, a sudden drop in the amount of sodium in the core would immediately intensify the chain reaction, which could be embarrassing.

Figure 23.4 is a schematic diagram of a breeder reactor generating plant. A second closed loop of sodium coolant is interposed between the primary loop (through the core) and the steam generator, because of the high radioactivity of the primary loop.

The future of the breeder reactor is one of the hottest items in the nuclear debate. The world's first two commercial breeder reactors went into operation in the Soviet Union in 1973 and 1980. Several demonstration plants are in operation abroad, and several new commercial plants are planned, primarily in France and the Soviet Union. The planned U.S. demonstration project on the Clinch River in Tennessee has been an object of intense debate for years.

Technological Problems of Nuclear Power

Nuclear power is among the most controversial items on the American scene. It is controversial because the entire energy crisis is controversial; because it seems to some to be an extreme form of big technology; because some observers claim that important yet unresolved questions stem from nuclear power; because it is a radically new energy source involving forces that have never before been released on Earth; and because of Hiroshima and Nagasaki and a half-conscious dread of anything labeled *nuclear.*

Thought Question 1
Is an ingrained fear such as this necessarily harmful?

As we discuss these problems, keep in mind that every energy technology, especially every centralized technology for producing electric power, has serious problems. For example, in Chapter Thirteen we discussed the problem of global atmospheric warming caused by carbon dioxide, a problem shared by coal, oil, and gas, but not by nuclear power. As we will see, the main centralized electric power alternative to nuclear today is coal, and coal has many drawbacks in addition to the CO_2 problem. Thus if society opts for additional centralized electric power, then the relevant question becomes "How do the problems of nuclear power compare with those of other power sources such as coal?" rather than the simple (and simple minded) question, "Does nuclear power have problems?"

1. Dwindling Uranium Resources

We have already seen that, without the breeder reactor, uranium supplies will run out (more precisely, will become too expensive to mine as a result of depletion of all the rich ores) in 40 or 50 years. This might be long enough to get us over the hump to some future solar-based or fusion-based economy, but if fission is to be a long-term solution, we must turn to breeder reactors.

2. Thermal Pollution

Every steam-electric generating plant, whether powered by gas, oil, coal, fission, or fusion, must have a means of cooling the back end of the turbine. If a lake or river is used for this purpose, its temperature is raised, possibly harming the environment. The solution, for the fission plant as for the fossil plant, is the cooling tower (Chapter Eleven).

3. Normal Release of Radioactivity

We saw in Chapter Twenty that the average radiation received per person per year from the normal operation of nuclear power plants is a negligibly small fraction (much less than 1 percent) of the radiation received from such sources as fallout from nuclear weapons tests, cosmic rays, natural radiation from rocks and soil, and medical diagnoses. Despite the publicity given to this problem, the release of radioactivity to the environment during normal operation does not appear to be a significant drawback of nuclear power.

4. Accidents during Operation

If a reactor were demolished by an earthquake, an airplane crash, or a terrorist group the results could be serious. Radioactivity would be spread over a wide area. Reactors are designed for safe containment of radiation in the event of earthquakes, crashes, and explosions, but the possibility remains that some disruption may breach this containment.

Less remote is the possibility that an accident will result from a malfunction in the power plant itself. The most discussed malfunction is the *loss of coolant accident,* in which faulty pipes or a faulty pump stops the flow of water (or other coolant) through the core. The core would then immediately begin heating up. To help prevent this, the control rods are set to automatically fall into the core in the event of a loss of coolant, shutting off the fission chain reaction. The problem, however, is that radioactive decay of the fission products will heat the core even after fission ceases. Thus all reactors are provided with an **emergency cooling system** to flood the core with water after a loss of coolant. These emergency cooling systems are designed for high reliability, but it is difficult to conduct full-scale tests of emergency cooling. Thus questions remain about a possible failure of these systems under the extreme and unpredictable conditions likely to prevail when emergency cooling is actually needed.

If the emergency core-cooling system fails to do its job and the core temperature continues to rise, the fuel will eventually melt. The core could then melt down through the bottom of the concrete containment building, in which case large quantities of radioactivity could be released to the environment. The core might continue sinking downward into the ground until it hit the water table. Groundwater would turn to steam and rise into the atmosphere, carrying radioactive isotopes upward to produce a fallout cloud similar to those produced by nuclear weapons. This cloud could produce intensely radioactive fallout as far as 200 miles downwind from the reactor, killing tens of thousands of people.

This chain of events leading to a **meltdown** and massive radiation release is extremely unlikely for any particular reactor in any particular year. But the important question is: How likely is it that this string of events will occur at some U.S. reactor at some time during, say, the next 50 years? I believe it is fair to say that the experts disagree on this question and that any answer must have a wide range of possible error.

The most serious reactor accident to date occurred in 1979 at the Three Mile Island power plant near Harrisburg, Pennsylvania. (Fig. 11.8). The trouble began when a small malfunction in the secondary loop of water (Fig. 23.2) caused the pumps in the secondary loop to shut down, stopping the flow of water around this loop. Sensing trouble, the reactor automatically dropped its fission-quenching control rods. Even though shut down, the reactor continued to generate about 5 percent of its thermal energy through radioactive decay. Thus the primary loop continued to heat up, but with no place to dump its thermal energy because water was not circulating in the secondary loop. The emergency cooling system was brought into play to cool the reactor, but a malfunctioning valve initiated a series of events that eventually ruptured a water tank. Thousands of gallons of radioactive water flooded onto the containment building floor and overflowed into another building, outside the concrete containment building.

Fuel rods cracked and partially melted in the overheated core, causing chemical reactions that produced a large, hydrogen-filled gas bubble in the core. It was feared that the bubble might cause a meltdown by blocking the cooling water or by combining explosively with atmospheric oxygen. Either event could have produced a massive release of radiation. Finally, plant operators succeeded in removing the bubble and cooling the reactor.

It is important to note that nobody was killed at Three Mile Island, although many scientists believe that some people may die during the next 20 or 30 years from cancer produced by the radiation releases.

The Nuclear Regulatory Commission (NRC) has listed six major errors that contributed to the Three Mile Island accident: (1) failure to keep spare auxiliary pumps on line (in violation of NRC rules); (2) failure of a relief valve to close; (3) a faulty water-level indicator; (4) a radioactive-leak-prevention system that didn't work; (5–6) human errors committed by technicians who turned off the emergency and primary cooling pumps.

The damaged plant, an investment of several hundred million dollars, may never operate again.

Thought Question 2

Suppose it could be shown that, with the 75 reactors operating in the United States today, a catastrophic accident would occur once every 20 years and that such an accident would kill 5000 people. Would this piece of information, by itself, be sufficient for you to favor banning nuclear power? What if the rate were one such accident every 100 years? Every 1000 years? (For perspective on this question, it is worth noting that automobile accidents kill about 50,000 Americans every year.)

---

5. Fuel Handling

Uranium fuel must be mined, milled, enriched, and incorporated into fuel rods. Uranium ores emit a radioactive gas, radon 222, which can produce lung cancer. In uranium mines this problem is controlled by ventilation and enforcement of radiation limits. However, uranium ore must be milled to extract uranium, and the milling process produces large piles of residue, or tailings, which emit small quantities of radon 222.

The enrichment and fuel-rod-fabrication processes have implications for the proliferation of nuclear weapons, discussed below.

A typical fuel rod remains in the reactor core for several years, until the U-235 (or Pu-239) content drops too low or until radiation has damaged the rod too severely. The rod is then removed. Whereas new fuel rods are not particularly dangerous (uranium is not highly radioactive), the highly radioactive used rod is very dangerous. So it is stored in a water pool near the reactor for several months to allow the decay of many of the radioisotopes produced by the chain reaction. At this point, the rod may be disposed of (see Waste Disposal, below), or it may be shipped to a **reprocessing plant** where the unused uranium and plutonium is chemically extracted and the unwanted radioactive isotopes are concentrated into small volumes for eventual disposal. The extracted fuel is then recycled into new fuel elements.

Reprocessing involves obvious potential for release of radioactivity or for theft or sabotage. Reprocessing of nuclear fuel, particularly breeder reactor fuel, also has important implications for nuclear weapons proliferation (discussed below). Some observers have suggested that the implications for weapons proliferation are so severe that reprocessing should be abandoned in favor of a **once-through fuel cycle,** in which the fuel is used once and then disposed of. Other observers feel that reprocessing will produce little, if any, increase in proliferation.

Thought Question 3

What implications would a once-through policy have for the expected lifetime of uranium resources? What implications would it have for the breeder reactor?

Nuclear materials must be transported from the fabrication plant to the power plant and then to the disposal or reprocessing site. Containers and procedures must be designed so that radioactive materials won't be released in a train or truck accident and to prevent theft or sabotage.

6. Waste Disposal

A glance at the streets and vacant lots of your own community, or at the litter along most highways, might convince you that this country has never learned to properly dispose of its trash. Perhaps this is a legacy of our pioneer history, when wilderness provided a plentiful dumping ground. The situation is no different with nuclear trash: We have not yet found a place to dump it.

Carroll Wilson, the first general manager of the Atomic Energy Commission, identified the origins of the problem:

> Chemists and chemical engineers were not interested with waste. It was not glamorous; there were no careers; it was messy; nobody got brownie points for caring about nuclear waste. The Atomic Energy Commission neglected the problem. . . . The central point is that there was no real interest or profit in dealing with the back end of the fuel cycle.*

Suggestions have been made that the very radioactive **high-level wastes** be buried at the base of the Antarctic ice cap, sunk in deep seas, buried in salt mines, or even rocketed to the sun. Reports by the U.S. Geological Survey, the National Academy of Sciences, and others have called for further studies before any final disposal plan leaves the drawing boards and have stated that the basic scientific aspects (for example, the geological properties of underground salt formations) of nuclear waste disposal are poorly understood. On the other hand, there is a concensus that disposal can eventually be achieved safely, given sufficient time, money, and scientific study.

High-level wastes include both the intensely radioactive fission products of the chain reaction and the heavy isotopes such as plutonium and uranium. This material is found in used fuel rods from military plutonium-production reactors, naval power reactors, and civilian power reactors. These used rods are currently kept in temporary storage in covered tanks of water on reactor sites while the Department of Energy looks for a permanent dumping ground. This material must be isolated from the environment for a time that is estimated at as little as a few thousand years by some scientists and at as much as 100,000 years by other scientists. After this time, the wastes will have decayed sufficiently that isolation is no longer required. Each of the country's 75 nuclear power plants generates about 1 ton of these high-level wastes every year.

Low-level wastes, such as the impurities removed from the cooling water as it circulates through the reactor, are less radioactive and contain no long-lived isotopes. These wastes are currently placed in canisters and buried. The buried canisters have some probability (presumably low) of leaking into the ground. A typical plant generates about 10 tons of these wastes every year.

Many observers argue that the technical and safety problems of permanent waste disposal are not severe and that the only real problem is the political one of convincing

*From an article by Wilson in *The Bulletin of the Atomic Scientists*, June 1979.

people that disposal is feasible and safe. Others argue that all further reactor construction should cease until this problem has been solved. Regardless of whether waste disposal is a technical problem or a political problem, the fact remains that it is one of nuclear power's biggest headaches. Despite assurances that a permanent dumping ground for high-level wastes will eventually be found, the fact is that one does not yet actually exist. Meanwhile, the wastes continue to accumulate.

Thought Question 4

Do you think that most people would vote to set aside some 30 square miles of their home state for a high-level nuclear waste disposal site? (The disposal site would occupy only a few hundred acres. The rest of the land would be needed for a buffer zone.) How would you vote?

Advantages of Nuclear over Fossil Power

We have discussed several technological problems of nuclear power, but other ways of generating electric power have their problems, too.

Electrical energy today is generated primarily from four nonrenewable resources: natural gas, oil, coal, and uranium. One renewable resource, hydraulic, contributes a sizable fraction (about 12 percent) of our electric power and might contribute more in the future through the use of small dams. Other renewable resources, such as direct solar, biomass, and wind, are not widely used for electric power today but might contribute more in the future (Chapters Ten and Eleven).

For the remainder of this century most of our electric power will probably continue to be generated from the fossil fuels and uranium. Thus it is important to compare nuclear power with power generated by natural gas, oil, and coal.

Certainly gas and oil are not viable options, not for long at any rate. As we saw in Chapter Ten, these fuels cannot generate significant electric power for many years beyond the end of this century. Furthermore, the United States is already severely overdependent on the importation of foreign oil.

Coal, then, is the only fossil fuel alternative to nuclear power for electric generation during the remainder of this century. We saw in Chapter Ten that this country has sizable coal resources, enough for perhaps a few hundred years even if all power plants switched to coal. But coal has severe problems. Some environmental scientists consider the heating of the atmosphere by increased carbon dioxide the most serious environmental problem of our time. Carbon dioxide is produced by all fossil fuel plants, but not by nuclear power plants. The carbon dioxide problem was discussed at length in Chapter Thirteen.

We have seen that a nuclear plant accident could cause large numbers of deaths. On the other hand, coal-generated electricity has already caused a large number of deaths. During this century, coal mining accidents in the United States have killed some 100,000 miners. Although the number of mining fatalities has dropped from 2000 per year in the early part of the century to 100 per year today, coal mining may still deserve its reputation as the most dangerous industrial occupation. Black lung, a respiratory ailment caused by inhalation of coal dust, is at least a contributing cause of 3000 to 4000 deaths per year. It is estimated that as many as 125,000 miners may have this life-shortening disease. Uranium mining is at least as hazardous as coal mining, but uranium mining is carried out on a much smaller scale than coal mining and so causes a much smaller number of deaths.

A form of pollution known as *acid rain* has already poisoned the soil and "killed" hundreds of lakes throughout wide regions of the United States. Acid rain is caused in large part by the release of the oxides of sulfur and nitrogen from coal-fired generating plants. In addition, acid drainage from abandoned coal mines has rendered some 10,000 miles of Appalachian streams biologically sterile.

The surface mining, or *strip mining*, of coal imposes a heavy penalty on the environment. The notched profiles and flattened peaks of large sections of central Appalachia will probably rank among humankind's most enduring marks on the planet. Legislation now requires operators to restore the landscape at new strip mines, but it is doubtful that such restoration is possible in the more arid western coal-mining regions.

Thermal pollution, the direct heating of a cooling lake or river, is a problem for coal plants just as it is for nuclear plants.

A 1000-MW power plant fired with coal generates about 2000 tons of ash daily. Compare this with the 10 tons of low-level waste and 1 ton of high-level waste generated *per year* by a nuclear power plant! Many plants sell their ash, primarily for construction of roadbeds. However, ash contains a rich variety of heavy metals and other chemicals, many of which are health hazards. The United States Environmental Protection Agency may declare ash a hazardous waste, ineligible either for sale or for disposal by conventional open dumping. Nobody seems to know what, then, will be done with it.

Devices known as *scrubbers* (Chapter Eleven) are now required in new coal-fired power plants, to remove most of the sulfur emissions. But everything has to go somewhere, and the sulfur removed from the stack emissions doesn't just vanish. Instead, it shows up as a white, soupy sludge that has to be disposed of. A 1000-MW plant generates enough of this sludge daily to cover an acre to a depth of 1 yard. Like ash, the sludge is a toxic waste.

One unexpected consequence of burning coal—the 10,000 tons of coal burned every day by a typical coal-fired plant actually releases more radioactivity to the environment than a normal day's operation of a nuclear plant.

Where do you suppose this radioactivity comes from?

The answer is that all rocks, including coal, contain a certain amount of uranium and thorium. Both elements are slightly radioactive. Like the radiation released by nuclear plants in normal operation, this radiation is negligible in the sense that it is far less than the amount received from such natural sources as the soil and cosmic rays.

Is coal preferable to nuclear? This is one of the key questions of the energy debate, and one that this book does not presume to answer. However, the above discussion does perhaps show that if we continue to ask for ever-increasing amounts of electric power, then we will necessarily expose ourselves to many new risks regardless of whether the power comes from nuclear fuels or fossil fuels.

In the competition for the new power plant market coal has pulled far ahead. For example, in 1979 coal's share of total generating capacity increased from 44 to 47 percent, while nuclear's share dropped from 13 to 11 percent. Oil dropped from 17 to 14 percent, natural gas increased slightly from 14 to 15 percent, and hydroelectric held steady at 12 percent.

Reference 1 presents the case for nuclear power and against such alternatives as coal; Reference 2 presents the case for nonnuclear alternatives to the electric power problem.

Thought Question 5
Would you prefer that the new power plants (assuming they are needed) constructed during the rest of this century be coal plants, nuclear plants, or some mixture of the two?

More General Issues

The nuclear power debate hinges at least as much on general questions about world affairs, nuclear war, economics, politics, society, and life-styles as it does on narrower technical considerations. Let's look at a few of the broader issues.

1. Theft and Sabotage

During 1969–1975 there were 11 instances of attacks on nuclear installations around the world. In most instances, one or more bombs exploded and some physical damage occurred, although nobody was injured. In addition, there have been many instances of vandalism, threats, hoaxes, and arson at nuclear plants. Nuclear sabotage might be motivated by a desire for publicity, by vehement opposition to nuclear power, and so forth.

There have been no confirmed instances of nuclear theft to date. Because reactor-grade uranium cannot be used to make a nuclear bomb, there isn't much motivation to steal nuclear materials. But the plutonium breeder reactor could change this situation radically, because reactor-grade plutonium can be made into a weapon. About 5 kg are enough for a 20-kiloton bomb, the size of the Nagasaki bomb. The potential thieves might be organized criminals seeking a profit by selling weapons material, or they might be terrorists seeking weapons for blackmail or other purposes.

Thought Question 6
Can you think of other motivations for sabotage or theft at nuclear power plants?

Working from unclassified documents available in most research libraries, terrorists, organized criminals, or any government in the world could design and construct a workable plutonium bomb. A junior aerospace engineering student at Princeton University used readily available information to design (but not construct) such a bomb as a project for a physics class on arms control, just to prove that it could be done. Weapons specialists have examined his design and found that it would indeed work. The bomb would need 9 kg of plutonium, it would have one-third the power of the Hiroshima bomb, it could fit into a car trunk, and it would be capable of leveling one-fourth of Manhattan. The student's professor gave him an A on the design.

In addition to security problems at reactor sites, there are many possibilities for theft or sabotage at fuel processing plants, reprocessing plants, waste disposal sites, and during transportation of nuclear materials.

Although nuclear plants already take impressive precautions, increased security measures would obviously help: better fences with more alarms, computerized systems to keep track of personnel and nuclear material, television cameras and radiation detectors placed throughout nuclear facilities, armed guards at plants and on trucks and trains, and so on.

Thought Question 7

Some observers feel that such measures would amount to an objectionable militarization of our country's electric power plants. What is your opinion?

Another type of security measure involves spiking any potential bomb material with radioisotopes, to make it too radioactive for potential thieves to handle, or with a substance that would make the material useless as a bomb while retaining its effectiveness in a reactor. There is some doubt as to how successful such measures would be in reducing theft, and they would, in any case, do little to reduce sabotage.

2. Proliferation of Nuclear Weapons

The U.S.S.R. became the world's second nuclear power with a test explosion in 1949. Great Britain entered the ranks in 1952, followed by France in 1960 and China in 1964.

Israel probably became nuclear weapons state number six during the early 1970s, although their government remains noncommital about it. We have seen (Chapter Twenty-Two) that Israel may have prepared nuclear weapons for use in the 1973 Yom Kippur War. According to a 1980 report in the highly respected journal *Science*, "Israel is widely believed to have developed her own nuclear weapons from the research reactor constructed for her by the French at Dimona."* A Central Intelligence Agency document released in 1974 stated, "We believe Israel has produced nuclear weapons."†

In 1974, India exploded a nuclear device based on materials from a Canadian-supplied reactor. Thus India joined the United States, the Soviet Union, Great Britain, France, China, and Israel as nuclear weapons state number seven. India's device utilized Pu-239 obtained by reprocessing used fuel rods. India's A-bomb has spurred a nuclear program in neighboring Pakistan. Pakistan's program is based on experience gained during construction and operation of that country's three nuclear power reactors. Thus Pakistan may soon become nuclear power number eight. Libya has been seeking nuclear weapons from China, has made an open offer to buy such arms from anyone else, and is buying a nuclear power reactor from the Soviet Union despite the fact that Libya is among the world's leading oil producers. Oak Ridge National Laboratory estimates that a "bandit" country could steal spent fuel from conventional reactor cooling ponds, extract and purify the plutonium, and build a bomb in just 7 days if it had preexisting facilities to do the work; some 6 months would be required to build these facilities.

In 1981, Israel bombed an Iraqi research reactor which France had provided as part of Iraq's nuclear power program. In Israel's view, the reactor was part of a secret effort to develop nuclear weapons. Israel's suspicions were raised by the fact that the

Science, August 29, 1980, p. 1001.
†*Bulletin of the Atomic Scientists*, January 1980, p. 11–16.

reactor was fueled by high-enriched uranium rather than by the more common low-enriched fuel, and by the fact that oil-rich Iraq seemed to have no need for a nuclear power program. The 150 pounds of bomb-grade fuel could have been used directly to build about four fission bombs.

Clearly the problem of nuclear weapons proliferation is real. And clearly this problem is related to nuclear power reactors. I believe it is fair to say that most informed observers, whether they are pronuclear, antinuclear, or neutral, agree that nuclear weapons proliferation is the most serious and the most difficult of the problems of nuclear power.

Some years ago it was hoped that the 1968 **Nonproliferation Treaty** (NPT) would resolve this problem. This treaty prohibits the transfer of nuclear weapons from nuclear-weapons states to non–nuclear-weapons states, and it obligates non–nuclear-weapons states to prevent diversion of nuclear energy from peaceful uses to weapons uses. The effectiveness of this treaty has been somewhat less than astonishing. Many would judge the treaty to be a failure. Most countries with the incentive or capability to develop nuclear weapons have not signed it. These countries include Israel, Formosa, Pakistan, Brazil, South Korea, India, and many others. Libya has signed the NPT but is flagrantly disregarding it.

If you view the NPT from the perspective of the nonnuclear states, it is not hard to see the reason for its failure. The treaty is essentially a request by the nuclear powers that the nonnuclear powers refrain from developing weapons. Yet the arms race between the two superpowers goes on and on, with no end or effective limitation in sight.

With the Carter presidency, the United States began to give serious consideration to the links between nuclear weapons and nuclear reactors. We have seen that breeder reactors pose the greatest proliferation danger, because the plutonium fuel that would be used in and produced by such a reactor is directly usable in bombs. Reprocessing plants, even in the absence of breeder reactors, are also a hazard, because weapons-grade plutonium can be extracted from used fuel rods in such plants. Conventional reactors, operating on a once-through fuel cycle in which used fuel rods are simply discarded rather than recycled, offer the least risk.

During the Carter presidency, these facts were the basis for U.S. policy on reactors and proliferation. During this period, the U.S. reactor program moved away from the plutonium breeder concept, deferred development of the Clinch River demonstration breeder reactor, and deferred development of a major reprocessing facility at Barnwell, South Carolina. In addition, the United States brought pressure on such countries as France and Germany to refrain from selling reprocessing plants and enrichment plants to other countries.

There is much debate about the effectiveness of the measures taken to inhibit proliferation. In the absence of concrete disarmament steps by the two superpowers, it is clear that such measures can at best buy time against the day when a large number of nations will possess nuclear weapons. At the heart of the spread of nuclear weapons to non–nuclear-weapons states lies the problem of massive deployment of new weapons by the United States and the Soviet Union.

3. Economics: Will Nuclear Power Survive?

From 1965 to 1974 some 200 orders were placed for new nuclear plants in the United States, an average of 20 per year. This now appears to have been the boom period for nuclear power. From 1975 to 1981, orders fell off to a rate of zero to two per year, and many old orders were rescinded or delayed. Nuclear power, once hailed in the United States as the coming energy source, is beginning to look like a "future option whose time is past."*

The reason—economics. In the early 1970s the construction cost of new nuclear power plants increased nearly fourfold, a rate far exceeding the inflation rate. The reasons for the increase were inflation, rapidly increasing costs of all new construction, regulatory restrictions and uncertainties, and delays and expenses stemming from public opposition to nuclear power.

Since new oil- and gas-fired plants are no longer being built, nuclear power's main competitor is coal. Economically, the advantage of nuclear over coal is lower fuel costs: it's not cheap to mine and ship the 10,000 tons of coal, enough to fill 50 railroad cars, needed daily by a coal plant. On the other hand, the initial, or capital, cost of a nuclear plant has always been larger than the initial cost of a coal plant. In fact, nuclear power is one of the most capital-intensive of all industries. And it is precisely the capital costs that have been increasing the fastest.

The economic woes of nuclear power have been compounded by rapidly increasing interest rates. During a period of tight money, when interest rates are high, it is difficult to amass large amounts of capital. This situation favors the less capital-intensive option, namely coal.

The growing doubt in private industry about nuclear prospects is paralleled by government forecasts. The Department of Energy once predicted that over 1200 nuclear plants would be operating by the year 2000. By the mid-1970s the prediction was for 300 to 400 plants by the year 2000. Now the prediction is below 300. About 75 plants are operating today.

4. Do We Need the Power?

During most of this century, electric power has grown at a nearly uninterrupted rate of 7 percent per year. Except for a slowdown during the depression decade of the 1930s, electricity demand has grown by about 7 percent annually from 1900 to 1973. Following the 1973 Arab oil embargo, the growth in demand fell to only 1 percent in 1974 and 2 percent in 1975. Since that time the rate has hovered around 3 percent. These figures cause us to speculate that perhaps the United States could get along with much less electric power than we now consume, or at least with a much lower expansion rate.

The President's Council on Environmental Quality has conducted a survey of 44 technical energy studies conducted by various sources over the past few years. Their survey, *The Good News About Energy* (Ref. 3), concluded that "the United States can do well, indeed prosper, on much less energy than has been commonly supposed." According to the council's conclusions, "achieving low energy growth will not be easy or cheap, but it will be far easier and less costly than achieving high energy growth."

An energy project at the Harvard Business School has recently received attention for its arguments along these lines (Ref. 4). This analysis indicates that conservation can greatly reduce our need for both nuclear and coal power during the remainder

*However, other countries are still ordering reactors. Nine plants were ordered worldwide in 1978, and 13 were ordered in 1979.

of this century and that the renewable resources can be phased in fast enough to provide a significant fraction of our needs by the year 2000. There is not space here to discuss the many energy efficiencies and alternative energy resources discussed by Stobaugh and Yergin (Ref. 4) and by many other writers (Refs. 2 and 3, for example), although a few of the alternative energy options have been touched on in Chapters Ten and Eleven.

Checklist

fission reactor	loss of coolant accident
fusion reactor	emergency cooling system
fuel rods	core meltdown
coolant	Three Mile Island
moderator	reprocessing plant
control rods	once-through fuel cycle
heavy water	high-level wastes
breeder reactor	acid rain
blanket	strip mining
liquid sodium coolant	scrubbers
uranium resources	nuclear weapons proliferation
thermal pollution	Nonproliferation Treaty

Further Thought Questions

8. Some 50,000 people are killed on U.S. highways every year, and yet few people want to ban automobiles. Nuclear power plant accidents, on the other hand, have not yet killed anyone. Do these facts seem inconsistent with the public's fears of nuclear power plant accidents? Is society more willing to accept a large number of small catastrophes than a small number of large catastrophes? If so, why?

9. If a catastrophic nuclear power accident did occur, what might be the response of the American people? Would such an accident shut down the nuclear power industry? What might be the consequences of a sudden shutdown of the entire nuclear power industry?

10. What should we do with our nuclear waste? Should we stop building new reactors until this problem is solved?

11. Some people argue that the best way to control proliferation of nuclear weapons is for the United States to maintain its leadership in nuclear energy by selling reactors abroad and developing the breeder reactor. It is argued that this will give the United States leverage in controlling nuclear weapons development in other countries. Do you agree? What suggestions do you have for solving the problem of proliferation of nuclear weapons?

12. Many observers of nuclear power feel that the capital-costs problem may eventually doom the entire industry. Would this be a good thing?

13. All things considered, what do you recommend for the energy future for the United States?

Further Exercises

4. Suppose that the fuel put into a particular reactor is 3 percent enriched uranium. After producing power for a few years, will this used fuel still have an enrichment of 3 percent, or will the enrichment be greater than or less than 3 percent?

5. Used reactor fuel could be recycled by extracting the uranium and sending it to an enrichment plant in order to bring its enrichment back to the level required for reactor fuel. Describe another recycling scheme. *Hint:* Is there another fissionable isotope in the used rods?

6. Suppose that for each neutron absorbed by U-235 in the core of a breeder reactor, an average of 0.9 neutrons is absorbed by U-238 in the blanket. Then for each kilogram of U-235 consumed, how many kilograms of Pu-239 are produced? Is this reactor actually "breeding" excess fuel? (Reactors that create less fuel than they consume are known as converter reactors.)

7. Repeat Exercise 6 for the case that an average of 1.1 neutrons is absorbed by U-238 in the blanket.

8. The water that is heated in a reactor does not pass directly through the turbines. Instead, thermal energy is transferred to a secondary loop of water. Why is this?

9. List at least two advantages that fusion power would have over fission power.

10. Pu-239 is radioactive. It decays by alpha emission with a half-life of 24,000 years. What nucleus results from this decay? If Pu-239 is used as reactor fuel, will much of the plutonium decay during the lifetime of the fuel?

Recommended Reading

1. Petr Beckmann, *The Health Hazards of Not Going Nuclear,* The Golem Press, Boulder, Colo., 1976. A hard-hitting statement of the case for nuclear power.

2. Amory Lovins, *Soft Energy Paths,* Ballinger Publishing Company, Cambridge, Mass., 1977. Wide-ranging analysis of the energy crisis. Lovins is a critic of nuclear energy.

3. The President's Council on Environmental Quality, *The Good News About Energy,* Superintendent of Documents, U.S. Government Printing Office, Washington, D.C. A short (49 pages), nontechnical survey that summarizes recent studies of the nation's energy problems.

4. Robert Stobaugh and Daniel Yergin, editors, *Energy Future: Report of the Energy Project at the Harvard Business School,* Random House, New York, 1979. Maps out a realistic energy path for the United States.

24

THE UNIVERSE AND EINSTEIN'S GRAVITY

Your legs are about 1 m long. They will carry you at a few meters per second. Anything that deviates vastly from such normal human sizes and speeds is likely to be difficult to comprehend, even unimaginable. Thus our minds, developed over the ages to deal with human sizes and speeds, cannot easily cope with the world of the fast (Chapters Fourteen and Fifteen) or the world of the small (Chapters Sixteen through Nineteen).

This chapter is an introduction to the world of the large.

A galaxy is some 10^{18} m across. That's a billion billion times larger than human sizes. And the universe contains billions of such galaxies. The cosmos is such a big place that our minds are likely to boggle at the answers to the typical cosmological questions: What is the shape of the universe? Does it have an edge, and, if so, what is beyond it? How did it all begin? Has it existed forever, with no beginning? Will it all end, and if so, how?

The Layout of the Universe

When you look into the night sky nearly all the objects you see are stars. The stars are so far away that they appear only as tiny points of light, despite the fact that many of them are larger than our own star, the sun. Even the nearest known star other than the sun, a star named Alpha Centauri (actually a three-star system, the closest of which is known as Proxima Centauri), is so distant that it takes light 4.3 years to get here from there. If you recall that light can circle Earth in an eighth of a second, you can see that distances between stars are truly immense.

Astronomical distances are so large that it is inconvenient to measure them in meters or kilometers. One popular unit is the *light-year:* the distance light travels in 1 year. Note that the light-year is a distance, not a time.

Exercise 1

How many light-years distant is Alpha Centauri? When did this star emit the light by which it is viewed on Earth today?

The stars that you view from Earth are separated from us not only by great distances but also by great times. The more distant visible stars are several thousand light-years distant from Earth. The light by which you see those stars was emitted several thousand years ago, perhaps before the dawn of recorded history.

Besides the stars, the only other easily seen permanent objects in the night sky are the "wanderers" (Chapter Two): the moon and the planets. In addition to Earth, the planets visible to the naked eye are Mercury, Venus, Mars, Jupiter, and Saturn. Three other planets, the ones farthest from the sun, are visible in telescopes: Uranus,

Neptune, and Pluto. All these objects orbit the sun and are thus members of the **solar system.** Compared to the distances between stars, our solar system is quite small. Light crosses it in just 11 hours.

Not every object in the night sky is a star or a planet. For example, in the vicinity of the constellation Cassiopeia (five bright stars in the northern sky arranged like a W or M) lies a very faint object that appears a little larger than a star, like a tiny, luminous cloud. It is so dim that it is nearly impossible to see with the naked eye except on dark, clear nights well away from city lights. When viewed through a telescope this object doesn't look like a star at all; it looks like Figure 24.1, which is a telescopic photograph of the cloud near Cassiopeia.

FIGURE 24.1 Andromeda galaxy.

This object is a *galaxy*. It is much more distant than the visible stars. Whereas a typical visible star is a few hundred light-years away, the distance to this galaxy is 2 million light-years.

Exercise 2

When was the light in Figure 24.1 emitted? What was happening on Earth at that time? (See Chapter Twenty.)

The cosmos holds many galaxies. In the *observable universe,* that part of the universe within range of our most powerful telescopes, there are some 100 billion galaxies. They range out to a distance of a few billion light-years for the most distant galaxies observed. Figure 24.2 shows several more of these lovely objects.

Each major galaxy is a gigantic affair made of gas, dust, and billions of stars bound together by gravitational attraction. There are two popular shapes for galaxies: spiral and elliptical. *Spiral galaxies* are flat, disk-shaped objects with a spiral structure, like a giant, flattened pinwheel. The galaxies in Figure 24.1 and in Figure 24.2a, b, and c have this form. Some *elliptical galaxies,* Figure 24.2d, are egg shaped; others are spherical. A few galaxies, like that shown in Figure 24.2e, don't fit into either category.

FIGURE 24.2 Galaxies. (a) Spiral galaxy, viewed face-on. (Palomar Observatory, California Institute of Technology.) (Figures 24.2 b-e on next page.)

FIGURE 24.2 (*b*) Spiral galaxy, viewed at an angle.

FIGURE 24.2 (*c*) Spiral galaxy, viewed edge-on.

FIGURE 24.2 (d) Elliptical galaxy

FIGURE 24.2 (e) Irregular galaxy.

Our sun is one star in a galaxy. But, being part of it, we are too close to see our whole galaxy the way we see (with the help of telescopes) the other galaxies such as those shown in Figures 24.1 and 24.2. Our galaxy is called the Milky Way galaxy. It is a typical spiral galaxy, similar to the galaxies shown in Figure 24.1 and 24.2*a*, *b*, and *c*.

Some clear and starry night, when you are far from city lights, find the Milky Way. This luminous band across the sky is actually an aggregation of millions of stars forming a small portion of our galaxy. When you look at the Milky Way, you are looking in a direction along the plane of the disk of our galaxy. Our star, the sun, is located in this disk at about two-thirds of the distance out from the center to the edge (Fig. 24.3). Even with the aid of telescopes we can see only a short distance along the plane of our galaxy because dust in our galaxy obscures our vision. Thus astronomers have never seen the center of our own galaxy. However, we do know that the center lies far beyond the visible Milky Way, in the direction of the constellation Sagittarius. For instruction in locating this direction, consult Chapter Eight (Fig. 8.6).

Ours is a typical spiral galaxy. It's not a small affair. The disk is 100,000 light-years across and contains some 100 billion stars. The galaxy near Cassiopeia, known as Andromeda galaxy, is our closest major neighboring galaxy (Fig. 24.1).

To get some feel for the enormous scale of all this, let's imagine that the universe has been reduced in size until the observable universe is only 5000 kilometers

FIGURE 24.3 Our galaxy. The sun's location is shown.

across—the distance across the United States. On this scale, a typical galaxy such as ours is 50 meters across, and the distance between galaxies is a mere 1 kilometer (about half a mile). The distance from Earth to the nearby stars becomes 2 *millimeters.* The distance from Earth to the sun becomes 10^{-8} meters, about 100 times the size of a real-life atom. Our solar system, and even our entire galaxy, is a speck in the immensity of space!

Receding Galaxies

The universe is made mostly of hydrogen. What isn't made of hydrogen is made mostly of helium. Hydrogen atoms form about 75 percent of the mass in the universe, and most of the remaining 25 percent is helium. The remaining elements, from atomic number 3 (lithium) on up, form less than 2 percent of the mass of the universe.

Exercise 3
How do you suppose we know this? (Answer below.)

Astronomers have learned about the composition of the universe by attaching spectroscopes (devices that spread light out into its different frequencies—Chapter Eighteen) to the eyepieces of their telescopes and studying the spectra gathered from the far reaches of space.

Heavier elements, such as carbon, oxygen, nitrogen, and iron, are abundant on Earth even though they are rare in the universe because in Earth's atmosphere lighter atoms, such as hydrogen and helium, move much faster than do the heavier atoms. Most of the lighter atoms move at speeds exceeding the escape velocity from Earth, some 11 km/sec, so most of the lighter atoms escaped from Earth long ago, leaving only the heavier elements to form our planet.

Although most astronomical objects have a spectrum characteristic of hydrogen and helium and a smattering of other elements, the spectra of many galaxies are moved over, or shifted, toward longer wavelengths. The spectra from these galaxies look normal except that the spectral lines (the specific frequencies present in a spectrum) have all been shifted toward the red end of the visible region. This phenomenon is drawn schematically in Figure 24.4, where the hydrogen spectra from three different galaxies are compared with the spectrum from hydrogen in the laboratory. This phenomenon is called the **red shift,** because the shift is toward the long-wavelength, or red, end of the electromagnetic spectrum. Some galaxies are shifted in this way more than others. Figure 24.4 shows the spectra of the light from three different galaxies. In Figure 24.4*a* the light from the galaxy is not shifted at all; in Figure 24.4*b* the galaxy is shifted; and in Figure 24.4*c* the galaxy is shifted farther.

What could be the cause of this? Are other galaxies made of some new material, not found on Earth, that emits a shifted spectrum? Or does the light from some galaxies pass through regions of space that somehow alter the wavelengths?

The commonly accepted cause of the red shifts is simpler and more natural than either of these ideas. Have you ever noticed the sound of a car horn (or siren, or train whistle) as the source of the sound moved past you? You may have noticed that the pitch suddenly drops as the vehicle passes. The frequency is higher while the vehicle approaches, lower as it recedes. This effect occurs whenever the source of sound and

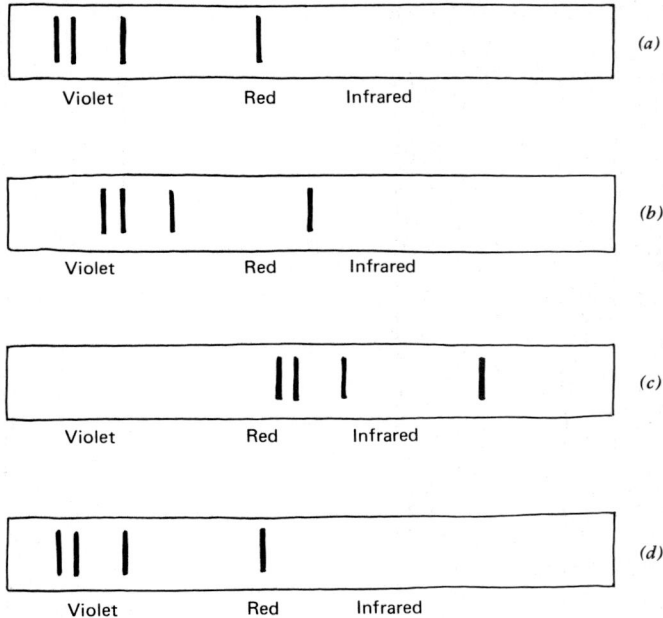

FIGURE 24.4 The red shift. Schematic drawings of three galactic hydrogen spectra, compared with the spectrum of hydrogen as seen in a laboratory. (*a*) Spectrum of a galaxy showing no red shift. (*b*) Spectrum of another galaxy, showing a red shift. (*c*) Spectrum of a third galaxy, showing a larger red shift. (*d*) Spectrum of hydrogen gas in a laboratory on Earth (compare Fig. 18.1).

the observer of the sound are moving toward or away from each other. It is called the **Doppler effect.** It occurs not only for sound, but for any wave phenomenon, including light.

Because you may have experienced the Doppler effect for sound, I'll explain it for sound waves rather than for light waves. Let's consider a stationary source of sound, such as a loudspeaker, and assume that the observer is moving toward this source. The explanation for a situation like that of the moving car horn, in which the observer is stationary and the source of the sound is moving, is similar, but more complicated.

Figure 24.5 shows several wave fronts, or high-pressure regions (crests), in a sound wave that is moving outward from a source of sound. These wave fronts move outward at the speed of sound. Suppose that the source is oscillating 500 times per second, so that 500 of these wave fronts are sent out each second. A stationary observer's eardrums would be hit by a wave front 500 times every second. But if the observer is moving toward the source, as shown in the figure, then the wave fronts will hit the observer's eardrum at a faster rate, say 520 times per second. This observer will hear an increased frequency, hence a higher-pitch sound.

If the source is moving toward a fixed observer, the result is similar: the observer hears a higher frequency. If, on the other hand, the source and the observer are getting farther apart instead of closer together (observer moving away from a fixed source or source moving away from a fixed observer), the observer hears a lower frequency.

Exercise 4

Now extend this reasoning to light waves. How would the light from a rapidly approaching object appear to a stationary observer? What if the light source is receding? (Answer below.)

The Doppler effect also occurs for light. If the source is moving toward the observer, higher frequency, or blue-shifted, light is observed. If the source is receding, lower-frequency, or red-shifted, light is observed.

Nearly every astronomer today agrees that the red shifts seen from some galaxies are a result of the Doppler effect. If so, the red-shifted galaxies must be moving away from us. In fact, these galaxies must be receding from us at a significant fraction of lightspeed, because a significant Doppler effect occurs only when the speed of the source is a significant fraction (1 percent or more) of the speed of the wave. The speed at which a galaxy is moving away from us, known as its *recessional speed,* can be calculated from its red shift. This calculation shows that some galaxies are receding from us at more than 20 percent of lightspeed.

Recall (Chapter Fifteen) that it takes tremendous amounts of energy to speed objects up to a significant fraction of lightspeed. What energy source could have hurled entire galaxies outward at 20 percent of this speed? Furthermore, why should so many galaxies be moving away from us? Is our galaxy special?

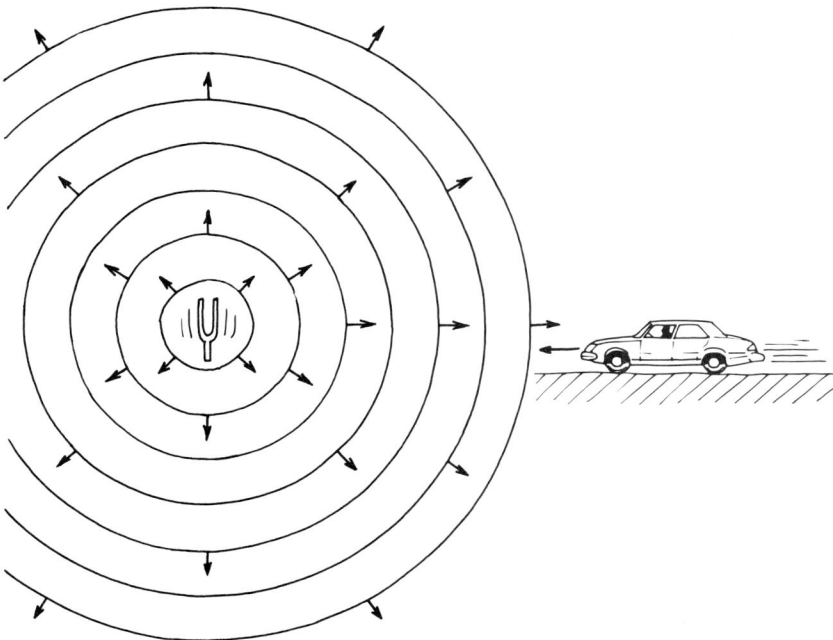

FIGURE 24.5 The Doppler effect.

Thought Question 1

Does it seem likely to you that our galaxy should be the one galaxy in the universe that the other galaxies are moving away from?

The answers to these questions begin to emerge when we consider the distances to the red-shifted galaxies. It's surprising, but true, that astronomers can determine these distances. That's no mean feat, when you stop to think about it. After all, you can't just stretch a tape measure from here to some other galaxy. The methods for determining these distances are fascinating, but we don't have space here to discuss them. Any astronomy text has details.

When the distances to various galaxies are compared with the recessional speeds of these galaxies (as determined by their red shifts), an interesting relationship emerges: The closer galaxies are receding slower, and the farther galaxies are receding faster. The farther away a galaxy is, the faster it is receding from us. This relationship turns out to be a proportionality: The recessional speed of any galaxy is proportional to its distance from us. For example, a galaxy 600 million light-years distant recedes at twice the speed of a galaxy 300 million light-years distant.

This relationship between speed and distance is just what we would expect if all the galaxies were moving away from each other. To illustrate, imagine that the universe is a loaf of raisin bread. The raisins are the galaxies. Now suppose that the bread is just beginning to rise, so that the entire loaf expands. As the load expands, all the raisins move away from each other. If you were a tiny ant standing on one of the raisins, you would see (if you could see through bread) all the other raisins moving away from you, with the farther ones moving away fastest. Furthermore, you would see this no matter which raisin you were standing on, because all the raisins are getting more distant from each other.

Thus the red shifts of the galaxies support the notion, now held by nearly every astronomer, that the entire universe is expanding, just as the bread expanded in our analogy. And the galaxies, being part of the universe, are all moving away from each other because of this expansion. Our galaxy is not in any privileged position—the situation would look the same from every galaxy. Whether you view the universe from this galaxy or from some other galaxy, you will see the distant galaxies moving away from you with the farthest moving fastest.

This raises another question: Why is the universe expanding?

The Big Bang

If the galaxies are rushing away from each other, then they must have been closer together in the past than they are now. With a knowledge of their present recessional speeds and their present separations, we can extrapolate backward in time to calculate their separations at any previous time. On the assumption that the galaxies have maintained roughly their present recessional speeds throughout their entire history, we find that all the galaxies were in the same place some 10 or 20 billion years ago.

So it seems that at some time in the distant past all the material in the universe was in one place. Then for unknown reasons this material began moving outward in all directions, and it has been moving outward ever since. This sounds like a description of a giant, universal explosion. In fact, we call it the

Big Bang Theory. Billions of years ago, our universe underwent a violent event, or *big bang,* that sent all matter and all radiation hurtling outward from a single point.

According to this theory, the universe as we presently know it was born in the big bang. The universe could have existed in some form before then, or perhaps it was created in the big bang.

Although most scientists accept this version of the birth of the universe, they disagree about the birth date. It's not easy to determine how far apart the galaxies were in the past, because it's not easy to determine how far apart they are at present. The problem is that it is extremely difficult to measure the distances to the more distant galaxies. Edwin Hubble,* around 1930, was the first to make an estimate of this sort. His estimate of the distances to other galaxies, combined with the recessional speeds as determined from the galaxies' red shifts, led him to date the creation of the universe at 2 billion years ago. More recent measurements have indicated that the distances between galaxies are larger than Hubble had thought, which has the effect of increasing the estimate of the age of the universe. Thus the age of the universe has been repeatedly revised upward until in 1974, Allen Sandage and Gustav Tammann put the figure at 16 to 19 billion years. This figure was widely accepted until 1979, when astronomers discovered a new, and presumably more accurate method of measuring distances to galaxies. Using this method, Gerard de Valcouleurs carried out a painstaking estimate of the age of the universe and found it to be between 9 and 11 billion years old. There is much disagreement today about which estimate is correct, if indeed either is correct. Perhaps the most honest conclusion that one can draw from all this is that the universe is some 10 or 20 billion years old (to be more accurate, 9 to 11 or 16 to 19 billion years old).

Thus throughout this chapter we will always use the rather awkward figure 10 or 20 billion years to represent the universe's age as estimated from its expansion rate.

Several other lines of evidence support this rough age estimate: Radioactive dating shows the oldest rocks of the solar system to be 4.6 billion years old, so our universe must be at least this old; certain clusters of stars in our galaxy have been determined to be between 9 and 15 billion years old; and the date of the formation of the first chemical elements has been estimated as either 7 billion or 18 billion years ago, depending on what assumptions are made as to the method of their formation.

Steven Weinberg's explosive book (sorry), *The First Three Minutes,* traces the history of the very early universe (Ref. 3). The principles of high-energy physics (Chapter Nineteen) indicate that during the first 3 minutes after creation the universe was so hot that no nucleus heavier than the hydrogen nucleus (a single proton!) could hold together. After 3 minutes the temperature dropped sufficiently that hydrogen 2, helium 3, and helium 4 nuclei could form and hold together. The calculations predict that by the end of the first 4 minutes the universe was made of about 25 percent helium 4 and 75 percent hydrogen 1, with small amounts of hydrogen 2 and helium 3. This mix of elements should have then persisted for many millions of years, when

*In 1924, Hubble presented the evidence that convinced the world that there are other galaxies besides our own.

the first galaxies and stars began forming. Now, this prediction can be tested by studying the composition of the oldest stars in our galaxy, stars that were presumably formed from the original stuff of the universe. We find that these stars are made almost entirely of hydrogen and helium, with the various isotopes occurring in the predicted proportions. This is one of the important pieces of evidence that supports the big bang theory.

Ponder this: If our telescopes could "see" 10 or 20 billion light-years away, what would they probably behold?

- *The answer: The big bang!*

Assuming of course that this event did actually occur. A telescope looking 10 or 20 billion light-years away would have to be looking backward in time 10 or 20 billion years because any light coming from that far away must have left its source that many years ago. Such a telescope would see the creation of the universe.

No telescope could actually capture the creation because the early universe was so hot and so dense that photons could not move freely through space. Photons could not come directly from the original **primeval fireball** to our telescopes.

But in a sense we should still be able to see the remnants of the creation. By the year 1 million (a million years after creation), the universe should have cooled to the point that photons could move freely through space. These primeval photons should still be traveling through the universe, and, in fact, they should be all around us. Because they come to us from 10 or 20 billion years in the past, and hence from 10 or 20 billion light-years away, they must be enormously red-shifted, just as distant galaxies are red-shifted. That is, the big bang is moving away from us (or, equivalently, we are moving away from it) as a result of the expansion of the universe. Another way of putting this is to say that the radiation from the big bang has cooled during its 10 or 20 billion year history, because redder photons (longer-wavelength photons) have lower energy. Thus the theory predicts that the cooled **remnant radiation** from the big bang should be all around us. Furthermore, the theory allows us to calculate the present temperature of this radiation: about 270° below 0 on the Celsius scale.

In 1964, three decades after the invention of the big bang idea, radio astronomers Arno Penzias and Robert Wilson discovered the predicted radiation, and at the predicted temperature! The story of this observation of the faint afterglow of the creation of the universe is one of the fascinating tales of science, and is recounted in Reference 4. This discovery is the most conclusive support to date for the big bang theory.

The General Theory of Relativity

Exercise 5

Of the four fundamental forces in nature, which one do you suppose has the most important effect on the overall structure of such astronomical objects as stars and galaxies? (Answer below).

Even though gravity is the weakest of the four basic subatomic forces, it shapes the universe. The shape and size of stars and the structure of galaxies is determined by this force, because gravity is the only one of the four forces that is able to exert a large effect over large distances. Both the strong and the weak nuclear force have too short a range to act between different stars or galaxies. The electric force is both strong and long-range, but in order for one star to influence another star electrically both stars would need to have a significant excess of positive or negative charge. Such buildups of charge don't occur because any imbalance of charge soon brings electric forces into play that neutralize the buildup, much as the charge buildup on a cloud is neutralized by a lightning stroke.

The gravitational force, on the other hand, is long range and cannot be neutralized as the electric force can. Whereas the electric force is attractive between unlike charges and repulsive between like charges, there is only one kind of gravitational force—attractive—so its effects always add up. Thus the strength of the gravitational force can be increased enormously by simply putting together large amounts of mass to produce a large cumulative effect.

Gravity rules the large-scale structure of the universe. It determines the general shape of the moons, planets, stars, and galaxies. It reaches from one galaxy to the next to influence the motion of these giant collections of stars. As we'll see below, it even reaches across the entire universe to determine the shape of the cosmos.

Apparently, if we want to understand the structure and evolution of the universe, we'll need to understand gravity. Three centuries ago, Newton invented a theory of gravity (Chapter Seven). Two centuries later, Einstein invented a different theory of gravity, a theory known as the **general theory of relativity.** Einstein's theory is the basis for our present understanding of **cosmology:** the study of the structure and evolution of the universe.

Recall that Einstein invented two theories. The special theory of relativity (Chapters Fourteen and Fifteen) relates the observations of two unaccelerated observers. The general theory of relativity (1915) removes this restriction and allows accelerated observers.

It's not difficult to see that general relativity might be related to gravity. For example, an upward-accelerating elevator squeezes its passengers toward the floor, giving them the impression that they are heavier than usual (Chapter Seven). The same effect could be produced by putting the elevator, at rest, on a different planet, where the force of gravity is stronger. Thus accelerations produce the same kinds of effects as gravity, and it is plausible that a theory of accelerated observers would be related to gravity.

Einstein followed this line of reasoning and made it into a basic principle of his theory. From the observation that accelerations often produce gravitylike effects, he made a far-reaching assumption known as

The Principle of Equivalence. The effects of acceleration are experimentally indistinguishable from the effects of gravity.

This idea has important implications for the behavior of light beams. As an illustration, suppose you are in an unaccelerated laboratory, at a place in the universe where gravity is not significant, and suppose that a light beam enters the lab through a hole in the wall. If the beam is directed horizontally as it enters, it will move horizontally straight across the lab (Fig. 24.6a). Now let's suppose the laboratory is accelerating upward. Then, even though the light beam enters horizontally, it will seem to you that the beam curves *downward* as it passes through the lab. This effect is caused by the upward acceleration of the lab. An outside observer, who is not

(a)

Flashlight

Direction
of
acceleration

(b)

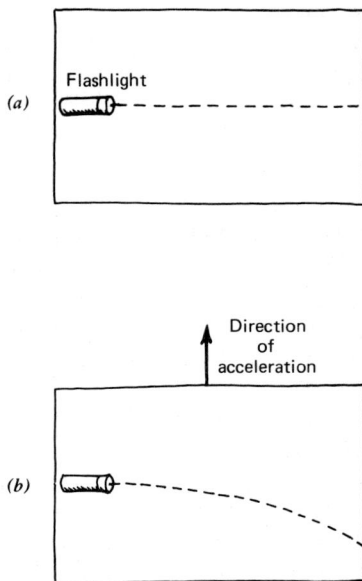

FIGURE 24.6 (a) A light beam crossing an unaccelerated laboratory. (b) A light beam crossing an accelerated
laboratory, as seen from inside the lab.

accelerating upward with the lab, would say that the beam keeps moving in a straight
line while the lab accelerates upward through the beam. But from your viewpoint,
inside the lab, the beam is curved downward because, relative to the lab, the beam
gets closer and closer to the floor (Fig. 24.6b).

Thus an observer in an accelerated lab will observe that light beams curve as
they cross the laboratory. But the principle of equivalence states that acceleration and
gravity produce identical effects. Because acceleration produces curved light beams,
we must also conclude that gravity causes light beams to curve.

Gravity causes little curvature of light beams of Earth, because Earth's gravity is
so weak. But a massive object like the sun produces a much larger effect, large
enough to be measured. This measurement was first made during a total eclipse of
the sun in 1919. During an eclipse, astronomers can photograph the stars in the
vicinity of the sun. Measurement of the positions of the stars seen close to the edge
of the sun showed that the light from these stars bends slightly as it passes near the
sun (Fig. 24.7). The degree of bending is precisely that predicted by Einstein's theory.
No other theory, including Newton's theory of gravity, could explain this result. It was
a triumph for Einstein's ideas.

Curved Space

Light beams are fairly basic in the universe. As we saw in Chapter Fourteen, the
very meaning of time is related to the behavior of light beams. As we will see shortly,
the meaning of straight is also related to light beams.

How would you go about determining whether a certain line is straight? First,
you would probably compare the line with the edge of a straight ruler. But how could
you be sure that the ruler is straight? Well, you would probably compare it with
another ruler. But maybe the other ruler is not straight!

Do you see the dilemma here? You can't use rulers to define *straightness*, because ultimately you must verify that the rulers themselves are straight.

The ultimate way to determine that your ruler is straight is to sight along it, or, to put it in terms of an experiment that could be carried out accurately in the laboratory, to send a light beam along the edge of the ruler. If the beam moves parallel to the edge, we say that the ruler is straight.

Thus light beams are the ultimate determiners of the physical meaning of *straight line*. But light beams themselves are curved by gravitational fields! What can this mean, when we use light beams to define straight lines?

FIGURE 24.7 The sun bends starlight. The dashed line is the path of a light beam. If light went in straight lines (solid line shown), the star would be hidden behind the sun.

In the general theory of relativity, this perplexing situation is handled by introducing the notion of **curved space.** The idea is that space itself can be curved by the presence of matter, and so everything *in* space (including light beams and rulers) must be curved also. Truly straight lines cannot exist in a curved space. Instead we must speak of the straightest possible lines. The path followed by any light beam is one of these straightest lines. That is, light beams don't really define straight lines, they define straightest lines. If a light beam moves parallel to the edge of your ruler, then your ruler is the straightest it can be in that region of space.

Unfortunately, it is impossible to visualize a curvature of space. The idea can be expressed in terms of mathematical relationships, but the idea cannot be visualized the way we can visualize a cat or a tree. The problem with visualizing curved space is that we are immersed in space, we are part of it, we cannot imagine standing outside of space in order to visualize its curvature.

The space around us is *three-dimensional*. By this, we mean that, starting from any agreed-upon point in space, the position of any other point can be located by giving three numbers, such as the distance north, the distance east, and the distance up. General relativity asserts that this three-dimensional space, the space we live in, is curved. Although we cannot visualize the meaning of this curved three-dimensional space, it is possible to visualize a certain analogy to it. I want to describe this analogy because it gives some insight into the structure of the curved universe in which we live.

The analogy is a curved, *two-dimensional* space.

First, let's consider an uncurved, or flat two-dimensional space. A chessboard, extended forever so as to have infinite surface area, is an example of such a space. As any point on the board can be located by giving just two numbers, such as the distance north and the distance east from some particular fixed point, the board is two-dimensional. And the board is obviously flat rather than curved. One way of defining the word *flat* here is to note that lines drawn on the board obey the principles of ordinary geometry, known as **Euclidean geometry.** For instance, an infinitely long straight line drawn on the board never intersects itself; the three interior angles of any triangle add up to the usual 180 degrees; and two short straight lines that start parallel to each other remain parallel no matter how far they are extended (Fig. 24.8). These widely accepted principles of geometry formed the basis of the work of the ancient Greek mathematician Euclid. Finally, it is reasonable to call the infinitely extended board a space because it has no edges and no boundaries. Thus our extended chessboard is a flat, two-dimensional space.

Now let's consider a curved two-dimensional space. A spherical surface is an example. The surface of a sphere is two-dimensional because any point on the surface can be located with just two numbers, latitude and longitude, for instance. But this surface is obviously curved rather than flat, like the chessboard. Finally, it is reasonable to refer to this spherical surface as a space because, like the infinitely extended chessboard, it has no boundary. For example, an ant crawling on the surface of a ball never comes to an edge. Thus our spherical surface is a curved, two-dimensional space.

Geometry on the surface of a sphere is interesting. For instance, there are no straight lines on this surface, because the space itself, the surface of the sphere, is curved. On the surface of a sphere, the best we can do is draw a straightest line. These straightest lines on the surface of a sphere are called *great circles*. On the

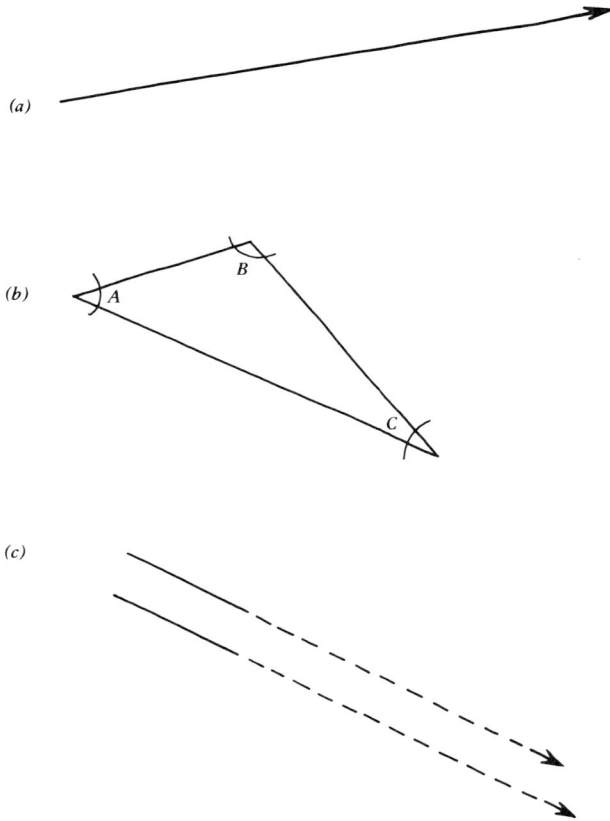

FIGURE 24.8 A few principles of Euclidean geometry (a)
A straight line never intersects itself, no matter how far it is
extended. (b) The three interior angles of a triangle (A, B,
and C in the diagram) add up to 180°. (c) Two short straight
lines that start parallel remain parallel no matter how far
they are extended.

spherical surface of Earth, the equator and the meridians of longitude are examples
of great circles. On the surface of a sphere, a long straight (i.e., straightest) line comes
around and meets itself; the three interior angles of any triangle add up to more than
180 degrees; and two extended "straight" lines, started parallel to each other,
eventually intersect (Fig. 24.9). Euclidean geometry doesn't work in curved spaces.
An interesting feature of the spherical surface is that, unlike the flat two-dimensional
space, this two-dimensional space is finite in extent. That is, its total surface area is
finite.

Still continuing our two-dimensional analogy, suppose you were magically
transformed into a two-dimensional creature who was aware of only two dimensions,
north and east, for example, and had no concept of up. How could you tell whether
your two-dimensional space was curved? For example, how could you tell whether
you lived on an infinitely extended chessboard or on the surface of a sphere?

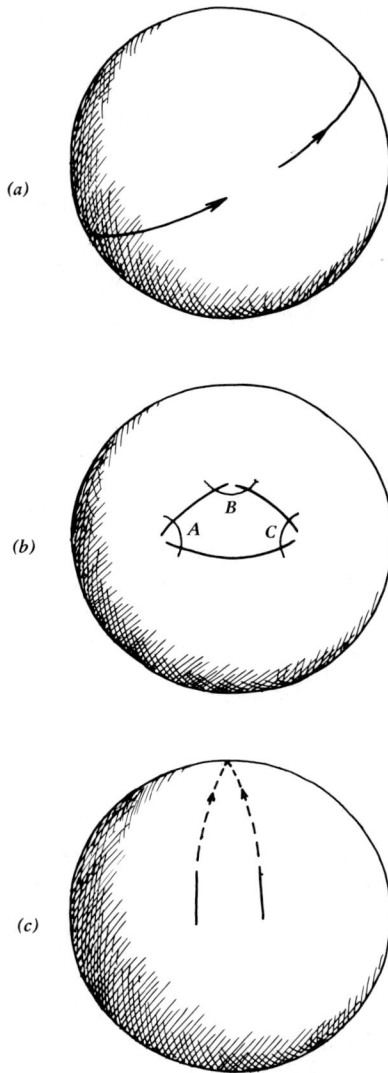

(a)

(b)

(c)

FIGURE 24.9 Euclidean geometry doesn't work on the surface of a sphere. Compare with Figure 24.8

Your first answer might be "just step back and look at the shape of the surface." But if you were a two-dimensional creature you couldn't step back because you couldn't step into a third dimension. You are immersed in the two-dimensional space and cannot visualize its curvature the way we three-dimensional creatures can visualize the curvature of a two-dimensional surface.

If you were a two-dimensional creature and you wanted to determine the shape of your two-dimensional universe, you would have to carry out geometry experiments. If Euclidean geometry worked, for instance if the interior angles of triangles always added up to 180 degrees and if two parallel lines remained parallel when extended, you would say that you were living in a flat two-dimensional space. Otherwise you would conclude that your space was curved.

Suppose you were a two-dimensional creature living on a spherical surface. You would find that even straight lines (the great circles) curve, so you would have to call them straightest lines. Extended straight (straightest) lines would curve around and meet themselves. Furthermore, the total extent (total area) of your universe would be finite, even though your universe had no edge.

So much for the two-dimensional analogy.

To return to our original question: What does it mean to assert that our three-dimensional space is curved? It means that, when we do geometry experiments, we will find deviations from the expected Euclidean results. We cannot step outside of our three-dimensional space, into some other dimension, in order to see the curvature of our own space. But if we find, for example, that straight lines (i.e., the paths of light beams) curve back on themselves, that the interior angles of triangles do not add to 180 degrees, or that two straight lines started parallel and then extended eventually meet, we can conclude that our space is curved.

The 1919 experiment to measure the positions of stars behind the edge of the sun was just such a geometry experiment. The result showed that even the straightest lines, namely light beams, are curved by gravity. Apparently our three-dimensional space is curved.

But it is not curved very much. The curvature of space is noticeable only over very large distances or near very massive objects. As we have seen, even an object as massive as the sun is able to curve space only slightly. It is only over distances comparable to the size of the universe or in intense gravitational fields such as are found near black holes that the curvature of space is pronounced.

The Shape of the Universe

The general theory of relativity gives certain quantitative relationships (equations) between masses and the curvature of space.* Matter is distributed throughout the universe in the form of galaxies, stars, and clouds of gas and other objects. If we knew precisely how this matter is distributed, we could use the equations of general relativity to predict the curvature of space throughout the entire universe. In other words, we could predict the shape of the universe.

Unfortunately, we don't know much about the distribution of matter in the universe. We don't even have a good idea of how much matter there is in the universe, much less how it is distributed. Thus the overall shape of the universe is unknown at present.

Despite our ignorance of the overall shape of the universe, we can still discuss the possible shapes that would be consistent with general relativity. I will just describe two of the several possible shapes. As one possibility, the large-scale shape of the universe could be simply an ordinary, flat, three-dimensional space. If this actually happens to be the case, then space is infinite in extent and has no large-scale curvature. Except for the smaller-scale "bumps" in space caused by individual stars and galaxies, Euclidean geometry would hold throughout the universe.

---

*I am simplifying the situation somewhat. According to the theory, masses affect not only the structure of space but also the structure of time; gravity can cause time to flow at different rates. So masses affect the structure of space *and* time. In the theory, space and time are combined into a single entity called space-time and thought of as a four-dimensional space. General relativity then prescribes the manner in which masses distort the geometry of space-time. This book neglects the role that time plays in all this and focuses only on the curvature of space.

A more interesting and more likely possibility is that our three-dimensional universe is curved in a manner analogous to a two-dimensional spherical surface. Don't try to visualize this! The best you or anyone else can do is to visualize the two-dimensional analogy to this three-dimensional shape, namely the two-dimensional surface of a sphere.

If the universe actually has this shape, then the universe is finite in extent, in other words, the total volume of the universe is finite, and yet, like the two-dimensional analogy, the universe has no edge. In this case, no matter how long we might travel, even if we traveled in a straight (straightest) line, we would never come to an edge. Instead, we would travel all the way around and come back to our starting point.

The Evolution of the Universe

When Einstein invented his theory in 1915 he found, to his surprise, that the equations of general relativity predict that the universe must be either expanding or contracting, that the universe cannot be static. In 1915 this prediction seemed absurd. The idea of an eternal and unchanging universe was so firmly fixed in the human mind that even Einstein didn't take his own prediction seriously. Instead, he added a "fudge factor" to his equations, a factor that allowed for a static universe. Einstein felt that it was arbitrary and inelegant to add this factor, but he thought that the facts of astronomy demanded it.

In 1927 the astronomer Edwin Hubble demonstrated, by studying the red shifts of distant galaxies, that the universe actually is expanding. Hubble's conclusion was in precise agreement with Einstein's original equations, without the despised extra factor. If Einstein had followed his own sense of what was fitting and beautiful, he would not have added the extra term and thus would have predicted as early as 1915 that the universe must be either expanding or contracting. Einstein called it "the biggest mistake of my life."

It is perhaps not coincidental that, whether we look at the large-scale, the small-scale, or the intermediate-scale, the rule of the universe seems to be ceaseless change. The world of the elementary particles (Chapter Nineteen) is one of chaotic creation and destruction, and the deeper we penetrate this world the greater the chaos becomes. Certainly at the human scale, birth, death and evolution are the order of the day. And now we find that even at the level of the apparently unchanging heavens the cosmos is always flowing, always changing, that, indeed, the principles of relativity do not allow a static universe.

Thought Question 2

One of the points made in *The Tao of Physics* (Chapter 17, Ref. 3) is that Eastern thinkers have always looked on change (energy) as the fundamental reality, whereas most Western thinkers look on objects (matter) as the fundamental reality. Thus Eastern thought is likely to say that motion, or change, brings objects into their brief existence. Western thought is likely to say that objects are prior to change and that motion exists only because of the objects that have that motion. Which view seems to be in closest accord with Newtonian physics? Which view seems in closest accord with post-Newtonian (twentieth-century) physics? Can you think of any aspects of present-day Western or Eastern culture that seem related to these different ways of viewing reality?

Modern cosmology envisions a curved universe whose shape is continually changing. It all started in a primeval fireball some 10 or 20 billion years ago. To be specific, let's suppose that the universe is actually shaped like the three-dimensional analogue of a spherical surface. This shape is sometimes called a **hypersphere.** Such a hyperspherical universe always has a finite volume, but no edge. But this volume has been increasing ever since the big bang. This is what we mean when we say the universe is expanding. In the first few seconds after the creation event, before the universe had expanded very far, the total volume of the universe (the total volume of all space) was very small. For example, at one time it was only 1 centimeter across. Space curved back on itself very tightly because an immense amount of mass was crowded into a small region and thus caused space to be highly curved. As time went on, the total volume of the universe became larger. Space expanded.

Thought Question 3

Is it possible to conceive of the universe when it was very small, say just 1 centimeter across? Remember that all space is inside this universe. Does it make sense to ask what was outside it? Does it make sense to ask what is outside our present universe?

You can view the creation and expansion of the universe as the three-dimensional analogue of the creation and expansion of a two-dimensional spherical surface, starting from a point. Picture the evolving universe as the three-dimensional analogue of the surface of an expanding balloon (Fig. 24.10). Initially, the unexpanded balloon occupied only a single point. Suddenly a cosmic breath of air, the big bang, started the balloon expanding. In the early stages the surface of the balloon (the total volume of the universe) was small, but as the expansion continued the surface increased in size. Picture the galaxies as grains of sand stuck to the surface of the balloon. The expansion of the balloon carries the galaxies farther and farther apart. As viewed by an observer on any single galaxy, the other galaxies are all receding and the farthest galaxies are receding fastest. Furthermore, every observer sees galaxies on all sides—the cosmos has no edge. And there is no center of the universe, hence no privileged galaxy at the center. There is a complete democracy of galaxies because every galaxy is on the surface of the sphere. It's a very Copernican idea: There are no privileged positions in the universe.

Do not view the universe as expanding *into* space. The big bang *created* space as well as time and matter. In the initial stages of its evolution, space was very small. Then it got bigger.

Don't be surprised if you have trouble visualizing all this. No one can visualize it, because we three-dimensional creatures are immersed in the space whose curvature we are trying to imagine. That is why we must use analogies to make any sense of it at all. As the philosopher put it (Chapter Seventeen), "the universe is queerer than we *can* suppose."

Speculations

What is the ultimate fate of the universe? Will it go on expanding forever? Not necessarily. The gravitational pull of all the matter in the universe must be slowing down the expansion, just as the gravitational pull of Earth slows down the upward motion of a ball. It is possible that the expansion will eventually grind to a halt and

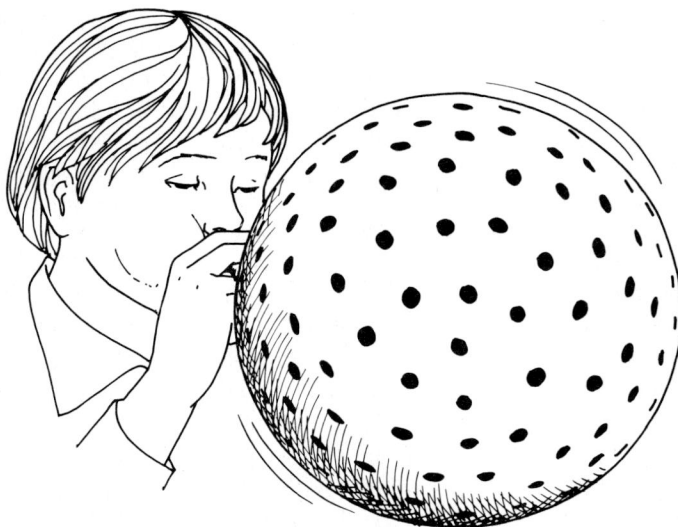

FIGURE 24.10 Two-dimensional analogy for the expansion of a hyperspherical universe. The dots represent a few of the billions of galaxies.

then turn around to become a contraction, just as an upward-moving ball can slow to a stop and then turn around and begin moving downward.

Will it turn around, or will it go on expanding forever? The answer depends on the amount of mass distributed throughout the universe. If there is sufficient mass per unit of volume, that is, if the universe is sufficiently dense, then the pull of gravity will be strong enough to stop the contraction and turn it around. A less dense universe would not exert enough gravitational force to stop the expansion. It is possible to estimate the average density of the observable universe, because we know roughly how massive an average galaxy is and we know roughly how many galaxies there are in the observable universe. Most such estimates to date have concluded that there is not nearly enough mass in the entire universe to stop the expansion. However, many scientists feel that a lot of mass remains to be discovered. For instance, the universe may contain superheavy black holes or giant gas clouds that have not yet been observed. In 1980 new results in high-energy physics (Chapter Nineteen) indicated that an elementary particle known as the neutrino may have rest mass, contrary to the previous belief that this type of particle has no rest mass. There are a lot of neutrinos in the universe. If they all have rest mass, then the universe contains much more mass than we had thought, perhaps enough to eventually reverse the expansion.

What would be the fate of a contracting cosmos? The universe would get smaller and smaller, rushing inward faster and faster as it collapsed. The collapse would heat the universe until it became a small, fiery ball of particles and photons. Finally, perhaps, it would contract into a single point—the ultimate gravitational collapse. A universal black hole.

And what then? Would the final collapse spell the end of space and time? Would there then be nothing? No thing?

Perhaps not. Many cosmologists feel that if the universe collapsed to some minimum size or to zero size, it would be reborn in a new creation event that would send everything hurtling outward once more.

This raises all sorts of fascinating possibilities. Perhaps our present expansion phase is just the latest in an unending series of expansions and contractions of the universe. Every 100 billion years or so such a universe would arise in a fiery rebirth out of the ashes of the old universe. Physicist John Wheeler has proposed that each successive universe would be radically different from its predecessors. Wheeler's idea is that in the tiny, atomic-size universe that might be attained at the end of each collapse, quantum uncertainties would play an important role. The structure of the entire universe would be subject to quantum uncertainties. Such fundamental quantities as the masses of the elementary particles and the strengths of the fundamental forces might then be entirely different in the next expansion phase. This would change the structure of reality in ways that are difficult to imagine.

Thus it is possible that over unending stretches of time, an infinity of different universes will pass into and out of existence, each one burning for a few brief aeons and then collapsing back into near nothingness.

In one such universe, in a galaxy called Milky Way, on a planet called Earth, you and I are privileged to hold such ideas in our minds.

Checklist

light-year	general theory of relativity
solar system	cosmology
galaxy	principle of equivalence
chemical composition of the universe	curvature of light beams
Doppler effect	curvature of space
red shift	Euclidian geometry
recessional speed	flat space
expansion of the universe	curved space
big bang theory	shape of the universe
primeval fireball	hypersphere
remnant radiation	contracting universe

Further Thought Questions

4. Physically, our star and our galaxy seem like insignificant specks in an immense universe of similar stars and galaxies. Does this circumstance make the idea that the universe was created especially for the human race any less plausible?

5. In theorizing about the birth of the universe, is science trying to answer a basically religious question? Should science be studying such questions?

6. Is space really curved, or is general relativity just a figment of scientists' imaginations?

Further Exercises

6. Even when viewed through the largest telescopes, all stars still appear as simply points of light, with no size or extension. But astronomers have obtained a great deal of information about various stars. For instance, their chemical compositions have been determined. How do you suppose such information was obtained? Why do astronomers use telescopes, since stars are too distant to appear as anything other than points of light?

7. Using the facts that lightspeed is 300,000 km/sec, and that light takes 8 minutes to get here from the sun, find the distance to the sun in kilometers.

8. How many kilometers are there in a light-year?

9. The giant planets Jupiter and Saturn are composed primarily of hydrogen and helium, and yet Earth has relatively little of these two elements. What can you conclude about the motions of atoms of hydrogen and helium in the atmospheres of these two planets?

10. The longitudinal lines running from the North Pole to the South Pole on globes are examples of great circles, or straightest lines, on the spherical surface of Earth. Is the equator a great circle? Are the other lines of latitude, such as the Tropic of Cancer, great circles?

11. Referring to the preceding exercise, imagine a triangle drawn on a globe in the following way: Beginning at the North Pole, follow the 0° longitudinal line down to the equator, then move westward along the equator until you come to the 90° west (that is 90° west of the 0° line) longitudinal line, then move back up on the 90° longitudinal line until you get back to the North Pole. Are the sides of this triangle straight (straightest) lines? How large is each of the three interior angles? Find the sum of these three angles (i.e., to what value do they add).

12. If you draw a relatively small triangle on Earth's surface, for instance one whose sides are each a kilometer long, and if each side is straight (i.e., straightest), to approximately what value will the three interior angles add? Will the precise value be a little larger or a little smaller than this approximate value?

Recommended Reading

1. Nigel Calder, *Einstein's Universe*, Penguin Books, New York, 1980. Based on the BBC and PBS television special marking Einstein's centenary in 1979.

2. William J. Kaufmann, III, *Relativity and Cosmology*, Harper and Row Publishers, New York, 1973. General relativity, black holes, white holes, quasars, the shape and creation of the universe.

3. Steven Weinberg, *The First Three Minutes*, Basic Books, Inc., New York, 1977. The title should be taken literally. Outstanding but demanding writing for intelligent nonscientists.

EPILOGUE

A PERSONAL VIEW OF PHYSICS AND THE MODERN AGE

> I call Heaven and Earth to record this day against you, that I have set before
> you life and death, blessing and cursing: therefore choose life, that both
> thou and thy seed may live.
>
> Deuteronomy 30:19

We live, on the one hand, in a time of unparalleled discovery. In the arts, technology, medicine, biology, government, agriculture, physics—on every side new ideas are bursting on the world with astonishing frequency. It is an exciting time of change and progress, a time of great potential. We stretch toward the stars, toward realms of the universe undreamed of in former centuries. It is the great age of exploration, a time of mapping not only our own planet, but the solar system and beyond. Closer to home, new energy sources replace human toil, medical discoveries extend human life, agricultural advances help feed a hungry world, mass education brings awareness and hope to increasing numbers of people. It is a time of tasting of the knowledge of nature's microscopic building blocks, the origin of life, the structure and history of the cosmos itself.

And yet we live at a time in which humankind may have become all too powerful. It is a century so captivated by the idea of progress that we have become oblivious to our heritage, oblivious to the destruction caused by our passing. It is a century in which science has cast doubt on traditional wisdom, a twilight time for the visions and the gods of ages past. It is a time of alienation from nature, from each other, perhaps even from our own hearts. Technology has so intervened between ourselves and our natural surroundings that many of us feel out of touch. Despite our wonderful new knowledge of the heavens, we search in vain for the stars among smog and city lights. We fly from New York to Paris in a few hours, yet we have difficulty finding an unspoiled field away from the din of civilization. Conversation in the home has largely been replaced by the dull throb of television. We have become addicted to an automobile that pollutes our environment while it depletes our resources, to an electrical technology that heats our atmosphere while it proliferates the knowledge and materials of nuclear weapons. It will be remembered as the century in which we learned to fashion the instruments of our own destruction, a time in which we truly tasted of the forbidden fruit.

What are we to make of all this? Is the twentieth century ushering in the millennium, or is it our death knell? The answer, I think, is that either is possible. What we do here on Earth in the next few decades, what you do and what I do, will determine which answer it is to be. We can continue driving our gas guzzlers, destroying the environment, consuming more electricity than we need, fashioning ever more deadly instruments of war, and we will have one outcome. Or we can work at making peace with our environment, with our neighbors, and especially within ourselves, and we will have another outcome.

In physics, the century was ushered in on two characteristic notes. The seeds of quantum mechanics were sown in 1900, and the theory of relativity was created in 1905. Relativity proclaimed that space and time, that solid framework within which Newton fashioned his physics, are different for different observers. No longer was the natural world entirely "out there," objective, objectlike. Quantum mechanics made a similar point in more radical fashion. This theory described a world altered by the very act of observation, a world thus rendered ultimately unknowable by science. With the new theories, the very idea of an independently existing natural world, the cornerstone of science since the time of the ancient Greeks, became suspect. For, according to the new physics, the natural world is unavoidably connected with the human mind: observer and observed, subject and object, are one.

The salvation of the modern age may lie in the attitudes of the new physics. Scientifically, the Newtonian age ended in 1900 with the birth of relativity and quantum physics. But, however post-Newtonian our physics may be, our culture remains firmly planted in the Newtonian model of an object-universe distinct from ourselves, a universe in which effects follow causes with clocklike precision. We imagine our object-universe to be composed of tiny parts, parts whose predictable motions determine the trajectories of the stars and of ourselves, parts that can be isolated, studied objectively, and then manipulated to our own advantage.

Thus when we are ill, depressed, or sleepless, we seek our the guilty organism or malfunctioning biochemical reaction that is causing our distress, and we manipulate it with drugs; when the weather doesn't satisfy us, we manipulate it with silver iodide crystals; if we desire to travel from Denver to Phoenix, we manipulate the mountains to fashion a place for our automobiles; if our national security is in doubt, we manipulate electrons in solids to build ever more fearsome weapons. We ask, "What can we do to the object-universe to make it behave as we desire?"

We have not taken account, in all this, of the theme of post-Newtonian physics: There are no true objects and no true parts in nature. The parts belong to each other, and the objects cannot be entirely distinguished from the subjects, that is, they cannot be separated from ourselves. In subduing and dissecting the universe, we subdue and dissect ourselves, increasingly to our own peril. The message of twentieth-century physics is that there is a ghost in the Newtonian world machine, and the ghost is ourselves.

Post-Newtonian physics suggests that we cannot master nature as we might master some alien creature, because we live *in* nature. Post-Newtonian physics suggests that our salvation lies not in mastering the universe, but rather in aiding the universe in its own pursuits, because we are inseparable from our environment.

Thus has the science of our own century rediscovered the mystery long contemplated by philosophers and mystics: Just as we three-dimensional creatures cannot picture a curved universe because we are woven into its very fabric, we can also never obtain a clear and complete picture of nature, for we are woven into the very fabric we seek to comprehend. Our dimensionality is too low. The division into subject and object is artificial, for we cannot stand outside the cosmos.

All this is not to suggest that we need to renounce the true benefits of modern science and technology, for if humankind and nature are one, then whatever truly benefits the one will benefit the other. But it is to suggest that we come more humbly to a natural and human world that passes all understanding.

Recommended Reading

1. Robert L. Heilbroner, *The Human Prospect,* W. W. Norton and Company, Inc., New York, 1975. A respected economist views our present predicament and future prospects.

2. Theodore Roszak, *The Making of a Counter Culture,* Doubleday and Company, Inc., Garden City, N.Y., 1968. Reflections on the technocratic society and its youthful opposition. Thoughtful questioning of our science-dominated culture.

3. E. F. Schumacher, *A Guide for the Perplexed,* Harper and Row, Publishers, New York, 1977. Shortly before he died, the author of *Small is Beautiful* (Chapter Eleven, Ref. 3) wrote this statement of the matters he considered most important. The book is a thoughtful and personal call for us to recapture the spiritual values which have been lost in these technological times.

4. Allen Wheelis, *The End of the Modern Age,* Harper and Row, Publishers, New York, 1973. The "modern age" means the Newtonian age, the time from Copernicus to Einstein. This book is a lyrical and scientifically literate study of the meaning of the transition to the post-Newtonian era.

EXERCISE SOLUTIONS

Only the solutions to the exercises interspersed with the reading material are given here. Solutions to the Further Exercises at the end of each chapter are not included.

Chapter 2

1. Retrograde motion occurs during the brief backward portion of the loops in the loop-the-loop motion (Fig. 2.3).
2. Yes. Put the two tacks "in the same place," that is, use only one tack.

Chapter 3

2. $6 + 12 + 6 = 24$.
3. Fortunately, air pressure is also pushing outward from inside your body.
4. 10×22 tons $= 220$ tons. As she descends to 300 ft the pressure inside her body also builds up to 10 atm.
5. No, because there is no outside air to press the soda up the straw.
6. The outside air presses the lid down tightly against the jar. You can remove the lid easily by using a utensil to lift a corner of the lid, allowing air to enter the jar.
7. They fall at the same rate, but note that rocks fall much slower on the moon than they do on Earth because of the decreased pull of gravity on the moon.
8. 5 km is 5,000 m, or 500,000 cm. 5 miles is 26,400 ft, or 316,800 in.
9. A millionth of a second, a thousandth of a meter, a thousandth of a second, 1,000 W, 1,000,000 W.
10. Thumbnail, 1 or 2 cm; thumb, 6 or 7 cm; foot, 25 or 30 cm; arm, a little less than a meter; body, about 1.5 to 2 m; football field, a little less than 100 m; marathon, 43 km; around Earth, 40,000 km; to sun, about 150 million km.
11. Similarities: elliptical orbits; several objects orbiting a central object; most of the mass is in the central object. Differences: the solar system is held together by the force of gravity, whereas the electrons are held into the nucleus by electrical forces; an atom is much smaller (!) than the solar system; the different electrons are identical, whereas the different planets are not identical.
12. Your muscles expand the lung cavity, increasing its volume and thus momentarily decreasing the air pressure inside the cavity below atmospheric pressure. Air is then pushed in from behind, through your nose or mouth, to fill the partial vacuum.

Chapter 5

1. Aristotle would say that the balloon got bigger because we added something to it, namely, some Fire.

2. Aristotle would say that it rises because sufficient fire has been added to overcome the natural downward tendency of the rubber covering, which is made of Earth.

5. Aristotle would say that there was no outside influence to maintain the book in its unnatural state of motion, so the book returned to its natural state of rest. Galileo would say that an outside influence, called friction, acted on the book to slow it down.

6. Speed $= \dfrac{\text{distance traveled}}{\text{time of travel}} = \dfrac{150 \text{ km}}{5 \text{ hr}} = 30 \dfrac{\text{km}}{\text{hr}}$ (about 19 mph)

7. (a) You flew over two time-zone boundaries. (b) 40 km/hr; 500 km/hr; 20 km/hr. (c) Total distance = 2100 km, total time = 8 hours, so average speed = 2100/8 = 262.5 km/hr (about 160 mph).

8. They have the same speeds (12 km/hr), but different velocities (12 km/hr north and 12 km/hr south).

9. The speed is constant but the velocity is not, because the direction changes.

10. 1 hour and 20 min (80 min). The speed is constant but the velocity is not.

11.

a	b	c	d	e	f	g	h	i	j
No	No	Yes	No	No	No	Yes	No	No	No
No	No	No	No	Yes	Yes	No	No	No	No
No	No	No	Yes	No	No	No	Yes	Yes	Yes
No	No	Yes	Yes	Yes	Yes	Yes	Yes	Yes	Yes

13. Accelerated in cases (b), (c), (d), and (e). Unaccelerated in case (a).

14. 2 m/sec².

15. 10 m/sec; 20 m/sec; 5 m/sec.

16. 5 m; 20 − 5 = 15 m; 45 − 20 = 25 m.

Chapter 6

1. Rover's weight is 50 lb. In space, mass = 20 kg and weight = 0.

2. On the moon, because the pull of gravity is weaker.

3. They have the same mass and the same weight, but the ton of feathers is larger.

4. On the moon.

5. Recall, from Chapter Five, that the acceleration of an object falling to Earth is 10 m/sec². (a) 0.4 × 10 = 4 m/sec². (b) 2 × 4 = 8 m/sec. (c) 2 × 10 = 20 m/sec.

6. (a) Brick. (b) The brick, because it picked up the most speed. (c) 0.

7. Alphonse the younger has half the mass so he must have twice the acceleration, because acceleration is inversely proportional to mass. Thus his acceleration is 6 g's or 60 m/sec².

9. 50,000 − 10,000 = 40,000 N, directed upward.

10. 0 net force, so the rocket would just sit there, not pressing down on the launch pad but not lifting off either.

11. The car has 0 acceleration. Thus, according to Newton 2, it must have 0 net force on it.

12. We need a forward drive force to balance the backward forces caused by air resistance and friction in order that the net force will be 0.

13. Starting: Because of the near lack of friction the tire has a difficult time pushing backward on the ice; the tire slips. According to Newton 3, the ice doesn't provide much, if any, forward force on the tires. Stopping: There is little backward frictional force by the road

405
Exercise Solutions

on the tires when the brakes are set. So, according to Newton 2, the acceleration (deacceleration in this case) is small.

14. Neither propeller nor jet airplanes can operate in the absence of air, because they both must "push backward" against the air in order to move forward.

Chapter 7

1. (a) Yes. (b) Yes, there must be other forces to balance the gravitational force. (c) The tree branch pulls upward on the apple, and the table pushes upward on the book. (d) The net force is 0 because both objects are unaccelerated.

3. Magnetic forces can only be caused by magnetic objects; electric forces can only be exerted by charged objects. There are other possibilities, such as the force that a sticky object, such as a wax ball, exerts on other objects.

4. (a) Doubles the force. (b) Quadruples the force, that is, makes it four times as large. (c) The force drops to one-ninth of its original value. (d) The force increases to four times its original value.

6. $\frac{1}{4}$ × normal weight, because the distance has doubled (from 6,000 to 12,000 km); $\frac{1}{9}$ × normal weight; $\frac{1}{16}$ × normal weight.

11. Distance around Earth is 6.28 × 6,000 or about 40,000 km. At 8 km/sec, the time required is 40,000/8 = 5,000 sec, or about 80 min.

12. (a) The required speed is different because the acceleration of gravity is less than 10 m/sec² at higher altitudes and also because the circular orbit isn't curved as strongly. The net effect of these two factors turns out to be a reduced orbital speed. (b) Because the speed is different and the distance around the orbit is larger, we would expect the orbital time to be different. (c) About a month (27.3 days, actually), because this is the time for the moon to orbit Earth!

14. (a) Friction between tires and road. (b) Gravity. (c) The pull, or tension, of the string.

17. She will feel weightless whenever her space vehicle is coasting in space with all rocket engines off. She will actually *be* weightless at the point between Earth and moon at which the moon's gravity just balances Earth's gravity. The distance from Earth to this point is about 90 percent of the distance to the moon.

Chapter 8

1. (b)

2. Yes. The floor is exerting a force on Rex, toward the center of the circle, that is, toward the axis of the cylinder.

3. Yes. Exercise 2 tells us that the floor exerts an inward force on Rex, so Newton 3 tells us that Rex must exert an outward force on the floor.

4. To Rex it seems that a gravitational force is acting on him, directed toward the outside of the cylinder. Thus he would call the cylindrical surface the floor. The reason Rex feels this way is that this surface is pressing inward against him, and he is pressing outward against it. Because of his everyday experiences with gravity and floors on Earth, Rex feels that he is being pressed against the floor by a gravitational force.

6. 100 × 100 = 10,000. (Gravitational force is inversely proportional to the square of the distance.)

1. No work, because the wall didn't move.
2. No work, because no force was exerted on the meteoroid.
3. Newton-meters (or meter-newtons); foot-pounds.
4. 8000 ft · lb; 6000 ft · lb; no work while lift is stationary.
5. 2000 N × 1.2 m = 2400 N · m.
6. 3000 ft · lb; 4000 J; 3.9 Btu.
7. Work = force × distance = 200,000 N × 0.02 m = 4,000 J. Power = work/time = 4,000 J/0.001 sec = 4,000,000 W!
8. Work = power × time. The number of seconds in 1 week is 60 × 60 × 24 × 7 = 604,800 sec, or about 600,000 sec. So work = 2 W × 600,000 sec = 1,200,000 J!
9. Work = power × time. The power is 1,000 W, or 1,000 J/sec, and the time is 1 hour, or 3,600 sec. Thus work = 1,000 × 3,600 = 3,600,000 J. Since 4 J = 1 cal, this is equivalent to 900,000 cal, or 2,700,000 ft · lb.
10. 60,000 cal = 240,000 J = 180,000 ft · lb.
11. 100 hp = 55,000 ft · lb/sec. Work done = power × time = 55,000 ft · lb/sec × 10,000 sec = 550,000,000 ft · lb, or 5.5×10^8 ft · lb. Because the gasoline could do 5.5×10^8 ft · lb of work, this must have been the amount of energy originally in the gasoline.
12. Radiant energy from the sun → gravitational energy of evaporated water → gravitational energy of water behind dam.
13. Blender: electromagnetic → kinetic (rotating machinery) → thermal (heating of blended substance). Toaster: electromagnetic → thermal. Light bulb: electromagnetic → radiant (light) and thermal.
14. It would appear as kinetic energy of a faster-moving river (because the turbines slow the water) and as thermal energy of a slightly warmer river.
15. In both cases it's because the air molecules inside are moving faster and thus hit the walls harder.
16. At higher temperatures the molecules of the liquid are moving faster, so they can escape from the liquid more easily.
17. Total energy increases by 20 cal + 40 J. Putting these into a single set of units, 20 cal = 80 J, so the total energy increases by 120 J. This increase goes entirely into thermal energy once the motion of the contents caused by stirring has subsided. Of the increase, 80 J came from electric energy and 40 J came from chemical energy stored in your body.

Chapter 10

1. Firewood, solar, wind, water behind a dam.
2. (a) Chemical. (b) Chemical. (c) Radiant. (d) Nuclear. (e) Kinetic. (f) Gravitational. (g) Chemical.
3. Burning consists of combining carbon from the fuel with oxygen from the air.
4. All except uranium, which was in the earth to begin with.
5. (a) 3,000 m² × 200 W/m² = 600,000 W. (b) 600,000/10,000 = 60 people.
6. (a) 200 J. (b) 200 $\frac{\text{joules}}{\text{sec}}$ × 60 sec = 12,000 J. (c) 12,000/600 = 20 m, because gravitational energy = weight × height.
7. $2 \times 10^{12}/200 = 10^{10}$ m² = 10,000 km².

8. In 1 sec, the energy produced is 6×10^6 N $\times$ 170 m = 1.02×10^9 J. So the power produced is 1.02×10^9 W.

9. 600.

Chapter 11

1. (a) Chemical to kinetic. (b) Elastic to kinetic. (c) Electric to kinetic. (d) Chemical to thermal.

3. 60 J exhaust; 40 percent efficiency.

4. 15 J.

5. Thermal energy of road and air, plus kinetic energy of car.

6. 9 + 36 = 45 kW. You need five times as much power as previously. The rate of consumption must now be 5 $\times$ 72 = 360 kW. 360 kW could light up *3600* 100-W bulbs!

7. Half as much, 4.5 kW. Half as much, 36 kW. Half as much pollution.

9. Cool the ocean, because the plant takes thermal energy from the ocean.

10. (a) 20 kW. (b) 16 kW. It could definitely heat the house. (c) 30 percent of 2kW = 0.6 kW, which is enough to run the lights. (c) 30 percent of 16 kW = 4.8 kW, which is enough to heat the house. (e) The smaller cell costs only 30 percent of the cost of the larger cell, so the total cost is $9,000, a saving of $11,000.

Chapter 12

2. The transmitted shape would be a stretched region of the Slinky.

5. 50 W $\times$ 300 sec = 15,000 J = 3,750 cal.

6. 40 W $\times$ 3,600 sec = 144,000 J. Since gravitational energy = weight $\times$ height, this could lift a 400-N person by 144,000/400 = 360 m.

7. Electric energy is consumed, and kinetic energy is produced.

8. Kinetic energy is consumed and electric energy is produced.

Chapter 13

1. The mass of a CO_2 molecule is $\frac{4}{3} + \frac{4}{3} + \frac{3}{3} = 11\frac{1}{3}$, or roughly four times the mass of one carbon atom. So 4 $\times$ 2000 = 8000 lb (4 tons) of CO_2 is produced.

2. CO_2 is a gas. It goes up the stack and into the air.

3. Options (b), (c), and (d) would help.

5. If all the ultraviolet from the sun got through the atmosphere, we would get about 50 times as much as we now get (since only 2 percent gets through at present), and a person would sunburn about 50 times as fast. Thus Caucasians would get a sunburn in about one-fiftieth of an hour, or about 1 minute!

Chapter 14

2. A rock moving north at 15 m/sec.

3. (a) A rock moving south at 17 m/sec. (b) A rock moving south at 23 m/sec.

4. (a) Because sound travels at 340 m/sec relative to the air, the sound wave moves at 340 m/sec relative to Velma. (b) 340 + 20 = 360 m/sec.

5. (a) 340 m/sec. (b) 340 − 20 = 320 m/sec.

6. Less. For example, if she were moving north at 340 m/sec, she would measure the speed of a northward-moving sound wave to be 0.!

7. 30 m/sec.

9. The speed of sound must be midway between 310 and 360, or 335 m/sec. Thus Velma must move at 25 m/sec relative to the air.

10. The penny will land behind your hand if you are speeding up, in front of your hand if you are slowing down, and to the left of your hand if you are turning a corner to the right.

Chapter 15

5. ⅔ of 70 is roughly 47 years old.

6. Again, 47 years old.

7. She must speed up after takeoff, turn around in order to begin heading back to Earth, and slow down to land on Earth.

9. Write out lightspeed as 300,000,000 m/sec. Multiply it by itself. The result is 9 followed by 16 zeros. In other words, 9×10^{16}.

10. Mass increase $= \dfrac{\text{energy increase}}{\text{lightspeed squared}} = \dfrac{90}{9 \times 10^{16}} = \dfrac{90}{90 \times 10^{15}} = \dfrac{1}{10^{15}} = 10^{-15}$ kg.

Chapter 16

1. Yes.

2. The curve would be lower (less radiation) and shifted toward the right (most of the radiation is infrared).

3. Radiant energy changes into electric energy (or, from another point of view, radiant energy turns into the kinetic energy of moving electrons).

5. State (a) has more energy because the charge is vibrating wider. The energy of the emitted radiation must equal the difference in energies of the initial state (a) and the final lower-energy state (b).

6. Shorter wavelength light has the higher frequency. Violet has the highest-energy photons, and red has the lowest-energy photons.

7. Gamma ray, because the individual photons have the most energy.

8. High-energy photons (ultraviolet photons, because these are the highest-energy photons in sunlight) are needed to break down the molecules. But most ultraviolet radiation doesn't penetrate the atmosphere, thus the chlorofluorocarbon molecules must rise above most of the atmosphere before they can be broken down by ultraviolet photons.

Chapter 17

1. 0. Three times as big at a as at c.

2. (a) A probability of ⅙ (about 17 percent) for each of the six outcomes. (b) No. (c) No.

3. (a) Most accurate: gamma ray. Least accurate: infrared. (b) Highest energy: gamma ray. Lowest energy: visible.

5. 200 m/sec.

6. 0.00004 m or 4×10^{-5} m.

7. Electron, proton, helium atom, water molecule, glucose molecule, grain of dust, baseball, automobile.

Chapter 18

2. Point *b*: 0.4 percent probability. Point *a*: 0.9 percent probability.

3. The speed of light, 3×10^8 m/sec. Thus its rest mass must be zero, because objects having nonzero rest mass cannot move at lightspeed (Chapter Fifteen).

4. Decreased, because the atom lost energy, and an object's mass is proportional to its energy (remember $E = mc^2$).

5. Highest frequency: 4 to 2. Longest wavelength: 4 to 3.

6. Highest frequency: 2 to 1. Longest wavelength: 4 to 3.

7. Ten. See Figure 18.6.

8. Highest frequency: 5 to 1. Lowest frequency: 5 to 4.

9. Largest: the line farthest to the left (deep violet). Smallest: Red.

10. Red line: state 3. The other lines correspond to jumps starting from states 4, 5, and 6. Figure 18.7 shows the first three of these lines.

11. Ultraviolet (higher energy than the five visible lines). Infrared (lower energy than the visible lines).

Chapter 19

1. Grain of dust, hydrogen atom, water molecule. As you will see later in this chapter, the proton might also be made of smaller pieces.

2. $mc^2 + mc^2 = 2\,mc^2$, where *m* stands for the mass of an electron.

3. $2\,mc^2$, where *m* stands for the mass of a proton.

4. Higher frequency, because they need to have more energy.

5. I wouldn't. I'd be annihilated.

6. Lots of gamma-ray (i.e., high-energy) photons.

Chapter 20

3. Remove three protons from the nucleus.

4. All contain one proton. They contain, respectively, zero, one, and two neutrons.

5. Radon, atomic number 86, mass number 222.

7. Nitrogen, atomic number 7, mass number 14.

8. Thorium, atomic number 90, mass number 234.

9. Protractinium, atomic number 91, mass number 234.

11. 50 kg; 25 kg; 6.25 kg ($\frac{1}{16}$ of 100 kg).

12. 5,000 years old. 4 half-lives, or 20,000 years old.

13. Decreases to ⅛ (12 percent) in 3 half-lives, 1/16 (6 percent) in 4 half-lives, 1/32 (3 percent) in 5 half-lives, 1/64 (1.7 percent) in 6 half-lives, 1/128 (0.8 percent) in 7 half-lives. So for a decrease to 10 percent, you must wait 3 or 4 half-lives, or 24 to 32 days. For a decrease to 1 percent, you must wait about 7 half-lives, or 56 days.

Chapter 21

1. Three nuclear particles (two protons and one neutron) on the left, three on the right.
3. 0.99 kg; 990 kg.
4. Atomic number 4, mass number 8, beryllium.
5. C-12.
6. N-13.
7. The result is C-13. Apparently one proton turned into a neutron (the opposite of ordinary beta decay).
8. N-14.
9. The cobalt nucleus is more massive.
10. The proton isn't a likely candidate because its positive charge causes it to be repelled by the nucleus. The electron doesn't seem likely to do the job because it is so light. More importantly, it happens that electrons do not produce the strong nuclear force that binds nuclei together.

Chapter 22

1. U-238 has three more nuclear particles than U-235, so the fractional difference in mass is about $3/235 = 1.3$ percent.
3. Easier. Plutonium and uranium are different elements, so chemical methods may be used to separate them.
4. Plutonium fission bomb.
5. 36 (6×6, because the area of a circle is proportional to the square of its radius).

Chapter 23

1. Millions of degrees, as in the H-bomb and the sun.
2. No. We saw in Chapter Twenty-Two that bomb-grade uranium must be at least 20 percent U-235.
3. Th-232 + n → Th-233; Th-233 → Pa-233 + beta; Pa-233 → U-233 + beta.

Chapter 24

1. 4.3 light-years. 4.3 years ago.
2. 2 million years ago. The earliest humanlike creatures had recently evolved.

EINSTEIN PHOTO ESSAY CREDITS

Page 208   *Top,* all Einstein Archive, Courtesy AIP Niels Bohr Library; *bottom left,* Lotte Jacobi; *bottom right,* Einstein Archive, Courtesy AIP Niels Bohr Library.

Page 209   *Top left,* Einstein Archive; *top right,* Lotte Jacobi; *bottom left,* Einstein Archive, Courtesy AIP Niels Bohr Library; *bottom right,* Lotte Jacobi.

Page 210   *Top left,* Einstein Archive, Courtesy AIP Niels Bohr Library; *top right,* Pittsburg Post-Gazette, Courtesy AIP Niels Bohr Library; *bottom left,* United Press International; *bottom right,* Einstein Archive.

Page 211   *Top left,* Ullstein Bilderdienst, Berlin; *top right,* Courtesy of the Archives, California Institute of Technology; *bottom left,* Courtesy AIP Niels Bohr Library; *bottom right,* Lotte Jacobi.

Page 212   *Top left and right,* from FDR Library, Courtesy of The Estate of Albert Einstein; *bottom left,* Ulli Steltzer, Courtesy AIP Niels Bohr Library.

Page 213   *Top left,* United Press International; *top right,* Ernst Haas/Magnum; *bottom left,* Copyright © H. Landshoff; *bottom right,* From *Herblock's Here and Now,* Simon & Schuster, 1955.

INDEX